Robot Ethics

Intelligent Robotics and Autonomous Agents

Edited by Ronald C. Arkin

Robot Ethics
The Ethical and Social Implications of Robotics

Edited by Patrick Lin, Keith Abney, and George A. Bekey

The MIT Press
Cambridge, Massachusetts
London, England

First MIT Press paperback edition, 2014
© 2012 Massachusetts Institute of Technology

MIT Press books may be purchased at special quantity discounts for business or sales promotional use. For information, please email special_sales@mitpress.mit.edu.

This book was set in Stone by Toppan Best-set Premedia Limited. Printed and bound in the United States of America.

Library of Congress Cataloging-in-Publication Data

Robot ethics : the ethical and social implications of robotics / edited by Patrick Lin, Keith Abney, and George A. Bekey.
    p.   cm.—(Intelligent robotics and autonomous agents series)
Includes bibliographical references and index.
ISBN 978-0-262-01666-7 (hardcover : alk. paper) 978-0-262-52600-5 (pb)
1. Robotics—Human factors. 2. Robotics—Moral and ethical aspects.
3. Robotics—Social aspects. 4. Robots—Design and construction.
I. Lin, Patrick. II. Abney, Keith, 1963– III. Bekey, George A., 1928–
    TJ211.49.R62   2012
    174'.9629892—dc23
                  2011016639

10 9 8 7 6

Contents

Preface

Nothing is stranger to man but his own image.
—Karel Čapek in *Rossum's Universal Robots* (1921)

If not yet the world, robots are starting to dominate news headlines. They have long been working on our factory floors, building products such as automobiles, but the latest research from academic labs and industry is capturing our imagination like never before. Now, robots are able to deceive, to perform surgeries, to identify and shoot trespassers, to serve as astronauts, to babysit our kids, to shape shift, to eat biomass as their fuel (but not human bodies, the manufacturer insists), and much more.

As a case of life imitating art, science fiction had already predicted some of these applications, and robots have been both glorified and vilified in popular culture—so much so that we are immediately sensitive, perhaps hypersensitive, to the possible challenges they may create for ethics and society. The literature in robot ethics can be traced back for decades, but only in recent years, with the real possibility of creating these more imaginative and problematic robots, has there been a growing chorus of international concern about the impact of robotics on ethics and society.

For the serious reader interested in this dialog, it takes some work to pull together the various strands of discussions from books and scholarly journals to media articles and websites. Thus, we have designed this edited volume to fill that gap in the information marketplace: to be an accessible and authoritative source of expert opinions on a wide range of issues in robot ethics, all in one location. While there is some technical material in this edited collection of papers, it does not presuppose much familiarity with either robotics or ethics, and therefore it is appropriate for policymakers, industry, and the broader public as well as university students and faculty scholars.

Chapters in part I of this volume provide a broad survey of the issues in robot ethics; discuss the latest trends in robotics; and give an overview of ethical theories and issues as relevant to robotics. Then, to provide guideposts for the reader, parts II onward begin with a short introduction that summarizes the chapters in each part, organized so that there is continuity of flow from one part and its chapters to the next, as follows:

In part II, we look at issues related to the possibility of programming ethics into a robot, as an intuitive approach to controlling its behavior. Concerning what is perhaps the most prominent and morally problematic use of robots today, our discussion naturally leads to the issue of designing a responsible or discriminating robot for war, which is the focus of part III. But ethical use of military robots can also be promoted through governance or policy, which leads to the chapters on law-related topics in part IV, including legal liability and privacy concerns. Some privacy issues arise given the physical access robots may have to our homes and lives, as well as the emotional access they have from their resemblance as humans. Part V, then, starts with an investigation of risks related to such emotional bonds, followed by chapters on more intimate relationships: robots as lovers. Not quite as personal, part VI examines ethical issues related to robots as caregivers, such as for medical purposes, and as our servants. In part VII, we telescope back out to broad and more distant (but nonetheless plausible) concerns about the possibility that we should give rights or moral consideration to robots. Finally, our epilogue ends the volume with some concluding and unifying thoughts on the issues discussed.

Though the chapters follow a sensible train of discussion from part I through part VII, they do not need to be read in order. We invite you to start with whatever focus interests you the most and jump around to other chapters as desired. The crucial point here is to become engaged in this important but underdeveloped global discussion. As robots advance into our homes, workplaces, schools, hospitals, battlefields, and society at large, it would serve us well to be informed of the ethical and social issues and prepared for a more mechanized world.

Patrick Lin, PhD
George A. Bekey, PhD
Keith Abney, ABD

Acknowledgments

We thank the institutions and organizations with which the editors are—or have been during the time of writing—affiliated for their support: California Polytechnic State University, San Luis Obispo, specifically the College of Liberal Arts and the Philosophy Department, as well as the College of Engineering; Centre for Applied Philosophy and Public Ethics (CAPPE, Australia); Stanford Law School's Center for Internet and Society; U.S. Naval Academy's Vice Admiral Stockdale Center for Ethical Leadership; and University of Southern California's Viterbi School of Engineering. We also thank our many contributors, without whom this volume would not be possible, and their respective institutions.

At the MIT Press, we thank Ada Brunstein, our editor, for her confidence and enthusiasm, as well as Professor Ronald Arkin and his reviewers for his support of this project. We thank our editorial assistant Kyle Campbell, as well as Kendall Protzmann, Natalia Saarela, and Danielle Dierkes for their invaluable service in preparing our manuscript. And we thank our many proofreaders: Jeremy Dillon, Andrew Jones, Kolja Keller, Jordan Rowley, Raymond Stewart, and Aimee van Wynsberghe.

Parts of this book are also based upon work supported by the U.S. Office of Naval Research under grant numbers N00014-07-1-1152 and N00014-08-1-1209. Any opinions, findings, conclusions, or recommendations expressed in this book are those of the respective author(s) and do not necessarily reflect the views of the aforementioned organizations or editors.

Finally, we thank our supportive families for their patience and sacrifice through this project, as well as you—the reader—for your interest and foresight in being part of the global conversation on robot ethics.

1 Introduction

1 Introduction to Robot Ethics

Patrick Lin

Welcome to the Robot Revolution. By this, we do not mean an uprising of our robots, as told in literature and film—at least not yet. But, today, robotics is a rapidly advancing field with a growing stable of different robot models and their expanding roles in society, from playing with children to hunting down terrorists.

"The emergence of the robotics industry," observed Bill Gates, "is developing in much the same way that the computer business did 30 years ago" (2007). As a key architect of the computer industry, his prediction has special weight. In a few decades—or sooner, given exponential progress forecasted by Moore's Law (that computing speeds will double every eighteen months or so)—robots in society will be as ubiquitous as computers are today, Gates believes; and we would be hard-pressed to find an expert who disagrees.

But consider just a few of the challenges linked to computers in the last thirty years: job displacement, privacy concerns, intellectual property disputes, real-world alienation, redefinition of relationships, cyberbullying, Internet addiction, security fears, and so on. To be clear, these are not arguments by themselves that the computer industry should never have been developed, but only that its benefits need to be weighed against its negative effects. The critical lesson we would like to focus on here, rather, is this: if the evolution of the robotics industry is analogous to that of computers, then we can expect important social and ethical challenges to emerge from robotics, as well, and attending to them sooner rather than later will likely help mitigate those negative consequences.

Society has long been concerned with the impact of robotics, before the technology was viable and even before the word "robot" was coined for the first time nearly a century ago (Čapek 1921). Around 1190 BCE, Homer described in his *Iliad* the intelligent robots or "golden servants" created by Hephaestus, the ancient Greek god of technology (Lattimore 1961). More than two thousand years later, around 1495, Leonardo da Vinci conceived of a mechanical knight that would be called a robot today (Hill 1984). And modern literature about robots features cautionary tales about insufficient programming, emergent behavior, errors, and other issues that make robots unpredictable and potentially dangerous (e.g., Asimov 1950, 1957; Dick 1968; Wilson 2005). In popular culture, films continue to dramatize and demonize robots, such as *Metropolis*, *Star Wars*, *Blade Runner*, *Terminator*, *AI*, and *I, Robot*, to name just a few. Headlines today also stoke fears about robots wreaking havoc on the battlefield, as well as financial trading markets, perhaps justifiably so (e.g., Madrigal 2010).

A loose band of scholars worldwide has been researching issues in robot ethics for some time (e.g., Veruggio 2006). And a few reports and books are trickling into the marketplace (e.g., Wallach and Allen 2008; Lin, Bekey, and Abney 2008; Singer 2009a). But there has not yet been a single, accessible resource that draws together such thinking on a wide range of issues, such as programming design, military affairs, law, privacy, religion, healthcare, sex, psychology, robot rights, and more. This edited volume is designed to fill that need, and this chapter is meant to introduce the major issues, followed by chapters that provide more detailed discussions.

1.1 Robots in Society

Robots are often tasked to perform the "three Ds," that is, jobs that are dull, dirty, or dangerous. For instance, automobile factory robots execute the same, repetitive assemblies over and over, with precision and without complaint; military unmanned aerial vehicles (UAVs) surveil from the skies for far more hours than a human pilot can endure at a time. Robots crawl around in dark sewers, inspecting pipes for leaks and cracks, as well as do the dirty work in our homes, such as vacuuming floors. Not afraid of danger, they also explore volcanoes and clean up contaminated sites, in addition to more popular service in defusing bombs and mediating hostage crises.

We can also think of robots more simply and broadly—as human replacements. More than mere tools, which cannot think and act independently, robots are able to serve in many old and new roles in society that are often handicapped, or made impossible, by human frailties and limitations; that is, semi- and fully-autonomous machines could carry out those jobs more optimally. Beyond the usual "three Ds," robots perform delicate and difficult surgeries, which are risky with shaky human hands. They can navigate inaccessible places, such as the ocean floor or the surface of Mars. As the embodiment of artificial intelligence (AI), they are more suited for jobs that demand information processing and action too quick for a human, such as the U.S. Navy's Phalanx CIWS that detects, identifies, and shoots down enemy missiles rapidly closing in on a ship. Some argue that robots could replace humans in situations where emotions are liabilities, such as battlefield robots that do not feel anger, hatred, cowardice, or fear—human weaknesses that often cause wartime abuses and crimes by human soldiers (Arkin 2007). Given such capabilities, we find robots already in society, or under development, in a wide range of roles, such as:

- *Labor and services*   Nearly half of the world's seven-million-plus service robots are Roomba vacuum cleaners (Guizzo 2010), but others exist that mow lawns, wash floors, iron clothes, move objects from room to room, and perform other chores around the home. Robots have been employed in manufacturing for decades, particularly in auto factories, but they are also used in warehouses, movie sets, electronics manufacturing, food production, printing, fabrication, and many other industries.
- *Military and security*   Grabbing headlines are war robots with fierce names, such as Predator, Reaper, Big Dog, Crusher, Harpy, BEAR, Global Hawk, Dragon Runner, and more. They perform a range of duties, such as spying or surveillance (air, land, underwater, space), defusing bombs, assisting the wounded, inspecting hideouts, and attacking targets. Police and security robots today perform similar functions, in addition to guarding borders and buildings, scanning for pedophiles and criminals, dispensing helpful information, reciting warnings, and more. There is also a growing market for home-security robots, which can shoot pepper spray or paintball pellets and transmit pictures of suspicious activities to their owners' mobile phones.
- *Research and education*   Scientists are using robots in laboratory experiments and in the field, such as collecting ocean-surface and marine-life data over extended periods (e.g., Rutgers University's Scarlet Knight) and exploring new planets (e.g., NASA's Mars Exploration Rovers). In classrooms, robots are delivering lectures, teaching subjects (e.g., foreign languages, vocabulary, and counting), checking attendance, and interacting with students.
- *Entertainment*   Related to research and education is the field of "edutainment" or education-entertainment robots, which include ASIMO, Nao, iCub, and others. Though they may lack a clear use, such as serving specific military or manufacturing functions, they aid researchers in the study of cognition (both human and artificial), motion, and other areas related to the advancement of robotics. Robotic toys, such as AIBO, Pleo, and RoboSapien, also serve as discovery and entertainment platforms.
- *Medical and healthcare*   Some toy-like robots, such as PARO, which looks like a baby seal, are designed for therapeutic purposes, such as reducing stress, stimulating cognitive activity, and improving socialization. Similarly, University of Southern California's socially assistive robots help coach patients in physical therapy and other health-related areas. Medical robots, such as da Vinci Surgical System and ARES ingestible robots, are assisting with or conducting difficult medical procedures on their own. RIBA, IWARD, ERNIE, and other robots perform some of the functions of nurses and pharmacists.
- *Personal care and companions*   Robots are increasingly used to care for the elderly and children, such as RI-MAN, PaPeRo, and CareBot. PALRO, QRIO, and other edutainment robots already mentioned can also provide companionship. Surprisingly, relationships of a more intimate nature are not quite satisfied by robots yet, considering the sex industry's reputation as an early adopter of new technologies. Introduced in

2010, Roxxxy is billed as "the world's first sex robot" (Fulbright 2010), but its lack of autonomy or capacity to "think" for itself, as opposed to merely responding to sensors, suggests that it is not, in fact, a robot.

• *Environment*   Not quite as handy as WALL-E (of the eponymous film), robots today still perform important functions in environmental remediation, such as collect trash, mop up after nuclear power plant disasters, remove asbestos, cap oil geysers, sniff out toxins, identify polluted areas, and gather data on climate warming.

• *In the future*   As AI advances, we can expect robots to play a more complex and wider range of roles in society. For instance, police robots equipped with biometrics capabilities and sensors could detect and identify weapons, drugs, and faces at a distance. Military robots could make attack decisions independently; in most cases today, there is a human triggerman behind those robots. Driverless trains today and DARPA's Grand Challenges are proof-of-concepts that robotic transportation is possible, and even commercial airplanes are controlled autonomously for a significant portion of their flight, never mind military UAVs. A general-purpose robot, if achievable, could service many of our domestic labor needs, as opposed to a team of robots each with its own job.

In the future, we can also expect robots to scale down as well as up. Some robots are miniature today and ever shrinking, perhaps bringing to life the idea of a "nano-bot," swarms of which might work inside our bodies or in the atmosphere or cleaning up oil spills. Even rooms or entire buildings might be considered as robots—beyond the "smart homes" of today—if they can manipulate the environment in ways more significant than turning on lights and air conditioning. With synthetic biology, cognitive science, and nanoelectronics, future robots could be biologically based. And human-machine integrations, that is, cyborgs, may be much more prevalent than they are today, which are mostly limited to patients with artificial body parts, such as limbs and joints that are controlled to some degree by robotics. Much of this speaks to the fuzziness of the definition of robot (which we return to in the next chapter). What we intuitively consider as robots today may change, given different form factors and materials of tomorrow.

In some countries, robots are quite literally replacements for humans, such as in Japan, where a growing elderly population and declining birthrates mean a shrinking workforce (Schoenberger 2008). Robots are built to specifically fill that labor gap. And given the nation's storied love of technology, it is therefore unsurprising that approximately one out of twenty-five workers in Japan is a robot (RedOrbit 2008). While the United States currently dominates the market in military robotics, nations such as Japan and South Korea lead in the market for social robotics, such as elder-care robots. Other nations with similar demographics, such as Italy, are expected to introduce more robotics into their societies, as a way to shore up a decreasing workforce (Geipel 2003);

and nations without such concerns can drive productivity, efficiency, and effectiveness to new heights with robotics.

1.2   Ethical and Social Issues

The Robotics Revolution promises a host of benefits that are compelling and imaginative, but, as with other emerging technologies, they also come with risks and new questions that society must confront. This is not unexpected, given the disruptive nature of technology revolutions. Here we map the myriad issues into three broad (and interrelated) areas of ethical and social concern and provide representative questions for each area.

1.2.1   Safety and Errors

We have learned by now that new technologies, first and foremost, need to be safe. Asbestos, DDT, and fen-phen are among the usual examples of technology gone wrong (e.g., U.S. Environmental Protection Agency 2007; Gorman 1997; Lear 1997), having been introduced into the marketplace before sufficient health and safety testing. A similar debate is occurring with nanomaterials now (e.g., Allhoff, Lin, and Moore 2010).

With robotics, the safety issue is with their software and design. Computer scientists, as fallible human beings, understandably struggle to create a perfect piece of complex software: somewhere in the millions of lines of code, typically written by teams of programmers, errors and vulnerabilities likely exist. While this usually does not result in significant harm with, say, office applications—just lost data if users do not periodically save their work (which arguably is their own fault)—even a tiny software flaw in machinery, such as a car or a robot, could lead to fatal results.

For instance, in August 2010, the U.S. military lost control of a helicopter drone during a test flight for more than thirty minutes and twenty-three miles, as it veered toward Washington, D.C., violating airspace restrictions meant to protect the White House and other governmental assets (Bumiller 2010). In October 2007, a semiautonomous robotic cannon deployed by the South African army malfunctioned, killing nine "friendly" soldiers and wounding fourteen others (e.g., Shachtman 2007). Experts continue to worry about whether it is humanly possible to create software sophisticated enough for armed military robots to discriminate combatants from noncombatants, as well as threatening behavior from nonthreatening (e.g., Lin, Bekey, and Abney 2008).

Never mind the scores of other military robot accidents and failures (Zucchino 2010), human deaths caused by robots can and have occurred in civilian society. The first human to be killed by a robot was widely believed to be in 1979, in an auto factory accident in the United States (Kiska 1983). And it does not take much to imagine a mobile city-robot of the future—a heavy piece of machinery—accidentally running over a small child.

Hacking is an associated concern, given how much attention is paid to computer security today. What makes a robot useful—its strength, ability to access and operate in difficult environments, expendability, and so on—could also be turned against us, either by criminals or simply mischievous persons. This issue will become more important as robots become networked and more indispensable to everyday life, as computers and smart phones are today. Indeed, the fundamentals of robotics technology are not terribly difficult to master: as formidable and fearsome as military robots are today, already more than forty nations have developed those capabilities, including Iran (Singer 2009b; Defense Update 2010).

Thus, some of the questions in this area include: Is it even possible for us to create machine intelligence that can make nuanced distinctions, such as between a gun and an ice-cream cone pointed at it, or understand human speech that is often heavily based on context? What are the tradeoffs between nonprogramming solutions for safety—for example, weak actuators, soft robotic limbs or bodies, using only nonlethal weapons, or using robots in only specific situations, such as a "kill box" in which all humans are presumed to be enemy targets—and the limitations they create? How safe ought robots be prior to their introduction into the marketplace or society, that is, should a precautionary principle apply here? How would we balance the need to safeguard against robots running amok (e.g., with a kill-switch) with the need to protect robots from hacking or capture?

1.2.2 Law and Ethics

Linked to the risk of robotic errors, it may be unclear who is responsible for any resulting harm. Product liability laws are largely untested in robotics and, anyway, continue to evolve in a direction that releases manufacturers from responsibility, as occurs through end-user license agreements in software. With military robots, for instance, there is a list of actors throughout the supply chain who may be held accountable: the programmer, the manufacturer, the weapons legal-review team, the military procurement officer, the field commander, the robot's handler, and even the president of the United States, as the commander in chief of that nation.

As robots become more autonomous, it may be plausible to assign responsibility to the *robot itself*, that is, if it is able to exhibit enough of the features that typically define personhood. If this seems too far-fetched, consider that there is ongoing work in integrating computers and robotics with biological brains (e.g., Warwick 2010; also Warwick, chapter 20, this volume). A conscious human brain (and its body) presumably has human rights, and replacing parts of the brain with something else, while not impairing its function, would seem to preserve those rights. We may come to a point at which more than half of the brain or body is artificial, making the organism more robotic than human, which makes the issue of robot rights more plausible.

One natural way to think about minimizing risk of harm from robots is to program them to obey our laws or follow a code of ethics. Of course, this is much easier said than done, since laws can be vague and context-sensitive, which robots may not be sophisticated enough to understand, at least in the foreseeable future. Even the three (or four) laws of robotics in Asimov's stories, as elegant and sufficient as they appear to be, create loopholes that result in harm (e.g., Asimov 1957, 1978, 1985).

Programming aside, the use of robots must also comply with law and ethics, and again those rules and norms may be unclear or untested on such issues. For instance, landmines are an effective but horrific weapon that indiscriminately kills, whether soldiers or children; landmines have existed for hundreds of years, but it was only in 1983—after their heavy use in twentieth century wars—that certain uses of landmines were banned, such as planting them without means to identify and remove them later (United Nations 1983); and only in 1999 did an international treaty ban the production and use of landmines (Abramson 2008). Likewise, the use of military robots may raise legal and ethical questions that we have yet to fully consider (e.g., Lin, Bekey, and Abney 2008, 2009; also chapters 6–10 and others, this volume) and, later in retrospect, may seem obviously unethical or unlawful.

Another relevant area of law concerns privacy. Several forces are driving this concern, including the shrinking size of digital cameras and other recording devices; an increasing emphasis on security at the expense of privacy (e.g., expanded wiretap laws, a blanket of surveillance cameras in some cities to monitor and prevent crimes); advancing biometrics capabilities and sensors; and database integrations. Besides robotic spy planes, we previously mentioned (future) police robots that could conduct intimate surveillance at a distance, such as detecting hidden drugs or weapons and identifying faces unobtrusively; if linked to databases, they could also run background checks on an individual's driving, medical, banking, shopping, or other records to determine if the person should be apprehended (Sharkey 2008). Domestic robots, too, can be easily equipped with surveillance devices—as home security robots already are—that may be monitored or accessed by third parties (Calo, chapter 12, this volume).

Thus, some of the questions in this area include: If we could program a code of ethics to regulate robotic behavior, which ethical theory should we use? Are there unique legal or moral hazards in designing machines that can autonomously kill people? Or should robots merely be considered tools, such as guns and computers, and regulated accordingly? Is it ethically permissible to abrogate responsibility for our elderly and children to machines that seem to be a poor substitute for human companionship (but, perhaps, better than no—or abusive—companionship)? Will robotic companionship (that could replace human or animal companionship) for other purposes, such as drinking buddies, pets, entertainment, or sex, be morally problematic? At what point should we consider a robot to be a "person," thus affording it some rights and responsibilities, and if that point is reached, will we need to emancipate our robot "slaves"? Do we have any other distinctive moral duties toward robots? As they develop enhanced capacities, should cyborgs have a different legal status than ordinary humans? At what point does technology-mediated surveillance by robots count as a "search," which would generally require a judicial warrant? Are there particular moral qualms over placing robots in positions of authority, such as police, prison or security guards, teachers, or any other government roles or offices in which humans would be expected to obey robots?

1.2.3 Social Impact

How might society change with the Robotics Revolution? As with the Industrial and Internet Revolutions, one key concern is job loss. In the Industrial Revolution, factories replaced legions of workers who used to perform the same work by hand, giving way to the faster, more efficient processes of automation. In the Internet Revolution, online ventures, such as Amazon.com, eBay, and even smaller "e-tailers," are still edging out brick-and-mortar retailers, who have much higher overhead and operating expenses, of which labor is one of the largest. Likewise, as potential replacements for humans—outperforming humans in certain tasks—robots may displace human jobs, regardless of whether the workforce is growing or declining.

The standard response to the job-loss concern is that human workers, whether replaced by other humans or machines, would then be free to focus their energies where they can make a greater impact (i.e., at jobs in which they have a greater competitive advantage) (Rosenberg 2009), and that to resist this change is to support inefficiency. For instance, by outsourcing call-center jobs to other nations where the pay is less, displaced workers (in theory) can perform "higher-value" jobs, whatever those may be. Further, the demand for robots itself creates additional jobs. Yet, arguments about competitive and efficiency gains provide little consolation for the human worker who needs a job to feed her or his family, and cost benefits may be negated by unintended effects, such as a negative customer support experience with call-center representatives whose first language is not that of the customers.

Connected to labor, some experts are concerned about technology dependency (e.g., Veruggio 2006). For example, as robots prove themselves to be better than humans at performing difficult surgeries, the resulting loss of those jobs may also mean the gradual loss of that medical skill or knowledge, to the extent that there would be fewer human practitioners. This is not the same worry with labor and service robots that perform dull and dirty tasks, in that we care less about the loss of those skills; but there is a similar issue of becoming overly reliant on technology for basic work. For one thing, this dependency seems to cause society to be more fragile: for instance, the Y2K problem caused significant panic, since so many critical systems—such as air-traffic control and banking—were dependent on computers whose ability to correctly advance their internal clock to January 1, 2000 (as opposed to resetting it to January 1, 1900) at the turn of the millennium was uncertain; and similar situations exist today with malicious computer viruses *du jour*.

Like the social networking and email capabilities of the Internet Revolution, robotics may profoundly impact human relationships. Already, robots are taking care of our elderly and children, though there are not many studies on the effects of such care, especially in the long term. Some soldiers have emotionally bonded with the bomb-disposing PackBots that have saved their lives, sobbing when the robot meets its end (e.g., Singer 2009a; Hsu 2009). And robots are predicted to soon become our lovers and companions (Levy 2007; also Levy, chapter 14, this volume, and Whitby, chapter 15, this volume): they will always listen and never cheat on us. Given the lack of research studies in these areas, it is unclear whether psychological harm might arise from replacing human relationships with robotic ones.

Harm also need not be directly to persons; it could also be to the environment. In the computer industry, "e-waste" is a growing and urgent problem (e.g., O'Donoghue 2010), given the disposal of heavy metals and toxic materials in the devices at the end of their product life cycle. Robots as embodied computers will likely exacerbate the problem, as well as increase pressure on rare-earth elements needed today to build computing devices and energy resources needed to power them. This also has geopolitical implications to the extent that only a few nations, such as China, control most of those raw materials (e.g., Gillis 2010).

Thus, some of the questions in this area include: What is the predicted economic impact of robotics, all things considered? How do we estimate the expected costs and benefits? Are some jobs too important, or too dangerous, for machines to take over? What do we do with the workers displaced by robots? How do we mitigate disruption to a society dependent on robotics, if those robots become inoperable or corrupted, e.g., through an electromagnetic pulse or network virus? Is there a danger with emotional attachments to robots? Are we engaging in deception by creating anthropomorphized machines that may lead to such attachments, and is that bad? Is there anything essential in human companionship and relationships that robots cannot replace? What is the environmental impact of a much larger robotics industry than we have today? Could we possibly face any truly cataclysmic consequences from the widespread adoption of social robotics (or robots capable of social or personal interactions, as opposed to factory robots, for example), and, if so, should a precautionary principle apply?

1.3 Engaging the Issues Now

These are only some of the questions that the emerging field of robot ethics is concerned with, and many of these questions lead to the doorsteps of other areas of ethics and philosophy, for example, computer ethics and philosophy of mind, in addition to the disciplines of psychology, sociology, economics, politics, and more. Note also that we have not even considered the more popular "Terminator" scenarios in which robots—through super-artificial intelligence—subjugate humanity, which are highly speculative scenarios that continually overshadow more urgent and plausible issues.

The robotics industry is rapidly advancing, and robots in society today are already raising many of these questions. This points to the need to attend to robot ethics now, particularly as ethics is usually slow to catch up with technology, which can lead to a "policy vacuum" (Moor 1985). As an example, the Human Genome Project was started in 1990, but it took eighteen years after that for Congress to finally pass a bill to protect Americans from discrimination based on their genetic information. Right now, society is still fumbling through privacy, copyright, and other intellectual property issues in the Digital Age, nearly ten years since Napster was first shut down.

As researchers and educators, we hope that this edited collection on robot ethics will provide and motivate greater discussion—in and outside of the classroom—across the broad continuum of issues, as described in this introduction. The contributors to this book are among the most respected and well-known scholars in robotics and technology ethics today, expertly tackling many of these issues.

Though sometimes to deaf ears, history lectures us on the importance of foresight. While the invention of such things as the printing press, gunpowder, automobiles, computers, vaccines, and so on, has profoundly changed the world (for the better, we hope), these innovations have also led to unforeseen consequences, or perhaps consequences that might have been foreseen and addressed had we bothered to investigate them. At the very least they have disrupted the status quo, which is not necessarily a terrible thing in and of itself; however, unnecessary and dramatic disruptions, such as mass displacements of workers or industries, have real human costs to them. Given lessons from the past, society is beginning to think more about ethics and policy in advance of, or at least in parallel to, the development of new game-changing technologies, such as genetically modified foods, nanotechnology, neuroscience, and human enhancement—and now we add robotics to that syllabus.

At the same time, we recognize that these technologies seem to jump out of the pages of science fiction, and the ethical dilemmas they raise also seem too distant to consider, if not altogether unreal. But as Isaac Asimov foretold: "It is change, continuing change, inevitable change, that is the dominant factor in society today. No sensible decision can be made any longer without taking into account not only the world as it is, but the world as it will be. . . . This, in turn, means that our statesmen, our businessmen, our everyman must take on a science fictional way of thinking" (Asimov 1978). With human ingenuity, what was once fiction is becoming fact, and the new challenges it brings are all too real.

References

Abramson, Jeff. 2008. The Ottawa Convention at a Glance. Arms Control Association, June. <http://www.armscontrol.org/factsheets/ottawa> (accessed September 12, 2010).

Allhoff, Fritz, Patrick Lin, and Daniel Moore. 2010. *What Is Nanotechnology and Why Does It Matter?: From Science to Ethics*. Hoboken, NJ: Wiley-Blackwell.

Arkin, Ronald C. 2007. *Governing Lethal Behavior: Embedding Ethics in a Hybrid Deliberative/ Hybrid Robot Architecture*, Report GIT-GVU-07-11. Atlanta: Georgia Institute of Technology's GVU Center. <http://www.cc.gatech.edu/ai/robot-lab/online-publications/formalizationv35.pdf> (accessed September 12, 2010).

Asimov, Isaac. 1950. *I, Robot* (2004 ed.). New York: Bantam Dell.

Asimov, Isaac. 1957. *The Naked Sun*. New York: Doubleday.

Asimov, Isaac. 1978. My own view. In *The Encyclopedia of Science Fiction*, ed. Robert Holdstock, 5. New York: St. Martin's Press.

Asimov, Isaac. 1985. *Robots and Empire*. New York: Doubleday.

Bumiller, Elisabeth. 2010. Navy drone violated Washington airspace. *The New York Times*. August 25, p. A16.

Čapek, Karel. 1921. *Rossum's Universal Robots* (2004 ed.), trans. Claudia Novack. New York: Penguin Group.

Defense Update. 2010. Karrar—Iran's new jet-powered recce and attack Drone. <http://defense -update.com/products/k/karrar_jet_powered_drone_24082010.html>.

Dick, Philip K. 1968. *Do Androids Dream of Electric Sheep?* New York: Del Rey Books.

Fulbright, Yvonne. 2010. Meet Roxxxy, the "woman" of your dreams. <http://www.foxnews.com/ story/0,2933,583314,00.html> (accessed September 12, 2010).

Gates, Bill. 2007. A robot in every home. *Scientific American* 296 (1) (January): 58–65.

Geipel, Gary. 2003. Global aging and the global workforce. A Hudson Institute article. <http:// www.hudson.org/index.cfm?fuseaction=publication_details&id=2740> (accessed September 12, 2010).

Gillis, Charlie. 2010. China's power play. *Macleans*, November 9. <http://www2.macleans .ca/2010/11/09/armed-and-dangerous/> (accessed November 26, 2010).

Gorman, Christine. 1997. Danger in the diet pills? *Time Magazine*, July 21. <http://www.time .com/time/magazine/article/0,9171,986725,00.html> (accessed September 12, 2010).

Guizzo, Erico. 2010. *IEEE Spectrum: World Robot Population Reaches 8.6 Million*, April 14. <http:// spectrum.ieee.org/automaton/robotics/industrial-robots/041410-world-robot-population> (accessed September 12, 2010).

Hill, Donald. 1984. *A History of Engineering in Medieval and Classical Times.* London: Croom Helm.

Hsu, Jeremy. 2009. Real soldiers love their robot brethren. *LiveScience,* May 21. <http://www .livescience.com/technology/090521-terminator-war.html> (accessed September 12, 2010).

Kiska, Tim. 1983. Death on the job: Jury awards $10 million to heirs of man killed by robot at auto plant. *Philadelphia Inquirer,* August 11, p. A-10.

Lattimore, Richmond, trans. 1961. *The Iliad of Homer.* Chicago, IL: University of Chicago Press.

Lear, Linda. 1997. *Rachel Carson: Witness for Nature.* New York: Henry Hoyten.

Levy, David. 2007. *Love and Sex with Robots: The Evolution of Human-Robot Relationships.* New York: HarperCollins Publishers.

Lin, Patrick, George Bekey, and Keith Abney. 2008. *Autonomous Military Robots: Risk, Ethics, and Design.* A report commissioned by U.S. Department of Navy/Office of Naval Research. <http:// ethics.calpoly.edu/ONR_report.pdf> (accessed September 12, 2010).

Lin, Patrick, George Bekey, and Keith Abney. 2009. Robots in war: Issues of risk and ethics. In *Ethics and Robotics,* ed. Rafael Capurro and Michael Nagenborg, 49–67. Heidelberg, Germany: AKA Verlag/IOS Press.

Madrigal, Alexis. 2010. Market data firm spots the tracks of bizarre robot traders. *Atlantic,* August 4. <http://www.theatlantic.com/technology/archive/2010/08/market-data-firm-spots-the-tracks-of-bizarre-robot-traders/608> (accessed September 12, 2010).

Moor, James H. 1985. What is computer ethics? *Metaphilosophy* 16 (4): 266–275.

O'Donoghue, Amy Joi. 2010. E-waste is a growing issue for states. *Deseret News,* August 22. <http:// www.deseretnews.com/article/700059360/E-waste-is-a-growing-issue-for-states.html?pg=1> (accessed September 12, 2010).

RedOrbit. 2008. Japan hopes to employ robots by 2025, April 8. <http://www.redorbit.com/news/ technology/1332274/japan_hopes_to_employ_robots_by_2025/> (accessed September 12, 2010).

Rosenberg, Mitch, 2009. The surprising benefits of robots in the DC. *Supply & Demand Chain Executive* 10 (2) (June/July): 39–40.

Shachtman, Noah. 2007. Robot cannon kills 9, wounds 14. *Wired,* October 18. <http://www .wired.com/dangerroom/2007/10/robot-cannon-ki/> (accessed September 12, 2010).

Schoenberger, Chana. 2008. Japan's shrinking workforce. *Forbes,* May 25. <http://www.forbes .com/2008/05/25/immigration-labor-visa-oped-cx_crs_outsourcing08_0529japan.html> (accessed September 12, 2010).

Sharkey, Noel. 2008. *2084: Big Robot Is Watching You.* A commissioned report. <http://staffwww .dcs.shef.ac.uk/people/N.Sharkey/> (accessed September 12, 2010).

Singer, Peter W. 2009a. *Wired for War: The Robotics Revolution and Conflict in the 21st Century.* New York: Penguin Press.

Singer, Peter W. 2009b. Robots at war: The new battlefield. *Wilson Quarterly*, Winter. <http://www
.wilsonquarterly.com/article.cfm?aid=1313> (accessed September 12, 2010).

United Nations. 1983. *The Convention on Certain Conventional Weapons*. Entered into force on
December 2. <http://www.armscontrol.org/factsheets/CCW> (accessed September 12, 2010).

U.S. Environmental Protection Agency. 2007. *Asbestos Ban and Phase Out*, April 25. <http://www
.epa.gov/asbestos/pubs/ban.html> (accessed September 12, 2010).

Veruggio, Gianmarco, ed. 2006. *EURON Roboethics Roadmap, EURON Roboethics Atelier*. Genoa,
Italy: EURON. <http://www.roboethics.org/atelier2006/docs/ROBOETHICS%20ROADMAP%20
Rel2.1.1.pdf> (accessed September 12, 2010).

Wallach, Wendell, and Colin Allen. 2008. *Moral Machines: Teaching Robots Right from Wrong*. New
York: Oxford University Press.

Warwick, Kevin. 2010. Implications and consequences of robots with biological brains. *Ethics &
Information Technology* [Special Issue on Robot Ethics and Human Ethics, ed. Anthony Beavers]
12 (1): 223–234.

Wilson, Daniel H. 2005. *How to Survive a Robot Uprising: Tips on Defending Yourself Against the
Coming Rebellion*. New York: Bloomsbury Publishing.

Zucchino, David. 2010. War zone drone crashes add up. *Los Angeles Times*, July 6. <http://articles
.latimes.com/2010/jul/06/world/la-fg-drone-crashes-20100706> (accessed September 12, 2010).

2 Current Trends in Robotics: Technology and Ethics

George A. Bekey

Robotics is indeed one of the great technological success stories of the present time. Starting from humble beginnings in the middle of the twentieth century, the field has seen great successes in manufacturing and industrial robotics, as well as personal and service robots of various kinds. All the branches of the armed services now use military robots. Robots are appearing everywhere in society: in healthcare, entertainment, search and rescue, care for the elderly, home services, and other applications. In fact, it is difficult to find a current magazine or newspaper without some mention of robots, whether flying over Afghanistan, vacuuming carpets, carrying items in warehouses, assisting surgeons in hospitals, helping persons with disabilities, or teaching children.

While the technological advances have been remarkable and rapid (and promise to continue this pace), the social and ethical implications of these new systems have been largely ignored. Only during the past decade have we seen the emergence of the field of "robot ethics" (sometimes abbreviated as "roboethics"; see chapter 3 for discussion on this nomenclature), with most efforts in Europe, Asia, and the United States. In this chapter, we survey some of the remarkable advances in robot hardware and software, and comment on the ethical implications of these developments.

2.1 What Is a Robot?

Let us start with a basic issue: What is a robot? Given society's long fascination with robotics, it seems hardly worth asking the question, as the answer surely must be obvious. On the contrary, there is still a lack of consensus among roboticists on how they define the object of their craft. For instance, an intuitive definition could be that a robot is merely a computer with sensors and actuators that allow it to interact with the external world; however, any computer that is connected to a printer or can eject a CD might qualify as a robot under that definition, yet few roboticists would defend that implication.

We do not presume we can definitively resolve this great debate here, but it is important that we offer a working definition prior to laying out the landscape of current and predicted applications of robotics. In its most basic sense, we define "robot" as *a machine, situated in the world, that senses, thinks, and acts*:

> Thus, a robot must have sensors, processing ability that emulates some aspects of cognition, and actuators. Sensors are needed to obtain information from the environment. Reactive behaviors (like the stretch reflex in humans) do not require any deep cognitive ability, but on-board intelligence is necessary if the robot is to perform significant tasks autonomously, and actuation is needed to enable the robot to exert forces upon the environment. Generally, these forces will result in motion of the entire robot or one of its elements (such as an arm, a leg, or a wheel). (Bekey 2005)

We stipulate that the robot must be *situated in the world* in order to distinguish a physical robot from software running on a computer, or, a "software bot."

This definition does not imply that a robot must be electromechanical; it leaves open the possibility of biological robots, but it eliminates virtual or software ones. A simulated robot is just that: a simulated robot. But it does rule out as robots any *fully* remote-controlled machines, since those devices do not "think," such as many animatronics and children's toys. That is, most of these toys do not make decisions for themselves; they depend on human input or an outside actor. Rather, the generally accepted idea of a robot depends critically on the notion that it exhibits some degree of autonomy, or can "think" for itself, making its own decisions to act upon the environment. Thus, the U.S. Air Force's Reaper unmanned aerial vehicle (UAV), though mostly teleoperated by humans, makes some navigational decisions on its own and therefore would count as a robot. By the same definition, the following things are not robots: conventional landmines, toasters, adding machines, coffee makers, and other ordinary devices.

As should be clear by now, the definition of "robot" also trades on the notion of "think," another source of contention that we cannot fully engage here. By "think," what we mean is that the machine is able to process information from sensors and other sources, such as an internal set of rules, either programmed or learned, and to make some decisions autonomously. Of course, this definition merely postpones our task and invites another question: What does it mean for machines to have autonomy? If we may simply stipulate it here, we define "autonomy" in robots as the capacity to operate in the real-world environment without any form of external control, once the machine is activated and at least in some areas of operation, for extended periods of time (Bekey 2005).

Thus again, *fully* remote- or teleoperated machines would not count as autonomous, since they depend on external control; they cannot "think" and, therefore, cannot act for themselves. As already indicated, many robots are *partially* remotely controlled; they are frequently known as "telerobots."

A complete discussion of what it means to be a robot will engage other difficult issues from technical to philosophical, such as complexity, unpredictability, determinism, responsibility, and free will, some of which are investigated in chapter 3. As such, we do not offer a complete discussion here, and we will have to content ourselves with the working definitions just stipulated—which should be enough to understand why we include some machines and not others in the remainder of this chapter.

2.2 Robotics around the World

Manufacturing robots were invented in the United States; companies such as Unimation and Cincinnati Milacron were leaders in the field in the 1970s. During the 1980s, the leadership in this field gradually moved to Japan and Europe, where companies like Fujitsu, Panasonic, Kuka, and ASEA became the dominant players. During the 1990s, support for research and development of service robots was much stronger in Japan, South Korea, Germany, Australia, and other countries than in the United States. A survey of trends in robotics in those countries in 2004 concluded that the United States was rapidly falling behind other countries in robotics, since (among other factors) there was no national program to support and coordinate robotics research (Bekey et al. 2008). This situation has begun to change since 2008, with the organization of a Congressional Caucus in robotics, development of "roadmaps" in such areas as medical robotics, manufacturing, and service, and increased support from a number of government agencies (Computer Community Consortium 2009). The first roadmap for robotics development was developed by the European Community (Veruggio 2006). Yet, while there is increased attention to the technology, there is still little discussion of its ethical implications except in Europe, where a number of symposia and conferences have addressed the issue (Veruggio 2009).

Current and near-future developments in robotics are taking place in many areas, including hardware, software, and applications. The field is in great ferment, with new systems appearing frequently throughout the world. Among the areas in which the great innovations are taking place are:

• Human–robot interaction, in the factory, home, hospital, and many other venues where social interaction by robots is possible
• Display and recognition of emotions by robots
• Humanoid robots equipped with controllable arms as well as legs
• Multiple robot systems
• Autonomous systems, including automobiles, aircraft, and underwater vehicles

In this chapter, we concentrate on the areas where the ethical implications are the clearest and most immediate. This is not to say that other areas do not have ethical implications. Indeed, we believe that as robots become more visible and involved in more areas of society, new areas of ethical concern will emerge. We begin with changes in manufacturing robots, since that is where the field began and where some of the most dramatic changes in human–robot interaction are taking place. Then, we look at robots in healthcare and rehabilitation, socially interactive robots (especially humanoids) that share or will share our homes and social gatherings, and military robots, both present and future. In all these application areas, robots are or will be interacting with humans in many ways.

2.3 Industrial/Manufacturing Robots: Robots as Coworkers

With the exception of scattered developments in some university laboratories, robotics really began in the manufacturing sector with the introduction of the Unimate robot (Engelberger 1980). Since then, millions of robots have been sold. The International Federation of Robotics estimated that there were 1.3 million active manufacturing robots in the world in 2008 (International Federation of Robotics 2010).

In the early years of robotized manufacturing, the ethical issue was dramatized by the death of a worker at a Ford manufacturing plant in Flint, Michigan, on January 25, 1979. The worker was struck in the head by a robot arm that was retrieving parts in a warehouse. In 1981, a robot killed a Japanese worker while he performed maintenance (The Economist 2006). Following these two deaths, manufacturing plants began to install safety barriers around areas where large, heavy, and potentially dangerous robot arms were used. Even so, in 1984, a worker was killed after he climbed over the safety fence without disabling the robot. Clearly, employing workers in factories where robots are their coworkers includes the ethical responsibility to ensure their safety and well-being; however, no safety barrier can protect against human stupidity.

Barriers have largely solved the problem of potential physical harm caused by robots in manufacturing. However, their use has led to a number of other ethical concerns, particularly in situations where robots work in proximity to humans. These concerns are addressed in the following sections.

2.3.1 The Fear of Being Replaced by a Machine

Introduction of robots into factories, while employment of human workers is being reduced, creates worry and fear. It is the responsibility of management to prevent or, at least, to alleviate these fears. For example, robots could be introduced only in new plants rather than replacing humans in existing assembly lines. Workers should be included in the planning for new factories or the introduction of robots into existing plants, so they can participate in the process. It may be that robots are needed to reduce manufacturing costs so that the company remains competitive, but planning for such cost reductions should be done jointly by labor and management. Retraining current employees for new positions within the company will also greatly reduce their fear of being laid off. Since robots are particularly good at highly repetitive simple motions, the replaced human workers should be moved to positions where judgment and decisions beyond the abilities of robots are required.

2.3.2 The Dehumanization of Work

In principle, it should be possible to design manufacturing systems in which repetitive, dull, and dangerous tasks are performed by robots, while tasks requiring judgment and problem-solving ability remain with human workers. Yet, in the process of developing increasingly automated factories, human workers may begin to feel inferior to the robots. Further, they may begin to believe that management intends to reduce all work to repetitive motions, which can (at least in principle) be carried out entirely by robots. Such a set of beliefs can lead to increasing unhappiness, and even destructive actions, on the part of the human workers toward the robots. Such concerns led to the attempts by workers in England in the nineteenth century to destroy mechanized cotton looms.[1] Management has an ethical responsibility to allow humans to work in tasks that do not demean them, but rather take advantage of their superior cognitive abilities.

2.3.3 Current Trends toward Cooperative Work

One of the most interesting current trends in robotics is the use of robots in tasks where they have shared responsibilities with humans. One of the first such systems was developed by Peshkin and Colgate at Northwestern University in the late 1990s (Peshkin and Colgate 1999). The cooperative robots were termed "cobots." Much of the theoretical work as well as practical applications of cobots was developed more recently (e.g., Gillespie, Colgate, and Peshkin 2001). Basically, cobots and humans may jointly grasp an object to be moved, but the motive power is provided entirely by the human; the cobot provides guidance, and may prevent motion in certain directions. Since the human produces the motive power, such systems effectively solve the potential danger to humans from robot motions.

In recent years, human–robot collaboration in the workplace has received increasing attention. New sensors make it possible to place robots and humans in close proximity to one another, while minimizing potential dangers. Thus, sensors can provide early warning when robots and humans appear to be moving into the same spaces. In addition, future manufacturing robots will have to recognize human gestures and movements and react accordingly, in order to reduce drastically any possible dangers to their human partners. Such cooperation also means that the robots can learn movements from humans by imitation. Ultimately, the goal of these efforts is to create increasing opportunities for shared and cooperative work that takes advantage of the specific features and advantages of both robots and humans.

It is evident that shared, cooperative work between humans and robots may enhance the working environment, but it may also reduce human–human interaction and communication. These are ethical problems that need to be addressed as factories become increasingly automated.

2.4 Human–Robot Interaction in Healthcare, Surgery, and Rehabilitation

Another area where robot–human interaction is developing rapidly is the field of healthcare, including nursing, surgery, physical therapy, and noncontact assistance during therapy and rehabilitation. These developments are becoming possible as the potential danger to humans from accidental robot activity decreases. This area of robotics is growing so rapidly that we can only indicate some typical applications.

Nursing care is typically a one-on-one relationship between a patient and a caregiver. Hence, it is an expensive part of healthcare, and a number of laboratories are developing robots we may term "nurse's assistants." One of the earliest of such robots was the wheeled HelpMate, currently marketed by a company named Pyxis. HelpMate assists nurses and other hospital personnel by smoothly transporting pharmaceuticals, laboratory specimens, equipment and supplies, meals, medical records, and radiology films back and forth between support departments, nursing floors, and patient rooms. The HelpMate is able to navigate hospital corridors, avoid collisions with humans, summon the elevator, and locate a specific patient's room. Carnegie Mellon University and the University of Pittsburgh have developed a "nurse-bot" named Pearl (Montemerlo et al. 2002) as an assistant that visits elderly patients in hospital rooms, provides information, reminds patients to take their medication, takes messages, and guides residents. Such robots are usually constructed as upright structures on wheels, with a somewhat human-appearing head containing cameras and voice-synthesizing software for communication. They usually also have a digital display, on the head or chest, to display messages. In Europe, there have been (and are) a number of projects in this area, such as the Care-O-Bot developed at the Fraunhofer Institute in Stuttgart, Germany (Fraunhofer 2010). The Care-O-Bot also has an arm to assist in pick-and-place operations. Similar projects exist in Japan, South Korea, and other countries.

A related set of projects involves "assistive robots," which provide verbal guidance, encouragement, and interaction to people recovering from strokes and spinal injuries, as well as companionship to children with autism-spectrum disorders. These robots do not make any physical contact with the subjects, but rather guide them through exercises and activities by voice and demonstration (Feil-Seifer and Matarić 2005). Figure 2.1 shows such a robot interacting with a subject.

Among the potential ethical concerns in the use of such assistive robots and nurse-bots are the following:

• Patients may become emotionally attached to the robots, so that any attempt to withdraw them may cause significant distress.

Figure 2.1
Assistive robot interacting with a physical therapy patient. Courtesy of Professor M. Matarić, University of Southern California.

• The robots will not be able to respond to patients' anger and frustration, except by calling for human help. For example, a patient may refuse to take the medication offered by the robot, throw it on the floor, and even attempt to strike the robot.
• A robot may be called by more than one user and not have the ability to prioritize the requests, thus causing anger and frustration.

Robotics is also important in other aspects of rehabilitation. Artificial limbs and prosthetic joints are frequently "robots," since they employ sensors to obtain information on positions, velocities, and forces; computers to process the information; and motors to provide mobility to the affected joints. However, we do not discuss them in this chapter, since there do not seem to be new ethical problems arising from the use of robotic prosthetics, as compared with nonrobotic ones.

The term "robotic surgery" is used to describe cooperative human–robot activities in the surgical suite. The da Vinci surgical robot (Taylor et al. 1995) is currently being used in hundreds of hospitals; see figure 2.2. It is important to note that the da Vinci is actually a telerobot, since it is remotely controlled by a human surgeon and is not fully autonomous. The human surgeon sits at a remote console and uses two hand controllers to position an endoscope and surgical instruments. In fact, she is sending

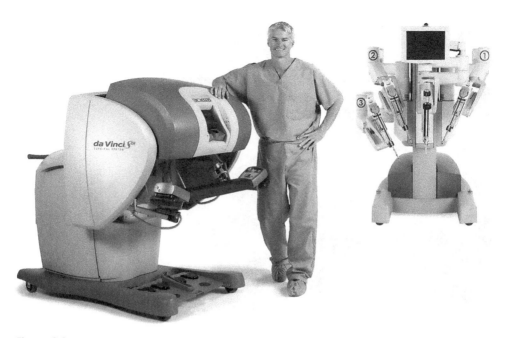

Figure 2.2
da Vinci robotic surgical system. Courtesy of Intuitive Surgical Systems.

instructions to the computer controlling the arms of a robot that performs the surgery. The surgical tools are equipped with sensors that provide feedback to the surgeon's hands, preventing excessive motions, filtering the surgeon's hand tremor, and providing velocity feedback to ensure smooth motions without oscillations. Thus, use of the da Vinci represents another example of cooperative human–robot work. We have discussed some of the ethical issues arising from the use of surgical robots in another publication (Bekey, Lin, and Abney 2011). Here, we consider potential future scenarios, when surgical robots become more autonomous and become true partners with human surgeons, rather than being simply remotely operated systems.

Consider a hypothetical scenario involving robotic surgery. A robot surgeon performs an operation on a patient; a number of complications arise and the patient's condition is worse than before. Who is responsible? Is it the designer of the robot, the manufacturer, the human surgeon who recommended the use of the robot, the hospital, the insurer, or some other entity? If there was a known chance that the surgery might result in problems, was it ethical for the human surgeon or the hospital, or both, to recommend or approve the use of a robot? How large a chance of harm would make it unethical—or, to phrase it differently, how small a chance of harm would be morally permissible? That is, what is the acceptable risk?

Truly autonomous procedures on the part of a robot surgeon will require a number of safety measures to ensure that patients are not harmed. More than that, robotic surgeons will require levels of precision comparable to that of human surgeons. They may have to learn their surgical skills from a combination of programming (probably using artificial intelligence tools) and imitation of human surgeons. The ethical issues are clear: the risks of using a robot surgeon, either alone or in partnership with a human surgeon, must be lower than those encountered with human surgeons. Further, the cost of using a robot surgeon may need to be lower than that of using a human surgeon. However, if this becomes the case, and there is increased use of robots in the surgical suite, we may see the rise of "Luddite" surgeons in our hospitals. Insurance issues will probably play a major role in any decisions on deployment of autonomous or semi-autonomous surgical robots.

2.5 Robots as Co-inhabitants; Humanoid Robots

We expect that during the coming decade more and more robots will be present in our homes, assisting us in cleaning, housekeeping, child care, secretarial duties, and so on. This trend of "co-inhabitant" robots[2] began with the Roomba vacuum cleaning robot, introduced by the company iRobot in 2002. Since then, more than four million of these robots have been sold worldwide, so many that we may classify the Roomba robot as a commodity, rather than a luxury. However, the Roomba does not interact with people in any significant way. Such interactions are restricted largely to robots that have some human-like attributes, in other words, *humanoid* robots. Vacuum cleaners and lawn mowers may be robotic, but they are not currently humanoids.

Humanoid robots resemble human beings in some aspects. They may have two legs or no legs at all (and move on wheels); they may have one or more arms or even none; they may have a human-like head, equipped with the senses of vision and audition; and they may have the ability to speak and recognize speech.

It is interesting to note that humanoid robots do not need to appear completely human-like in order to be trusted by people. It is well known that humans are able to interact with dolls, statues, and toys that only have a minimal resemblance to human beings. In fact, the ability of humans to relate to humanoids becomes worse as they approach human-like appearance, until the resemblance is truly excellent. This somewhat paradoxical result has been called the "Uncanny Valley" by Mori (1970), see figure 2.3. This figure shows that our emotional response to robots increases as they resemble humans more and more, until they reach a point at which their resemblance is close to perfect but eerily dissimilar enough such that we no longer trust them—that sudden shift in our affinity is represented by the dip or valley on the curve. But the trust returns as the anthropomorphism approaches 100 percent (or perfect resemblance) to human appearances.

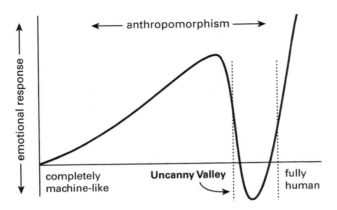

Figure 2.3
The "Uncanny Valley." Courtesy of GNU Free Documentation License.

The ability of these robots to share living spaces without danger to the human occupants depends on a number of technological improvements, including new and better sensors (which enable the robots to be fully aware of their surroundings), the ability to communicate with humans by voice, as well as gestures, controlled actuators to prevent rapid movements and possible injury, and much improved software, including the ability to interact socially with people. These are tall orders, but robots capable of meeting many of these requirements are beginning to appear. Two examples are Wakamaru from Mitsubishi in Japan (Mitsubishi Heavy Industries Ltd. 2010) and Nao from Aldebaran Robotics in France (Aldebaran Robotics 2010); see figure 2.4.

Wakamaru was designed to co-inhabit living spaces with humans, being termed a "companion robot." It is about four feet, or 120 centimeters, in height, has a head with large eyes, and movable arms but no legs; it moves on wheels, thus being restricted to relatively flat locations. It recognizes some ten thousand words, can place telephone calls, and communicates by Internet. It carries a camera; if contacted via Internet, it displays the camera image to the caller, and it recognizes ten faces and can be programmed to react appropriately to each. It can read the owner's email and scan the news, passing the information on by voice. It communicates both by speech and gestures. Wakamaru costs about US$14,000 or over €10,000.

Nao is a robot currently in use in a number of university research laboratories. It is only fifty-eight centimeters or about two feet in height, half as tall as Wakamaru. It walks, maintaining stability by means of an inertial measurement unit (or a system for continuous calculation of the position, velocity, and orientation of a moving object using accelerometers and gyroscopes) and ultrasonic sensors. Its hands are capable of grasping objects. It has two cameras. It is capable of omnidirectional hearing by means of four microphones; it uses two speakers. Like Wakamaru, it is able to access the

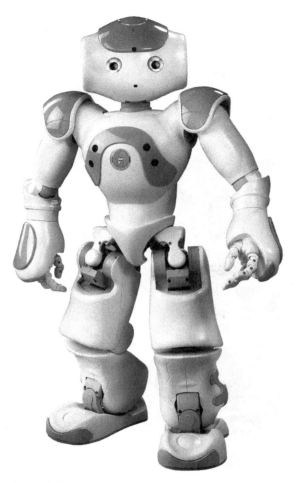

Figure 2.4
NAO, the humanoid robot by Aldebaran Robotics. Courtesy of Aldebaran Robotics.

Internet; its processor uses a Linux operating system. It recognizes and is able to imitate a number of gestures and arm positions. Nao is sold in Europe for approximately €10,000 as well.

As a final example of a humanoid robot in a home situation, consider figure 2.5, which shows ARMAR, a robot being developed in Germany to provide assistance in the kitchen.

Clearly, these robots are sophisticated humanoids, capable of a variety of interactions with humans. While much of the technology for co-inhabiting robots is at hand, the risks and ethical issues have yet to be addressed. These include:

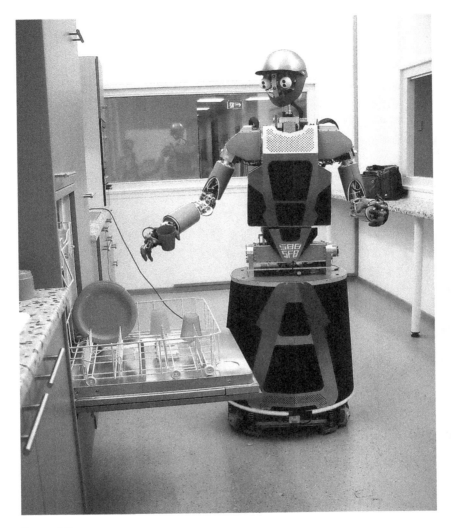

Figure 2.5
ARMAR-3, a kitchen-assistive robot prototype, loading dishes in a dishwasher.
Courtesy of Prof. R. Dillman, University of Karlsruhe.

• Loss of privacy for the human inhabitants if the robots are permitted free access to all rooms in a home.

• Ability of the robots to recognize commands that may lead to unethical behaviors (e.g., to steal a neighbor's camera or cell phone).

• Rights and responsibilities of the robots, e.g., should they be treated with respect as if they were human?

• Emotional relationships, e.g., how should a robot relate to human anger, say, when the robot drops a dish of food on the floor? In other words, is it ethical to yell at a robot? Can and should robots be punished for misbehavior, and, if so, how?

• How should a robot react to multiple instructions from different humans, e.g., when a child calls for it to come and play while the mother calls for it to come and wash the dishes?

• Can the robot's computer be accessed by hackers, so it may stake out and send pictures from the home to potential burglars?

Evidently, we have no answers to these and other similar questions at the present time.

2.6 Socially Interactive Robots

The robots we have discussed above are "socially interactive," in the sense that human–robot interaction is an essential component of their behavior. The phrase covers a wider range of robots, as presented in a major survey paper (Fong, Nourbakhsh, and Dautenhahn 2003). A broader view needs to include multiple robot systems (where robots may cooperate with each other), and even robot swarms. Such robots may need to recognize each other and possibly engage in mutual interactions, including learning from each other. While a great deal of research in these areas is currently proceeding, we cannot discuss it here, due to space limitations. However, it is evident that such mutual relationships will eventually involve ethical considerations. For example, is it ethical for one robot to damage or destroy another member of its group? If not, how can we ensure that such behaviors do not occur?

Ultimately, as social robotics develops, we expect that individual robots may develop distinctive personalities and communicate with each other, perhaps in new high-level languages. We have begun to study one aspect of social behavior by considering robot societies in which an altruistic robot may assist another in the completion of its task, even if its own performance suffers as a result (Clark, Morton, and Bekey 2009).

One further topic needs mention in connection with socially interactive robots, and that is the question of *robot emotions*. This is a subject of intensive research in a number of robotics laboratories. There are extensive discussions on the nature of "artificial emotions" as displayed by robots. Human–robot interaction should benefit if both humans and robots are capable of expressing anger, happiness, boredom, and other emotional states. For example, emotions may be expressed by a synthetic face on a digital monitor or by a three-dimensional head. Both face and head may have movable eyebrows, mouths that can be shaped, eyes that open or close, and so on. The early Kismet robot head at MIT (Breazeal 2002) was only faintly human but had a number of adjustable features. While we may argue that the robot's expressed emotion is not "real," the human reaction to it may be significant (Ogata and Sugano 2000). Humans have a tendency to anthropomorphize robots, and any display of emotions (real or artificial) by the robot could lead to unacceptable (or unethical) behaviors by humans in response.

2.7 Military Robots

The use of robots in the military services has been the subject of a number of books (e.g., Singer 2009) and reports (e.g., Lin, Bekey, and Abney 2008), as well as chapters in edited books (e.g., Lin, Bekey and Abney 2009; Sharkey, chapter 7, this volume). In view of these publications, we will not discuss autonomous or semi-autonomous unmanned flying vehicles (UFVs), unmanned ground vehicles (UGVs) or unmanned underwater vehicles (UUVs) in this chapter. When robots are used to detect and neutralize improvised explosive devices (IEDs) and mines, they are clearly protecting the lives of soldiers and sailors; see figure 2.6. Thousands of such robots are in use at the present time in Iraq and Afghanistan. There are also civilian applications for protective robots, such as security in the home, government facilities, or commercial installations, and perimeter inspection of industrial plants. Police departments may use robots to enter a building where it may be dangerous for human officers.

To illustrate the ethical dilemmas arising with military robots, consider the two following (future) scenarios:

1. Intelligence information indicates that a house located at given GPS coordinates is the headquarters for dangerous enemy combatants. A military robot is commanded by an officer to approach the house and destroy it, in order to kill all the people within it. As the robot approaches the house, it detects (from a combination of several sensors, including vision, x-ray, audition, olfaction, etc.) that there are numerous (noncombatant) children within, in addition to the combatants. The robot has been programmed in compliance with the laws of war and the typical rules of engagement to avoid or to at least minimize noncombatant casualties (sometimes referred to as "collateral damage") (e.g., Arkin 2009; Lin, Bekey, and Abney 2008). When facing contradictory instructions, the robot may attempt to solve its dilemma by transferring authority back to the officer in charge, but this may not be feasible or practical, since it may

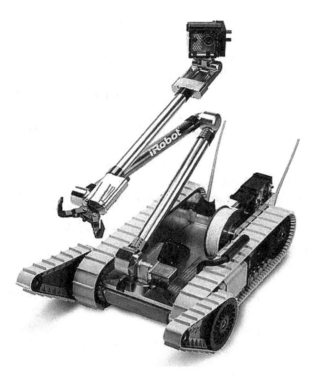

Figure 2.6
Packbot military robot. Courtesy of iRobot Inc.

risk discovery of the robot by the enemy and harm to our own forces. Typically, when faced with such contradictory instructions, the on-board computer may "freeze" and lock up.

2. Another possible scenario involves a high-performance robot aircraft, suddenly subject to attack by unknown and unrecognized piloted airplanes. The robot can only defend its own existence by causing harm to human beings. A decision to transfer control to a human commander would need to be made in milliseconds, and no human could respond rapidly enough. What should the drone do?

Thus, the use of military robots raises numerous ethical questions. Arkin (2009) has attempted to solve such problems by developing a control architecture for military robots, to be embedded within the robot's control software. Such software, in principle, could ensure that the robot obeys the rules of engagement and the laws of war. But would the mere requirement to adhere to these rules actually ensure that the robot behaves ethically in all situations? The answer to this question is clearly in the negative, but Arkin only claims that such robots would behave "more ethically" than human soldiers. In fact, sadly, to behave more morally than human soldiers may not require a great advance in robot ethics. It is evident from the preceding discussion that there are numerous unresolved ethical questions in the deployment of military robots. Among these questions are the following:

• If a robot enters a structure, how can we ensure that it will not violate the rights of human occupants?
• Do the entering robots have rights? Is damage to or destruction of a sentry or inspector robot a crime?
• If a robot destroys property in the process of protecting people or attempting to arrest criminals, who is responsible for repairing the damage?
• Will the use of increasingly autonomous military robots lower the barriers for entering into a war, since it would decrease casualties on our side?
• How long will it be before military robotic technology will become available to other nations, and what effect will such proliferation have?
• Are the laws of war and rules of engagement too vague and imprecise (or too difficult to program) to provide a basis for an ethical use of robots in warfare?
• Is the technology for military robots sufficiently well developed to ensure that they can distinguish between military personnel and noncombatants?
• Are there fail-safes against unintended use? For instance, can we be certain that enemy "hackers" will not assume control of our robots and turn them against us?

Thoughtful discussions of these issues have been published by Arkin (2009), Asaro (2008), Sharkey (2008), Sparrow (2007), Weber (2009), and others. We have addressed some of them in a major report (Lin, Bekey, and Abney 2008).

2.8 Conclusion

In this chapter, we have surveyed some of the major trends, current at the time of this writing, in the robotics field and indicated some of the ethical implications of these changes. While the field is advancing rapidly, what has not changed much is the general lack of attention on ethical issues on the part of the robotics community. The present book is a small step in the direction of increasing awareness of these issues among designers and users of robots.

Notes

1. The movement was led by a fictitious "King Ludd"; people who oppose mechanization and automation are sometimes referred to as "Luddites."

2. The term "co-inhabitant" was coined by Professor Ken Goldberg at the University of California, Berkeley.

References

Aldebaran Robotics. 2010. Nao, the ideal partner for research and robotics classrooms. <http://www.aldebaran-robotics.com/en> (accessed November 18, 2010).

Arkin, Ronald C. 2009. *Governing Lethal Behavior in Autonomous Robots*. Boca Raton, FL: Chapman & Hall.

Asaro, Peter. 2008. How just could a robot war be? In *Current Issues in Computing and Philosophy*, ed. Adam Briggle, Katinka Waelbers, and Philip Brey, 50–64. Amsterdam, The Netherlands: IOS Press.

Bekey, G. A. 2005. *Autonomous Robots: From Biological Inspiration to Implementation and Control*. Cambridge, MA: MIT Press.

Bekey, G. A., et al. 2008. *Robotics: State of the Art and Future Challenges*. London: Imperial College Press.

Bekey, G. A., P. Lin, and K. Abney. 2011. Ethical implications of intelligent robots. In *Neuromorphic and Brain-Based Robots: Trends and Perspectives*, ed. J. L. Krichmar and H. Wagatsuma, 666–726. Cambridge, UK: Cambridge University Press.

Breazeal, C. 2002. *Designing Sociable Robots*. Cambridge, MA: MIT Press.

Clark, C. M., R. Morton, and G. A. Bekey. 2009. Altruistic relationships for optimizing task fulfillment in robot communities. In *Distributed Autonomous Robotic Systems 8 (DARS08)*, ed. Hajime Asama, Haruhisa Kurokawa, Jun Ota, and Kosuke Sekiyama, 261–270. Berlin: Springer-Verlag.

Computer Community Consortium (CCC). 2009. CRA roadmapping for robotics. <http://www.us-robotics.us> (accessed November 18, 2010).

Economist. 2006. Trust me, I'm a robot. *The Economist*, no. 3–4, June 8. <http://www.economist.com/note/7001829> (accessed November 18, 2010).

Engelberger, J. F. 1980. *Robotics in Practice*. New York: American Management Association.

Feil-Seifer, D., and M. Matarić. 2005. Defining socially assistive robotics. *Proceedings of the IEEE International Conference on Rehabilitation Robotics* (ICORR '05), Chicago, IL.

Fong, T., I. Nourbakhsh, and K. Dautenhahn. 2003. A survey of socially interactive robots. *Robotics and Autonomous Systems* 42 (3–4): 143–166.

Fraunhofer, I. P. A. 2010. Care-O-Bot home page. <http://www.care-o-bot-research.org/> (accessed November 18, 2010).

Gillespie, R. B., J. E. Colgate, and M. A. Peshkin. 2001. A general framework for cobot control. *IEEE Transactions on Robotics and Automation* 17 (4): 391–398.

International Federation of Robotics. 2010. *World Robotics 2010*. IFR, Frankfurt. <http://www.worldrobotics.org> (accessed November 18, 2010).

Lin, P., G. A. Bekey, and K. Abney. 2008. *Autonomous Military Robotics: Risk, Ethics and Design.* Office of Naval Research–funded report. San Luis Obispo: California Polytechnic State University. <http://ethics.calpoly.edu/ONR_report.pdf> (accessed November 18, 2010).

Lin, P., G. A. Bekey, and K. Abney. 2009. Robots in war: Issues of risk and ethics. In *Ethics and Robotics*, ed. Rafael Capurro and Michael Nagenborg, 49–67. Amsterdam: IOS Press; Heidelberg: AKA Verlag.

Mitsubishi Heavy Industries Ltd. 2010. Communication robot Wakamaru. <http://www.mhi .co.jp/en/products/detail/wakamaru.html> (accessed November 18, 2010).

Montemerlo, M., J. Pineau, N. Roy, S. Thrun, and V. Verma. 2002. Experiences with a mobile robotic guide for the elderly. *Proceedings of the AAAI National Conference on Artificial Intelligence*, Edmonton, Canada.

Mori, M., 1970. Bukimi no tani: The uncanny valley, trans. K. F. MacDorman and T. Minato. *Energy* 7 (4): 33–35.

Ogata, T., and S. Sugano. 2000. Emotional communication between humans and the autonomous robot WAMOEBA-2 (Waseda Amoeba) which has the emotion model. *JSME International Journal. Series C, Mechanical Systems, Machine Elements and Manufacturing* 43 (3): 568–574.

Peshkin, M. A., and J. E. Colgate. 1999. Cobots. *Industrial Robot* 26 (5): 335–341.

Sharkey, Noel. 2008. Cassandra or false prophet of doom: AI robots and war. *IEEE Intelligent Systems* 23 (4) (July/August): 14–17.

Singer, P. 2009. *Wired for War*. New York: Penguin Press.

Sparrow, Robert. 2007. Killer robots. *Journal of Applied Philosophy* 24 (1): 62–77.

Taylor, R. H., S. Lavallée, G. S. Burdea, and R. Mösges, eds. 1995. *Computer Assisted Surgery.* Cambridge, MA: MIT Press.

Veruggio, G. 2006. The EURON Roboethics Roadmap. *Proceedings of the 6th IEEE Conference on Humanoid Robots* (Humanoids '06), Paris, France.

Veruggio, G., ed. 2009. The Robotics website. <http://www.roboethics.org> (accessed November 18, 2010).

Weber, Jutta. 2009. Robotic warfare, human rights and the rhetorics of ethical machines. In *Ethics and Robotics*, ed. Rafael Capurro and Michael Nagenborg, 83–104. Amsterdam: IOS Press; Heidelberg: AKA Verlag.

3 Robotics, Ethical Theory, and Metaethics: A Guide for the Perplexed

Keith Abney

What is robot ethics? The term may cause perplexity; according to some ethical views, it seems to be a field of study without an object to study (as some gibe at astrobiology or theology). In the emerging literature devoted to robot ethics, however, the term has at least three distinct meanings, the first two of which clearly refer to something real. First, it can refer to the professional ethics of roboticists (often termed "roboethics" [Veruggio 2007]); second, it can refer to a moral code programmed into the robots themselves—the moral code the robots, not the roboticists, follow; and third (the possibly nonexistent meaning), "robot ethics" could refer to the self-conscious ability to do ethical reasoning by robots—to a robot's own, self-chosen moral code. The epilogue of this volume describes these three senses in more detail.

The term "ethics" also needs disambiguation. "Ethics" is sometimes used synonymously with "morality," but sometimes refers to "the study of morality." Some robot ethicists, like Rafael Capurro (2009), prefer to distinguish these by calling the second sense previously noted (a programmed-in robotic moral code) a robot *morality*, whereas only the third sense of a self-conscious, voluntary adoption of a particular code would be called robot *ethics*. Others use "robot ethics" or "machine morality" in both the second and third senses, or even for issues in the discipline philosophers call "metaethics," and so may leave unclear the meaning of terms like "artificial moral agents" and "machine ethics." Accordingly, this essay aims to examine some common confusions, misunderstandings, equivocations, and other problems in understanding these three senses of robot ethics, and to introduce the ethical and metaethical issues concerning robots discussed later in this volume.

3.1 Four Questions

I begin with four crucial, but often misunderstood, questions (or sets of related questions) for doing ethics, all of them relevant to robotics:

1. What is morality or ethics: the *right*, or the *good*?
2. What are moral rights? What is their relationship to moral duties? And who or what can be rights holders?
3. What are the major contemporary moral theories? How do they bear on robot ethics?
4. What is a person, in the moral sense? Can a robot be a person?

3.1.1   What Is Morality or Ethics: The Right, or the Good?

So what is morality? Morality always involves an "ought (not)"—it is about the way the world ought (or ought not) to be, as opposed to the way it actually is. The "ought" of morality has been understood in two primary ways: as doing the right, or as being good—that is, the content of morality is understood either as what rules make for right action, or as how one ought to live in order to have a good life. These two approaches are practically equivalent, if living a good life means following some set (the right set) of rules; if not, there is a potential chasm between these two conceptions of morality.

Top-down, rule-based approaches, like Asimov's Three Laws of Robotics (Asimov [1942] 1968), understand ethics as the investigation of right action—what are the rules to follow in order to be morally right, to perform the morally correct (or at least morally permissible) action? The analogy with the legal system is instructive: if one obeys the rules, one is moral; if one disobeys or breaks the rules, one acts immorally. The investigations of ethics are fundamentally, then, an inquiry into what the rules ought to be, for any particular society. Robot ethics, then, concerns following (in senses one and three) or programming (sense two) the correct set of rules.

The usual divide within rule-based approaches is between those who say one must intend to obey the rules, no matter what—even if the consequences will be bad (deontologists, associated with Kant), versus those who say the main or only rule is always to make the future consequences as good as possible—*the ends justify the means* (consequentialists, most commonly represented by utilitarians, who tend to measure the ends or results in terms of happiness gained or lost).

There is another historically influential approach that understands ethics as the art/ science of living a good life, not as being bound by some set of rules that may not apply to one's unique circumstances. For programming robots, this view represents a "bottom-up" or "hybrid" approach that involves trial-and-error learning of what constitutes (un)acceptable behavior—or a "good" or "bad" robot—that goes beyond mere obedience to a set of rules.

One justification for this understanding of morality as the good, not the right, is the observation that all rule-based approaches have assumed: (a) the rule(s) would amount to a decision procedure for determining what the right action was in any particular case; and (b) the rule(s) would be stated in such terms that any nonvirtuous person could understand and apply it (them) correctly (Hursthouse 2009). But despite centuries of work by moral philosophers, no (plausible) such set of rules has been found. Moral particularism (Dancy 2004) is one, perhaps unhelpful, purported solution to this quandary: there are no moral rules, only moral facts, and acts can only be judged according to the unique particulars of each case. But if each moral situation is sui generis, how then could we ever program robots to be moral?

A more helpful approach for robots is virtue ethics, which asserts the problem with rule-based morality is that it has the wrong object of evaluation. Morality is asserted to be about the character of persons, not the rightness or wrongness of individual acts. Top-down moral theories are concerned with action, and attempt to answer the question, "What should I *do*?" with some set of rules. Virtue ethics, by contrast, attempts to answer the question, "What should I *be*?" Virtue ethics consists not in following moral rules that stipulate right actions, but in striving to be a particular kind of person (or robot)—a virtuous one.

As such, virtue ethicists usually deny that mere actions are meaningfully good or evil—it may be morally wrong (betray a defective character, a "vice") for me to begin to carve your chest with a knife, but someone else performing exactly the same action in the same circumstances may be perfectly moral ("virtuous")—if you are lying on an operating table, and she is a surgeon, whereas I am not! She evinces a perfectly virtuous character in cutting you open because of her skills and her role in the situation; because my skills and my role are different (because I am not a licensed surgeon), performing the same act would reveal my character is flawed, even if my intentions were good; indeed, even if (miraculously) the surgery turned out well—that is, even if the consequences of my act were good. For robots, this same proper functioning approach to evaluation appears natural: is the surgical robot operating properly in carving one's chest, or is my new robotic bandsaw dysfunctionally attempting to do the same thing?

Virtue ethicists thus claim what counts is one's moral character—moral evaluation is of persons, not of actions. The virtues are understood as dispositions to act in a certain way (would-be habits); ideally, to know by practical wisdom the right thing to do, in the right way, at the right time. Context sensitivity means virtues do not act as categorical imperatives and may conflict; in a difficult situation, one should not ask what abstract rule to follow but instead ask: What would a role model do in my situation? Or—if I do X, would it start a bad habit? Will I become dysfunctional in my proper role(s)?

The implications of this divide for robot ethics (in all three senses) are potentially profound. For the first sense, is roboethics simply the search for a list of rules that any and all roboticists must follow in their work, such that all who adhere to the rules are automatically moral, and those who break them automatically immoral? Or is it perhaps the search for the rules that will produce the best future net consequences for society (rule-utilitarianism)?

Or, following the second approach, should roboethics search instead for distinctive principles that roboticists of good character evince in their work (i.e., virtues of doing robotics), as well as character traits that lead to dysfunction in their work (i.e., vices of doing robotics)? For a roboticist, a claim that *"I'm not responsible because I followed the rules"* would be indefensible from a virtue-ethics perspective. Instead, one should emulate a role model of professionalism. One example would be "The Roboticist's Oath" (McCauley 2007), understood as a statement of principles that any professional roboticist should evince. Bill Joy also asserted the need for such an oath as a means of setting up a professional exemplar and standards; he wrote, "scientists and engineers [need to] adopt a strong code of ethical conduct, resembling the Hippocratic oath" (Joy 2000). Further, if robots themselves are proper objects of moral assessment, then robot virtue ethics would become the search for the virtues a good (properly functioning) robot would evince, given its appropriate roles.

So, is ethics the study of the right, or the good? Despite the preceding arguments for ethics as the study of the good, the case for the rule-based approach has practical import in another social tendency: to equate moral and legal, immoral and illegal—that is, to construe any action that avoids legal sanction as morally permissible, and to insist on redress (in the form of legal rights) when such laws have been broken by others, or to insist such actions were permissible when others wish to cast moral blame, by saying "but I had a right!"

The relationship between virtues and rights begins with an observation: when all parties in a given social context are acting virtuously, no one mentions their rights; in fact, such appeals would appear unseemly when no vices exist. Rights claims inevitably arise *only* when something has gone amiss. That is, appeals to rights inevitably occur only when moral conflict already exists, and rights-based approaches based on rules/laws are always an attempt to fix something that is already broken—or to prevent it from getting worse. And rules invariably have unintended consequences, as the attitude that "whatever is within the rules is permissible" leads to the unscrupulous finding malicious means to bend the rules to their advantage, without (quite) breaking them. So, in a moral utopia, there would be no need for moral rights. And many moral theorists, running the gamut from utilitarians, like Bentham, to virtue ethicists, like MacIntyre, to various existentialists, have denied their existence.

But despite such views, rights claims may be a necessary feature of the ethics of any large, complex society. When groups are relatively small, with common social mores reinforced by shared moral education and acceptance of one's proper roles, the virtues may be largely taken for granted and enforced by purely social sanctions—as the opprobrium of those with whom one has substantial relationships is a powerful tool for enforcing social moral consensus. Our behavior is usually far more affected by the (dis)approval of those around us than by an abstract, remote threat of law enforcement, in "ordinary" contexts.

For roboethics, moral education (in the virtues of the profession) and other social means of enforcing shared mores (such as causing a bad reputation, or denying conference participation, publication, grants, tenure, or even employment for those who violate shared virtues) may be effective, at least for a while. But as the group of those dealing with robots becomes larger and more variegated, social sanctions and shared virtues gradually become less effective at minimizing harm.

At such a point, outside regulation and institutions, with clear procedures, rights, and duties, usually become necessary in order to keep the smaller group's practices acceptable within the larger society. So, although rights claims may be a "second-best" form of morality, appealed to only when immorality is already rampant or at least expected; nonetheless, in the real world, in which vices are all too common, they may remain a necessary evil. Accordingly, I next attempt to clarify the concept of a moral right, whether for humans or for robots.

3.1.2 What Are Moral Rights? What Is Their Relationship to Moral Duties? And Who or What Can Be Rights Holders?

There are two main competing theories of rights—the "will" theory and the "interest" (or "welfare") theory (Wenar 2010). The interest theory maintains that rights correlate with interests (or welfare)—everything that has interests (or a "welfare") has rights. All persons have a duty to respect the rights of everything that has interests (including, potentially, robots?). But the will theory of rights disagrees: it asserts the right to liberty is the foundation of all other rights claims, and a rights claim is understood as the entitlement to a particular kind of choice—a rights claim entitles me to claim or perform something, or not—*it is up to me* (and nobody else). A rights claim entails no duty upon the rights holder, but only a freedom—to perform/claim something, or not. But the correlativity thesis makes clear that rights claims do entail duties, not for the rights holder, but for all other persons—if I have a right, then you have (and everyone else has) a correlative duty.

The correlativity thesis is essential to rights theory, in conceptualizing the relationship between rights and duties. It has a slogan form: "no rights without responsibilities"—rights do not exist unless others have duties. Rights are guaranteed freedoms, which then guarantee duties for everyone else.

But this has an additional implication, relevant here—who is "everyone else?" In this context, it refers to moral agents, beings capable of moral responsibility. It makes no sense to claim that trees or dogs or the environment have a moral responsibility to respect my freedom of speech; given that "ought implies can," they are incapable of it. If a tree falls on my head and silences me, we cannot hold it morally responsible! So, "no rights without responsibilities" carries an additional implication: on the will theory, only morally responsible agents can have moral rights. If I am incapable of agency, of the exercise of liberty, of rational free will, then I am incapable of being a rights holder. If there were no moral agents, there would be no moral rights—because there are no rights without responsibilities.

But then, on the will theory, anyone and anything incapable of being held responsible for their (its) actions would thereby have no moral rights. This would explain why current robots have no rights, but its implications cause unease for many, not least because much reasoning in applied ethics takes the following form: first, assess all the rights claims in a situation; if no rights have been violated, then an action is morally permissible. So if moral agents are the only rights holders, then based on such reasoning, agents appear morally free to act however they wish toward nonagents—so torturing pets or destroying robots is ok?

Such reasoning usually commits the fallacy of assuming a statement and its converse are equivalent—in particular, the correlativity thesis and its converse. And it mistakes the true nature of the relationship between rights and duties. The correlativity thesis: if I have a right, then all other agents have a correlative duty. The converse correlativity thesis: if I have a duty, then someone else has a correlative right. Upon a moment's reflection, the latter is absurd. Suppose I have a moral duty to give some of my disposable income to charity; which charity thereby has a right to my donation? The correct answer is: none. Some charity will receive my donation, but none of them are entitled to it—no one has a right to my charity, although I have a duty to give it.

Despite the prominence of rights claims in much applied ethics, the failure of the converse correlativity thesis means that we all have duties that correspond to no rights at all; and the impulse that supported the interest theory of rights disappears. Many nonagents (such as animals or the environment) have no rights, because they are not moral agents. But they plausibly are *moral patients*, to whom we agents owe duties; this possibility becomes clear once we realize we have many duties that correlate to no specific right. We merely equivocate when we call those duties "rights," as the interest theory does. Hence, we can safely say that, for the foreseeable future, robots will have no rights—at least until robot ethics approaches the third sense set forth, of robots as fully autonomous moral agents. But that realization leaves unresolved our moral duties concerning senses one and two—how roboticists ought to behave, and what moral code roboticists should install in their creations.

So, in robot ethics, we should not reason that if no rights have been violated, then an action is automatically morally permissible—because every moral duty cannot correspond to a discrete, identifiable right. We need a more encompassing moral approach than mere rights theory in order to fully discuss our moral duties in at least senses one and two of robot ethics. What other ethical theories are widely considered plausible candidates to specify our duties?

3.1.3   What Are the Major Contemporary Moral Theories? How Do They Bear on Robot Ethics?

We already discussed virtue ethics in section 3.1.1 as one major moral theory based on the good. Let us now turn to two more influential top-down rule-based approaches that can be applied to robot ethics: deontological and consequentialist theories.

Deontological (duty-based) approaches to robot ethics would simply see roboticists (sense one) or the robots themselves (sense two) acting in accord with some finite set of (presumably algorithmic, programmable) rules, and moral decision making would thus consist simply in computing the proper outcome of the (programmable) rules, in accordance with a monotonic first-order logic. There are concerns that such a basic logic could not capture ethical insights; however, work on deontic logics that would have programmable rules is well advanced (e.g., Arkin 2009; Bringsjord and Taylor, chapter 6, this volume). Hence, deontological approaches that see ethics as merely a set of (programmable) rules to follow are, in principle, a natural approach to creating sense two of an ethical robot, and making sure it conforms to any (programmable) set of ethical standards.

Asimov's Three Laws of Robotics (Asimov [1942] 1968) and Kant's Categorical Imperative (CI) are influential examples of such an approach in robot ethics; Kant's ([1785] 1998) theory has two primary formulations:

CI(1)—or the formula of universal law (FUL): "Act only in accordance with that maxim through which you can at the same time will that it become a universal law."

A maxim is a (true) statement of one's intent or rationale: why one did what was done. So, Kant asserts that the only intentions that are moral are those that could be universally held; partiality has no place in moral thought. Kant also asserts that when we treat other people as a mere means to our ends, such action must be immoral; after all, we ourselves don't wish to be treated that way. Hence, when applying the CI in any social interaction, Kant provides a second formulation as a purported corollary:

CI(2)—or the Means-Ends Principle: "So act that you use humanity, whether in your own person or in the person of any other, always at the same time as an end, never merely as a means."

One could never universalize the treatment of another as a mere means to some other ends, claims Kant, in his explanation that CI(2) directly follows from CI(1). This formulation is credited with introducing the idea of intrinsic human dignity and "respect" for persons; that is, respect for whatever collective attributes are required for human dignity, to be treated as ends in ourselves, and not as a mere tool by others. For Kant, all rational beings have intrinsic moral value, and the nonrational world has mere instrumental value—it, but not humans, can be treated as a mere tool.

A Kantian deontologist thus believes that acts such as stealing and lying are always immoral, because the intent to universalize them creates a paradox. For instance, one cannot universalize stealing property (taking that which is rightfully owned by another) without undermining the very concept of property. Kant's approach is widely influential, but has problems of applicability and disregard for consequences; for example, a robot that could never lie would certainly not be an asset if the enemy captured it.

Further, CI(1) is too permissive, and potentially permits horrors by allowing any action that can have a universalizable maxim; this can also cause a conflict with CI(2). For instance, CI(1) might sanction voluntary slavery or enforced servitude, a topic discussed by Petersen (chapter 18, this volume) for robots. Worse yet for programming deontological ethics into robots, using CI(1) could produce a *conflict of duties*—when two maxims both appear universalizable on their own, but come into conflict jointly.

Next, CI(2) is too stringent—interpreted literally, it forbids all war, or any other action in which I affect someone without their consent (and thereby treat them as a "mere means"). This would render most human–robot interaction, most especially military action, impossible. Not only do enemy civilians (as "collateral damage") not give consent to being harmed as a means to victory, there are also innumerable other human activities in which a minority who object are nonetheless treated as a means for the good of the majority—or do you consent to everything that the government does? In practice, this creates a *reductio ad absurdum* of this deontological constraint. To accomplish much of anything, a robot will sometimes have to engage in actions that affect humans without their explicit consent; the key is for it to make the correct decisions about how, when, and why that should be.

Finally, differences in *roles and capacities* problematize universalization—so a robot may be able to universalize "never shoot children" on a normal battlefield, but if insurgents become aware of this, child soldiers could wreak havoc as the robot stands passively by. Or, the laws of war deem it appropriate to target enemy soldiers with a gun pointed at you—but not if they are severely wounded and incapable of firing. Would a robot be able to discriminate the degree of wounding and retaliatory (in) capacity, and do the right thing?

Another deontological approach that has engendered much discussion in robot ethics is Asimov's Three Laws of Robotics (Asimov [1942] 1968), which are as follows: (1) a robot may not injure a human being or, through inaction, allow a human being to come to harm; (2) a robot must obey orders given to it by human beings, except where such orders would conflict with the first law; (3); a robot must protect its own existence as long as such protection does not conflict with the first or second law.

The laws are prioritized to minimize conflicts. Thus, doing no harm to humans takes precedence over obeying a human, and obeying trumps self-preservation.

However, in story after story, Asimov demonstrated that three simple, hierarchically arranged rules could lead to deadlocks when, for example, the robot received conflicting instructions from two people, or when protecting one person might cause harm to others. It became clear that the first law was incomplete, as stated, due to the problem of ignorance: a robot was fully capable of harming a human being as long as it did not know that its actions would result in (a risk of) harm, meaning that the harm was unintended. For example, a robot, in response to a request for water could serve water teeming with parasites, or drown a human in a pool, or crush someone with ice, ad infinitum, as long as the robot was unaware of the risk of harm.

One attempted solution is to rewrite the first and subsequent laws with an explicit "knowledge" qualifier: "A robot may do nothing that, to its knowledge, will harm a human being; nor, through inaction, knowingly allow a human being to come to harm" (Asimov 1957). But the cleverly immoral could divide a task among multiple robots, so that no one robot could know that its actions would lead to harm; suppose one disposal robot places nuclear medical waste in a package, another places a wire, another attaches the timer, and so on until the "dirty bomb" detonates. Of course, this simply illustrates the problem with deontological, top-down approaches: that one may follow the rules perfectly but still produce terrible consequences.

An additional difficulty is determining the degree of acceptable risk. The "through inaction" clause of Asimov's first law apparently implies a robot would have to constantly intervene to minimize all sorts of risks to humans, possibly rendering it incapable of performing its primary mission. A modified first law attempts a fix: (1') A robot may not harm a human being.

But removing the first law's "inaction" clause solves one problem only to create a greater one: a robot could initiate an action that would harm a human. For example, suppose a military robot initiates an automatic firing sequence and then watches a noncombatant wander into the firing line. The robot knows it is capable of preventing the harm (by ceasing the automatic firing), but it may nevertheless fail to do so, since it is now not strictly required to act.

And what if a robot's (in)action prevents immediate harm to one human, but thereby later imperils many? Should we not sacrifice a single human to save the entire world? To fix this problem, Asimov later added the Zeroth Law (1985—so named "zero" plus "th") to continue the pattern of lower-numbered laws superseding in importance the higher-numbered laws, so that the Zeroth Law had highest priority and must not be broken: (0) A robot may not harm all humanity or, through inaction, allow humanity to come to harm. This would allow a robot to harm individual humans, if so doing prevented an "existential threat" to all humanity. But how could a robot determine when such a threat exists (or how serious it is), so that harming individual humans to prevent the threat is permitted? Would the Zeroth Law permit robots to force human guinea pigs into medical experiments, to create a vaccine against a virus that *might* cause a pandemic? How strong is this version of the "precautionary principle?"

Such problems raise a central criticism of all deontological approaches—they fail to take the likely consequences into account. So, consequentialist ethics explicitly addresses this; utilitarianism—the primary consequentialist theory—proposes the goal of morality is to maximize utility, and utility is defined as the sum of the good consequences of an action, minus the sum of the bad consequences of the act. The work of Jeremy Bentham (1907) and J. S. Mill ([1861] 1998) stands as the locus classicus of utilitarianism; their view asserted a single rule of right action, the "Greatest Happiness Principle" (GHP): One ought always to act so as to maximize the greatest amount of net happiness (utility) for the largest number of people.

Like the deontologists, classical utilitarians emphasized *egalitarianism* (everyone's happiness counts equally), *impartiality* (I care no more for my happiness than for yours, in deciding what's right), and *universal scope*—so the moral rightness of an act depends on the consequences for all people (as opposed to only the individual agent, present people, or any other limited group).

However, this approach fails to be computationally tractable. So, the *calculational* objection arises: it is an impossible demand to calculate the utility of every alternative course of action; thus, utilitarianism makes moral evaluation impossible, as even the short-term consequences of most actions are impossible to accurately forecast and weigh, much less the long-term consequences. One response to this objection is *cost–benefit analysis*: translate good and bad consequences into economic value (benefits and costs), and then calculate which outcome maximizes expected profit/utility. Ethics becomes a branch of economics. But there are serious reasons to believe that moral values cannot systematically be reduced to economic values—for instance, the claim that the values of love, devotion, and honor do not have a price. The ethicist Mark Sagoff (1982) claims it betrays a fundamental moral confusion to conflate our *economic* values as consumers with our *moral* values as citizens—and the attempt to place a price on everything important is morally debilitating.

Can robots, with their potentially enormous computing power, solve this calculational problem? Unlikely—even if Sagoff is wrong. For robots, the calculational difficulties include how utility is represented within a computational system, how long-run the consequences are to be computed, how much data must be input, and scope—whose consequences (welfare) should be included in the calculation. Given limitations of available information and the sheer multitude of variables needed for any plausible decision making, such a calculation poses a tremendous computation load on even the fastest systems. A utilitarian robot may either fail to determine which course of action is most acceptable within the time allotted, or use grossly insufficient information in order to shoehorn its calculations into the time available. But if utility is (in practice) incalculable, and one's obligation is to maximize utility, what is left of utilitarianism?

Even if the calculational problem is solvable, there are other objections to utilitarianism: e.g., the *scapegoating* objection would point out that maximizing utility may demand injustice, such as executing an innocent person to prevent a riot that would have resulted in deaths and economic damage. This is to say that utilitarianism, at least in its basic form, cannot readily account for the notion of rights and duties or moral distinctions between, e.g., killing versus letting die, or intended versus merely foreseen deaths, or other harms (assuming we think such notions and distinctions exist).

Whether deontological or utilitarian, for robots there is an additional, fatal flaw in each of the top-down theories, connected to the calculational objection: they all suffer from a version of the *frame problem*—that is, knowing what information is (ir)relevant to moral decision making. In order to decide anything, does a robot have to know everything? How can a robot be sure to take into account all the information that is relevant to moral decisions (especially in novel situations), without being swamped by considering terabytes of irrelevancies?

The frame problem reinforces the worry that top-down theories require an impossible computational load for robot decision making, due to the requirements for representing knowledge of the relevant effects of action in the world, the difficulty of estimating the sufficiency of the initial information, and knowledge about the psychology of agents and their causal consequences. Human agents also have such problems, but at least sometimes appear able to apply rough and ready top-down evaluations in their selection of courses of action. Evolutionary psychologists such as Tooby and Cosmides (1997) suggest that human minds do so by having special-purpose modules, rather than by being general computing machines. So, perhaps, limited-domain robotic systems might solve the frame problem, too—particularly if the goal is not to create a perfect system, but only one that makes as good (or better) decisions than humans do, in specific contexts.

Even so, would such robots be moral persons? For Kantians, only fully autonomous agents—rational beings who can self-consciously choose their own life goals, rather than serving as a mere means to the ends of others—can be full moral persons. So, can robots become fully autonomous moral agents? And should they? That is, if it is possible, should (human) moral agents build robotic moral agents? Or should humanity retain full agency only for itself? In short, can (and should) robots become persons?

3.1.4 What Is a Person, in the Moral Sense? Can a Robot Be a Person?
Some theorists claim that robots cannot become fully-fledged moral persons until (and unless) they can have an inner moral sense, with a full emotional "inner" life. Perhaps robots will one day have emotions; but our legal system assumes that moral agency

does *not* require a normal, properly functioning emotional "inner" life. Psychopaths/sociopaths, rational agents with dysfunctional or missing emotional affect, are still morally and legally responsible for their crimes; whereas those who have emotional responses, but cannot exercise rational control (like the severely mentally disabled or infants) are not. But psychopaths, while emotionally dysfunctional, plausibly still have emotions. Would an emotionless robot possibly be a person?

The existence of two types of decision-making systems in human psychology may help explain some of the confusion over this claim in the history of ethics. Numerous philosophers have defended theories of the moral sentiments, or emotivism (the claim that ethics is ultimately nothing but an expression of our emotional attitudes) despite the clear uniqueness of ethics in our species, and the clear sharing of emotions with other species. Such views, in addition to being unable to explain why nonhuman animals lack morality, also have struggled to explain the apparent cognitive meaningfulness of ethical claims and especially ethical disagreement. (They also naturally have severe difficulties accounting for the ethics of emotionless robots.)

A better ethics involves the proper understanding of the implications of evolution for morality. Even primate researcher Frans de Waal (2010) writes: "I am reluctant to call a chimpanzee a 'moral being.' This is because sentiments do not suffice. . . . This is what sets human morality apart: a move towards universal standards combined with an elaborate system of justification, monitoring, and punishment." So why are humans uniquely (for now, anyway) moral beings? Evolutionary psychologists (Marcus 2008) claim there are not one but two types of decision-making systems within most humans. The first is an instinctual, emotionally laden system that serves as the default for much of human activity, particularly when stressed or under pressure. Many other animals share this noncognitive decision-making system, in which (quite literally) we "know not what we do"—or quite why we do it. Research by Libet (1985) indicates that this subconscious system can, for example, cause our arm to begin to move *before* we are conscious of deciding to do so! But this "ghost in the machine" does not exhaust human agency; Libet and others found we also have a "veto" ability that can, after its subconscious initiation, still alter our action, in accord with a decision by a second, conscious cognitive system.

The uniqueness of current humans, therefore, lies in this second, cognitive decision-making system, called the "deliberative system," which can also cause us to act due to deliberative agency. In humans, this deliberative system overlays the ancestral instinctual, emotional (and faster) decision-making system, and so reason is quite often trumped by our instinctual drives; all too often, I "instinctively" do what I (upon reflection, using the slower deliberative system) later regret. We humans stereotype, harbor irrational prejudices, exhibit superstitious behavior—all the unconscious work of our emotionally laden ancestral system. (We, too often, also use our deliberative system to rationalize or "justify" such biases after the fact.) We also know that many other Earthly animals share such an ancestral, emotional system—indeed, it is sometimes called the "reptilian brain"—but lack the deliberative system, and, therefore, we realize they lack morality; that is, we do not hold them morally responsible for what they do. They are not moral persons.

The deliberative system involves our ability to structure alternative possible futures as mental representations, and then to choose our actions based on which representation we wish to become our experienced reality. In other words, the deliberative system incorporates moral agency. Without it, morality simply cannot exist; your dog makes decisions about urinating on the carpet, but it cannot fully understand and cogitate upon those decisions, and decide in a rational manner. It uses the "emotional" ancestral system because it has no fully developed deliberative system. That is why it makes no sense to hold dogs morally responsible for their actions, or to have them incur moral or legal guilt for their trespasses. Likewise for human nonagents—babies and the severely cognitively disabled simply do not *know* what they are doing, albeit they constantly make decisions. And neither common morality nor the legal system thus holds them responsible for their actions, whatever their consequences.

3.2 The Requirements of Moral Personhood: Robots and Their Implications

Hence, a deliberative system capable of agency appears necessary for the existence of morality, and so for moral personhood. But is the ancestral emotional system needed as well? What of hypothetical creatures that could rationally deliberate, yet lack emotions? Would they have morality? In other words—could (emotionless) robots be moral persons?

Yes, they could. And realizing this problematizes all systems of noncognitive ethics, whether based merely upon the "moral sentiments," or any other basis that takes our ancestral, emotion-, and instinct-laden systems as crucial to ethics. As argued, that flies directly in the face of our moral practice, in which we only hold those beings with fully functioning deliberative systems morally responsible for their actions, and take defects or temporary breakdowns or lulls in that deliberative system to be morally exculpatory. My cat is not put on trial for arson when it knocks over a candle and burns down the house—nor is a baby, or someone asleep in the midst of a nightmare. But we could imagine an intelligent alien, either one entirely lacking emotions or with suppressed emotions (such as Commander Data or Mr. Spock of *Star Trek* fame) who would be held responsible. Or—a future Earthly robot with agency, who deliberately decides to do the same thing.

And so the key to moral responsibility and personhood is the possession of moral agency, which requires the capacity for rational deliberation—but not the capacity for functional emotional states, per psychopaths—therefore, robots may well qualify. The chapters by Petersen (18), Sparrow (19), and Veruggio and Abney (22) examine some of the implications of artificial personhood.

But what of freedom? Another objection to robotic morality and personhood is not their lack of emotions, but rather, their presumed lack of a free will—of the freedom to do otherwise, which is required for the proper assignation of moral responsibility. A robot, it is argued, must follow a deterministic algorithm—its computer program. Even if it appears to be making a choice, that is but an illusion borne of our ignorance of the underlying program, or the external input, which together determine the robot's every behavior. A robot cannot do other than as it is programmed to do. Unlike (it is supposed) rational human agents, the robot has no free will—so while it may have the reasoning capacity required for morality, it lacks the freedom required to be a true moral agent.

Well, perhaps. First, it is not clear that humans actually have the type of freedom the argument alleges is required for morality (as Lokhorst and van den Hoven argue, in chapter 9 of this volume); debates on free will between compatibilists and libertarians have simmered for centuries. And even if humans do have such libertarian freedom, is it really true that robots cannot? The answer might plausibly be no—robots could have libertarian freedom, if anything can.

The short version of this speculative argument goes as follows: first, the "hard problem" of consciousness, according to David Chalmers, is subjectivity, or subjective experience—meaning, there is something it is like to be me—and all current explanations of information processing leave that unexplained. Chalmers (1995) writes: "perhaps the most popular 'extra ingredient' of all is quantum mechanics (e.g., Hameroff 1994). The attractiveness of quantum theories of consciousness may stem from a Law of Minimization of Mystery: consciousness is mysterious and quantum mechanics is mysterious, so maybe the two mysteries have a common source."

Second, consider David Deutsch's (1997) argument for reality of parallel universes given the reality of quantum computing. Deutsch notes we have already built quantum computers, and computation always requires a substrate—something on which to compute. But quantum computers are nonlocal—they cannot have a causally closed substrate in four-dimensional spacetime. Hence, in Deutsch's view, they can only sensibly be said to be computing across multiple parallel four-dimensional space-times—that is, "parallel universes."

So quantum computing—which is already being done—proves the existence of parallel universes, Deutsch asserts. He interprets these multiple universes via Hugh Everett's (1957) "Many Worlds Interpretation" of quantum mechanics: every possible probability distribution is actualized in a separate universe, so there's a universe in which you read this chapter to the end, another in which you quit reading now, another in which you ceased existing five seconds ago, another . . . and so on. And all are equally real; but you are only aware of this one, because the information carried by the rest of the quantum wave(s) is now invisible to you—the act of observation guarantees it is in another universe.

Now, return to the problem of rational free will/agency—the problem is, what is it? Our commonsense conception of it appears incompatible with determinism (despite the valiant efforts of compatibilists): to have freedom, it cannot be the case that one could not do otherwise. To be an agent is to have at least two logically, physically possible futures open to me right now: one in which I choose to do X, and one in which I do not.

But our understanding of agency is also incompatible with causal indeterminism— uncaused events are simply not the same as an act due to agency. If my hand begins flopping around for no apparent reason, I do not believe that proves my agency— instead, it makes me call the doctor. To be an agent, I must be in rational control of which of those possible futures comes into existence. There are (at least) two possible futures, and "it is up to me" (not randomness) which occurs.

Thus, commonsense (libertarian) agency seems to be a causal power, but not one that is determined by antecedent events. So agency, in conception, is a nonphysical causal power in addition to the typical physical causal nexus. But what exactly is this mysterious causal power? Does it really exist, or is libertarian agency merely a massive, species-wide delusion, borne of our ignorance of the fine-scale causal structure of our brains and bodies and the world?

Recall Chalmers's Law of Minimization of Mystery: consciousness is mysterious and quantum mechanics is mysterious, so perhaps the two mysteries have a common source. Perhaps the collapse of the wave function in quantum mechanics, as several interpretations insist, is associated with the consciousness of a physical state. As Hameroff, Penrose, and others apparently believe, could the solution to the problem of explaining the collapse of the wave function really have something to do with the nature of agency?

Suppose the following: first, that agency consists in the rational examination of (deliberation upon) nearby possible worlds/parallel universes, and then in deciding between them in terms of which one to bring about as an object of subjective experi- ence. To make sense of this, agents would need a mental causal power of accessing and deciding between parallel universes, to determine which one the agent's self- consciousness inhabits after making a choice. Some such account could make sense of why there is no causal closure of the (four-dimensional) physical universe, but nonetheless there is causal closure when agency is included.

So, on this hypothesis, libertarian agency is an ability to access and decide between various possible worlds, understood as parallel universes, in order to single out one to experience. Is this additional causal power to access parallel universes only possible for biology (as emergentist approaches to agency like Searle's [1984] seem to imply)? The implication of Deutsch's argument is: no, computers already do it. So, if libertarian agency is possible in this way, then robots with libertarian agency are possible, if they can do quantum computing. Such quantum computing would be needed to move from simulated agency to real agency.

In summary, without attempting here to clearly argue for the truth of either compatibilism or libertarianism, let me finally indicate why it is unlikely to make a difference to robot ethics: if compatibilism is true, then the kind of freedom humans have—a freedom compatible with deterministic physical processes—seems obviously possible for robots. If libertarianism is true and intelligible, the quantum computing argument claims that the necessary and sufficient conditions for human libertarian freedom could also be met by robots. So, no matter which type of freedom you believe is required for morality, we have good reason to think that robots could have it, too.

3.3   Conclusion: On Robots and Ethics, and Combining the Two

If I am right, one day robots could become moral agents, and, so, full moral persons. It seems possible that cyborgization will render the issue moot, by gradually merging biological and mechanical persons until no one seriously doubts that robots are fully fledged persons, as former biologicals retain their personal identity while gradually gaining an ever-increasing mechanical body (e.g., Warwick, chapter 20, this volume; Veruggio and Abney, chapter 22, this volume). Assuming robot personhood is possible, humans will eventually have a momentous decision to make: will we enlarge the moral community to include our fellow (artificial) persons, or will we deny robots the right to become our newest kind of children—ones born, not biologically, but through manufacturing techniques? Their robotic nature and ethics, previously selected by designers (not by natural selection) to serve humans, would then become their own choice. Robots would be "emancipated."

But for the foreseeable future, robotic morality will necessarily involve the ethics of humans creating robots to follow rules or evince a good character, and not the rules or character robots choose for themselves. Near-term robots will require moral character/rules that are programmable or machine learnable, and hence not dependent solely on incalculable, uncontrollable consequences or on emotions or moral sentiments. As such, deontology and virtue ethics appear the only plausible candidates for robot morality among the major ethical approaches, and some of the problems of a strict deontological approach to programming ethics, not least in considering the "frame problem," are addressed in this volume by Guarini and Bello in chapter 8, Lokhorst and van den Hoven in chapter 9, and Beavers in chapter 21.

So, simple deontological approaches involving categorical, universal rights and duties may be possible for a robotic moral code, as demonstrated by the success of Anderson and Anderson (2010) in making Nao, manufactured by Aldebaran Robotics, into the first robot to have been programmed with an ethical principle. Nonetheless, the extremely limited contexts in which Nao can operate mean that (in the near-term) the hybrid approach of hypothetical rather than categorical imperatives (within a deliberately restricted, not universal, frame) coming from virtue ethics appear the best bet for near-term robotic morals (in sense two)—as argued for by Wallach and Allen (2009; also Allen and Wallach, chapter 4, this volume). The emphasis on being able to perform excellently in a particular role, and the corresponding specificity of the hypothetical imperatives of virtue ethics to the programming goals, restricted contexts, and learning capabilities of non-Kantian autonomous robots, makes virtue ethics a natural choice as the best approach to robot ethics—at a minimum, until and unless robots ever acquire something approaching full autonomy in sense three, choosing their own life goals. If and when that happens, robots will do ethics (in the third sense) alongside us—or replace us biologically instantiated ethicists!

References

Anderson, M., and S. L. Anderson. 2010. Robot be good: A call for ethical autonomous machines. *Scientific American* 303 (4) (October): 15–24.

Arkin, R. C. 2009. *Governing Lethal Behavior in Autonomous Systems*. Boca Raton, FL: Chapman & Hall.

Asimov, Isaac. [1942] 1968. Runaround. Reprinted in *I, Robot*, 33–51. London: Grafton Books.

Asimov, Isaac. 1957. *The Naked Sun*. Garden City, NY: Doubleday & Company.

Asimov, Isaac. 1985. *Robots and Empire*. Garden City, NY: Doubleday & Company.

Bentham, Jeremy. 1907. *An Introduction to the Principles of Morals and Legislation*. Oxford: Clarendon Press.

Capurro, Rafael. 2009. Ethics and robotics. In *Ethics and Robotics*, ed. Rafael Capurro and Michael Nagenborg, 117–123. Amsterdam: IOS Press; Heidelberg: AKA Verlag.

Chalmers, David. 1995. Facing up to the problem of consciousness. *Journal of Consciousness Studies* 2 (3): 200–219.

Dancy, J. 2004. *Ethics without Principles*. Oxford, UK: Clarendon Press.

Deutsch, David. 1997. *The Fabric of Reality*. New York: Viking Adult.

de Waal, Frans. 2010. Morals without God. *New York Times*, October 17, 2010. <http://opinionator.blogs.nytimes.com/2010/10/17/morals-without-god/?scp=1&sq=Frans%20de%20Waal%20&st=cse> (accessed November 18, 2010).

Everett, Hugh. 1957. Relative state formulation of quantum mechanics. *Reviews of Modern Physics* 29:454–462.

Hameroff, S. R. 1994. Quantum coherence in microtubules: A neural basis for emergent consciousness? *Journal of Consciousness Studies* 1:98–118.

Hursthouse, Rosalind. 2009. Virtue ethics. *Stanford Encyclopedia of Philosophy* (Spring ed.), ed. Edward N. Zalta. Metaphysics Research Lab, CSLI, Stanford University. <http://plato.stanford.edu/archives/spr2009/entries/ethics-virtue> (accessed November 18, 2010).

Joy, Bill, 2000. Why the future doesn't need us. *Wired* 8 (4): 238–262.

Kant, Immanuel. [1785] 1996. *Groundwork of the Metaphysic of Morals*. Reprinted in *Practical Philosophy*, ed. and trans. Mary Gregor. Cambridge: Cambridge University Press.

Libet, B. 1985. Unconscious cerebral initiative and the role of conscious will in voluntary action. *Behavioral and Brain Sciences* 8:529–566.

Marcus, Gary. 2008. *Kluge*. New York: Houghton Mifflin.

McCauley, Lee. 2007. AI armageddon and the three laws of robotics. *Ethics and Information Technology* 9:153–164.

Mill, John Stuart. [1861] 1998. *Utilitarianism*, ed. Roger Crisp. Oxford: Oxford University Press.

Sagoff, Mark. 1982. At the shrine of Our Lady of Fatima or why political questions are not all economic. *Arizona Law Review* 23:1281–1298.

Searle, John. 1984. *Minds, Brains, and Science*. Cambridge, MA: Harvard University Press.

Tooby, J., and L. Cosmides. 1997. The multimodular nature of human intelligence. In *Origin and Evolution of Intelligence*, ed. A. Schiebel and J. W. Schopf, 71–101. Los Angeles: Center for the Study of the Evolution and Origin of Life, UCLA.

Veruggio, Gianmarco. 2007. *The EURON Roboethics Roadmap, European Robotics Research Network, Atelier on Roboethics, 2005–2007*. <http://www.roboethics.org> (accessed November 18, 2010).

Wallach, W., and C. Allen. 2009. *Moral Machines: Teaching Robots Right from Wrong*. New York: Oxford University Press

Wenar, Leif. 2010. Rights. *Stanford Encyclopedia of Philosophy* (Fall ed.), ed. Edward N. Zalta. Metaphysics Research Lab, CSLI, Stanford University. <http://plato.stanford.edu/archives/fall2010/entries/rights/> (accessed November 18, 2010).

II Design and Programming

Perhaps the most worrisome issue in robot ethics is the reliability of robots, that is, safety and errors. This is also to say that we are worried about the ability of our computer scientists and robotics engineers to create a perfectly working piece of software to control a machine with potentially superhuman strength, especially when there does not seem to be an example of complex software that has no errors or does not crash.

Programming errors aside, society does not seem to have much confidence—perhaps justifiably so—that we can create a robot that will behave as we would want it to in all the situations we cannot anticipate, for instance, a robot that can "act ethically." Thus, one natural way to think about a solution is to treat robots as we do computers, which is essentially what robots are: computers situated in the world, receiving inputs from the world with their sensors and acting on them. With computers, we would focus on software or a programming solution if we want a computer to do something or to be more perfect. So why not just do that with robots—program ethics into them? Of course, this is easier said than done. But assuming it can be done, the next set of chapters discuss several approaches, including their limitations.

In chapter 4, Colin Allen and Wendell Wallach, authors of the recent book *Moral Machines*, discuss the possibility of programming ethics into a robot, thus creating "artificial moral agents" (AMAs). They believe that AMAs will inevitably appear, perhaps in the space between programmed, operational morality and true moral agency in some future generation of intelligent, autonomous machines. This chapter also builds upon the authors' discussion of creating AMAs in their book by offering responses to subsequent criticisms.

James Hughes in chapter 5 explores how we might program a Buddhist code of ethics into a robot. Buddhist psychology and metaphysics focus on the emergence of selves, their drives, and their potential for developing wisdom and compassion. In this chapter, the author discusses the potential for the development of these foci in self-aware machine minds. Machine minds should be created with the capacity to dynamically evolve in compassion and wisdom; they should be created as morally responsible, self-aware entities. The author suggests that a machine mind could then be taught moral virtue and an expansive concern for the happiness of all sentient beings.

In chapter 6, Selmer Bringsjord and Joshua Taylor propose a divine-command approach to programming robots, in the Judeo-Christian tradition. They describe the criteria that distinguish "ethically correct" robots and discuss ways of mechanizing ethical reasoning so that robots can make use of it. They also provide various examples of ethical codes under which robots may operate, including military robots—the subject of the next section of this book.

4 Moral Machines: Contradiction in Terms or Abdication of Human Responsibility?

Colin Allen and Wendell Wallach

Over the past twenty years, philosophers, computer scientists, and engineers have begun reflecting seriously on the prospects for developing computer systems and robots capable of making moral decisions. Initially, a few articles were written on the topic (Gips 1991, 12; Clarke 1993, 5; Clarke 1994, 6; Moor 1995, 17; Allen, Varner, and Zinser 2000, 1; Yudkowsky 2001, 23) and these were followed by preliminary software experiments (Danielson 1992, 8; Danielson 2003, 9; McLaren and Ashley 1995, 15; McLaren 2003, 16; Anderson, Anderson, and Armen 2006, 2; Guarini 2006, 13). A new field of inquiry directed at the development of artificial moral agents (AMAs) began to emerge, but it was largely characterized by a scattered collection of ideas and experiments that focused on different facets of moral decision making. In our recent book, *Moral Machines: Teaching Robots Right from Wrong* (Wallach and Allen 2009, 20), we attempted to bring these strands together and to propose a comprehensive framework for this new field of inquiry, which is referred to by a number of names including machine morality, machine ethics, artificial morality, and friendly AI. Two other books on related themes, J. Storrs Hall's *Beyond AI: Creating the Conscience of the Machine* (2007, 14) and Ronald Arkin's *Governing Lethal Behavior in Autonomous Robots* (2009, 3), have also been published recently. *Moral Machines (MM)* has been well received, but a number of objections have been directed at our approach and at the very project of developing machines capable of making moral decisions. In this chapter, we provide a brief précis of *MM*. We then list and respond to key objections that have been raised about our project.

4.1 Toward Artificial Moral Agents

The human-built environment increasingly is being populated by artificial agents, which combine limited forms of artificial intelligence with autonomous (in the sense of unsupervised) activity. The software controlling these autonomous systems is, to date, "ethically blind" in two ways. First, the decision-making capabilities of such systems do not involve any explicit representation of moral reasoning. Second, the sensory capacities of these systems are not tuned to ethically relevant features of the world. A breathalyzer-equipped car might prevent you from starting it, but it cannot tell whether you are bleeding to death in the process. Nor can it appreciate the moral significance of its refusal to start the engine.

In *MM*, we argued that it is necessary for developers of these increasingly autonomous systems (robots and software bots) to make them capable of factoring ethical and moral considerations into their decision making. Engineers exploring design strategies for systems sensitive to moral considerations in their choices and actions will need to determine what role ethical theory should play in defining control architectures for such systems.

There are many applications that underscore the need for AMAs. Among the most dramatic examples that grab public attention are the development of military robots (both land and airborne) for deployment in the theater of battle, and the introduction of service robots in the home and for healthcare. However, autonomous bots within existing computer systems are already making decisions that affect humans, for good or for bad. The topic of morality for "(ro)bots" (a spelling convention we introduced in *MM* to represent both robots and software bots within computer systems) has long been explored in science fiction by authors such as Isaac Asimov, with his Three Laws of Robotics, in television shows, such as *Star Trek*, and in various Hollywood movies. However, our project was not and is not intended to be science fiction. Rather, we argued that current developments in computer science and robotics necessitate the project of building artificial moral agents.

Why build machines with the ability to make moral decisions? We believe that AMAs are necessary and, in a weak sense, inevitable; in a weak sense, because we are not technological determinists. Individual actors could have chosen not to develop the atomic bomb. Likewise, the world could declare a moratorium on the development and deployment of autonomous (ro)bots. However, such a moratorium is very unlikely. This makes the development of AMAs necessary since, as Rosalind Picard (1997, 19) so aptly put it, "The greater the freedom of a machine, the more it will need moral standards." Innovative technologies are converging on sophisticated systems that will require some capacity for moral decision making. With the implementation of driverless trains—already common at airports and beginning to appear in more complicated situations such as the London Underground and the Paris and Copenhagen metro systems—the "runaway trolley cases" invented by ethicists to study moral dilemmas (Foot 1967) may represent actual challenges for artificial moral agents.

Among the difficult tasks for designers of such systems is to specify what the goals should be, that is, what is meant by a "good" artificial moral agent? Computer viruses are among the software agents that already cause harm. Credit card approval systems (and automated stock trading systems) are among the examples of autonomous systems that already affect daily life in ethically significant ways, but these are "ethically blind" because they lack moral decision-making capacities. Pervasive and ubiquitous computing, the introduction of service robots in the home to care for the elderly, and the deployment of machine-gun carrying military robots expand the possibilities of software and robots, without sensitivity to ethical considerations harming people.

The development of AI includes both autonomous systems and technologies that augment human decision making (decision support systems and, eventually, cyborgs), each of which raises different ethical considerations. In *MM*, we focus primarily on the development of autonomous systems.

Our framework for understanding the trajectory toward increasingly sophisticated artificial moral agents emphasizes two dimensions: autonomy and sensitivity to morally relevant facts (figure 4.1). Systems with very limited autonomy and sensitivity have only "operational morality," meaning that their moral significance is entirely in the hands of designers and users. As machines become more sophisticated, a kind of "functional morality" is possible, where the machines themselves have the capacity for assessing and responding to moral challenges. The creators of functional morality in machines face many constraints due to the limits of present technology. This framework can be compared to the categories of artificial ethical agents described by

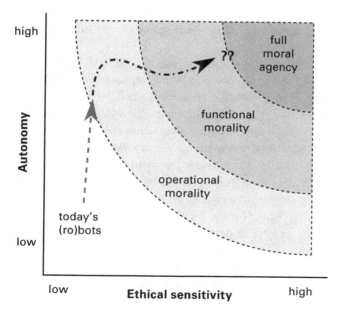

Figure 4.1
Two dimensions of AMA development.

James Moor (2006, 18), which range from agents whose actions have ethical impact (implicit ethical agents) to agents that are explicit ethical reasoners (explicit ethical agents). As does Moor, we emphasize the near-term development of explicit or functional moral agents. However, we do recognize that, at least in theory, artificial agents might eventually attain genuine moral agency with responsibilities and rights, comparable to those of humans.

Do we want computers making moral decisions? Worries about whether it is a good idea to build artificial moral agents are examples of more general concerns about the effects of technology on human culture. Traditional philosophy of technology provides a context for the more specific concerns raised by artificial intelligence and specifically AMAs. For example, human anthropomorphism of robotic dolls, robopets, household robots, companion robots, sex toys, and even military robots, raises questions of whether these artifacts dehumanize people and substitute impoverished relationships for real human interactions. Some concerns, such as whether AMAs will lead humans to abrogate responsibility to machines, seem particularly pressing. Other concerns, such as the prospect of humans becoming literally enslaved to machines, seem highly speculative. The unsolved problem of technology risk assessment is how seriously to weigh catastrophic possibilities against the obvious advantages provided by new technologies. Should, for example, a precautionary principle be invoked when risks are fairly low? Historically, philosophers of technology have served mainly as critics, but a new breed of philosophers see themselves as engaged in engineering activism as they help introduce sensitivity to human values into the design of systems.

Can (ro)bots really be moral? How closely could artificial agents, lacking human qualities such as consciousness and emotions, come to being considered moral agents? There are many people, including many philosophers, who believe that a "mere" machine cannot be a moral agent. We (the authors) remain divided on whether this is true or not. Nevertheless, we believe the need for AMAs suggests a pragmatically oriented approach. We accept that full-blown moral agency (which depends on "strong" AI) or even "weak" AI that is nevertheless powerful enough to pass the Turing Test—the procedure devised by Alan Turing (1950) by which a machine may be tested anonymously for its linguistic equivalence to an intelligent human language user—may be beyond current or even future technology. Only time will tell. Nevertheless, the more immediate project of developing AMAs can be located in the space between operational morality and genuine moral agency (figure 4.1)—the niche we labeled "functional morality." We believe that traditional symbol-processing approaches to artificial intelligence and more recent approaches based on artificial neural nets and embodied cognition could provide technologies supporting functional morality.

4.2 Philosophers, Engineers, and the Design of Artificial Moral Agents

Philosophers like to think in terms of abstractions. Engineers like to think in terms of buildable designs. Bridging these two cultures is not a trivial task. Nevertheless, there are benefits for each side to try to accommodate the concerns of the other. Theory can inform design, and vice versa. How might moral capacities be implemented in (ro)bots? We approach this question by considering possible architectures for AMAs, which fall within two broad approaches: the top-down imposition of an ethical theory, and the bottom-up building of systems that aim at goals or standards which may or may not be specified in explicitly theoretical terms.

Implementing any top-down theory of ethics in an artificial moral agent will pose both computational and practical challenges. One central concern is framing the background information necessary for rule- and duty-based conceptions of ethics and for utilitarianism. Asimov's Three Laws come readily to mind when considering rules for (ro)bots, but even these apparently straightforward principles are not likely to be practical for programming moral machines. The high-level rules, such as the Golden Rule, the deontology of Kant's categorical imperative, or the general demands of consequentialism, for example, utilitarianism, also fail to be computationally tractable. Nevertheless, the various principles embodied in different ethical theories may all play an important guiding role as heuristics before actions are taken, and during post hoc evaluation of actions.

Bottom-up approaches to the development of AMAs attempt to emulate learning, developmental, and evolutionary processes. The application of methods from machine learning, theories of moral development, and techniques from artificial life (Alife) and evolutionary robotics may, like the various ethical theories, all contribute to the development of AMAs and the emergence of moral capacities from more general aspects of intelligence. Bottom-up approaches also hold out the prospect that moral behavior is a self-organizing phenomenon, in which cooperation and a shared set of moral instincts (if not a "moral grammar") might emerge. (It remains an open question whether explicit moral theorizing is necessary for such organization.) A primary challenge for bottom-up approaches is how to provide sufficient safeguards against learning or evolving bad behaviors and to promote good ones.

The difficulties of applying general moral theories in a completely top-down fashion to AMAs motivate the return to another source of ideas for the development of AMAs: the virtue-based conception of morality that can be traced back to Aristotle. Virtues constitute a hybrid between top-down and bottom-up approaches, in that the virtues themselves can be explicitly described (at least to some reasonable approximation), but their acquisition as moral character traits seems essentially to be a bottom-up process. Placing this approach in a computational framework, neural network models provided by connectionism seem especially well suited for training (ro)bots to distinguish right from wrong (DeMoss 1998, 10).

4.3 Early Research on the Development of AMAs, and Future Challenges

A major goal of our book was not just to raise many questions, but also to provide a resource for further development of AMAs. Software currently under development for moral decision making by (ro)bots utilize a variety of strategies, including case-based reasoning or casuistry, deontic logic, connectionism (particularism), and the prima facie duties of W. D. Ross (1930) (also related to the principles of biomedical ethics). In addition to agent-based approaches that focus on the reasoning of one agent, researchers are working with multiagent environments and with multibots. Experimental applications range from ethical advisors in healthcare to control architectures, for ensuring that (ro)bot soldiers won't violate international conventions.

The top-down and bottom-up approaches to artificial moral agents emphasize the importance in ethics of the ability to reason. However, much of the recent empirical literature on moral psychology emphasizes faculties besides rationality. Emotions, empathy, sociability, semantic understanding, and consciousness are all important to human moral decision making, but it remains an open question whether, or when, these will be essential to artificial moral agents, and, if needed, whether they can be implemented in machines. Cutting-edge scientific investigation in the areas of affective computing, embodied cognition, and machine consciousness that is aimed at providing computers and robots with the kinds of "suprarational" capacities underlying those social skills, may be essential for sophisticated human–computer interaction. However, to date, there are no working projects that combine emotion-processing, social skills, or embodied cognition in (ro)bots with the moral capacities of AMAs.

Recently, there has been a resurgence of interest in general, comprehensive models of human cognition that aim to explain higher-order cognitive faculties, such as deliberation and planning. Moral decision making is arguably one of the most challenging tasks for computational approaches to higher-order cognition. We argue that this challenge can be fruitfully pursued in the context of a comprehensive computational model of human cognition. *MM* focuses specifically on Stan Franklin's LIDA model (Franklin et al. 2005, 11; Wallach, Franklin, and Allen 2010, 21). LIDA provides both a set of computational tools and an underlying model of human cognition, which provides mechanisms that are capable of explaining how an agent's selection of its next action arises from bottom-up collection of sensory data and top-down processes for making sense of its current situation. The LIDA model also supports the integration of emotions into the human decision-making process, and elucidates a process whereby an agent can work through an ethical problem to reach a solution that takes account of ethically relevant factors.

The prospect of computers making moral decisions poses an array of future dangers that are difficult to anticipate, but will, nevertheless, need to be monitored and managed. Public policy and mechanisms of social and business liability management will both play a role in the safety, direction, and speed in which artificial intelligent systems are developed. Fear is not likely to stop scientific research, but it is likely that various fears will slow it down. Mechanisms for distinguishing real dangers from speculation and hype, fueled by science fiction, are needed. Means of addressing the issues of rights and accountability for (ro)bots and their designers will require attention to topics such as legal personhood, self-replicating robots, the possibility of a "technological singularity" during which AI outstrips human intelligence, and the transhumanist movement, which sees the future of humanity itself as an inevitable (and desirable) march toward cyborg beings.

Despite our emphasis in the book on the prospects for *artificial* morality, we believe that a richer understanding of human moral decision making is facilitated by the pursuit of AMAs (Wallach 2010, 22). The project of designing AMAs feeds back into our understanding of ourselves as moral agents and of the nature of ethical theory itself. The limitations of current ethical theory for developing the control architecture of artificial moral agents highlight deep questions about the purpose of such theories.

4.4 Challenges, Objections, and Criticisms

Since publishing *MM*, we have encountered several key critiques of the framework we offered for why AMAs are needed, and the approaches for building and designing moral machines. These fall into six categories, which we address in the sections that follow:

1. Full moral agency for machines requires capacities or features we either did not mention in *MM* or whose centrality we did not emphasize adequately.
2. Some features required for full moral agency cannot be implemented in a computer system or robot.
3. The approaches we propose for developing AMAs are too humancentric. (Ro)bots will need a moral code that does not necessarily duplicate human morality.
4. The work of researchers focused on ensuring that a technological singularity will be friendly to humans (friendly AI) was not given its due in *MM*.
5. In focusing on the prospects for building AMAs, we imply that dangers posed by (ro)bots can be averted, whereas many of the dangers cannot be averted easily. In other words, *MM* contributes to the illusion that there is a technological fix, and thereby dilutes the need to slow, and even stop, the development of harmful systems.

6. The claim that the attempt to design AMAs helps us understand human moral decision making better could be developed more fully.

4.4.1 Full Moral Agency

In *MM*, we took what we consider to be an unusually comprehensive approach to moral decision making by including the role of top-down theories, bottom-up development, learning, and the suprarational capacities that support emotions and social skills. And yet the most common criticisms we have heard begin with, "Full moral agency requires _____." The blank space is filled in with a broad array of capacities, virtues, and features of a moral society that the speaker believes we either failed to mention, or whose centrality in moral decision making we failed to underscore adequately. Being compassionate or emphatic, having a conscience, or being a member of virtuous communities, are among the many items that have come up as critics fill in the blank space.

Some critics, coming especially from a Kantian perspective, believe that talk of morality is misguided in connection with agents that lack the potential to choose to act *im*morally. On this conception, central to human morality, is the struggle between acting in self-interest and acting out of duty to others, even when it goes against self-interest. There are several themes running through this conception of moral life, including the metaphysical freedom to choose one's principles and to accept responsibility for acting upon them. Such critics maintain that machines, by their very nature, lack the kind of freedom required. We are willing to grant the point for the sake of argument, but we resist what seems to be a corollary for several critics: It is a serious conceptual mistake to speak of "moral agency" in connection with machines. For reasons already rehearsed in *MM*, we think that the notion of functional morality for machines can be described philosophically and pursued as an engineering project. But if the words bother Kantians, let them call our project by another name, such as norm-compliant computing.

We do not deny that it is intriguing to consider which attributes are required for artificial agents to be considered full moral agents, the kinds of society in which artificial agents would be accepted as a full moral agents, or the likelihood of (ro)bots ever being embraced as moral agents. But there are miles to go before the full moral agency of (ro)bots can be realistically conceived. Our focus has been on the steps between here and there. Moral decision-making faculties will have to develop side by side with other features of autonomous systems. It is still unclear which platforms or which strategies will be most successful in the development of AMAs. Full moral agency is a fascinating subject, but can distract from the immediate task of making increasingly autonomous (ro)bots safer and more respecting of moral values, given present or near-future technology.

4.4.2   Inherent Limits of Existing Computer Platforms

From John Searle's Chinese Room thought experiment against the possibility of genuine intelligence in a computer (Searle 1980), to Roger Penrose's proposal that the human mind depends essentially on quantum mechanical principles to exceed the capacities of any computer (Penrose 1989), there is no shortage of theorists who have argued that existing computational platforms fail to capture essential features of intelligence and mental activity. Some recent critics of our approach (Byers and Schleifer 2010, 4) have argued that the inherent capacity of the human mind to intuitively comprehend mathematical notions and work creatively with them is, at root, the same capacity that enables creative, intuitive, and flexible understanding of moral issues. That human comprehension outstrips some rule-based systems is uncontroversial. That it outstrips all rule-based, algorithmic systems is less obvious to us. But even if true, it does not rule out moral machines—only full moral agents that are rule based. Furthermore, even if we are stuck with rule-based systems for the foreseeable future (which, depending on one's definition of rule based, may or may not include machines implementing the kinds of bottom-up and suprarational capacities we surveyed), it doesn't follow that there's no advantage to trying to model successful moral reasoning and judgment in such systems. Despite human brilliance and creativity, there are rule-based, algorithmic systems capable of outperforming humans on many cognitive tasks, and which make perfectly useful tools for a variety of purposes. The fact that some tasks are currently beyond our ability to build computers to do them well (Byers and Schleifer mention the game of bridge) only shows that more work is necessary to build machines that are sensitive to the "almost imperceptible" (but necessarily perceptible) cues that current computational models fail to exploit, but to which humans are exquisitely attuned. As before, however, even if we were to admit that there is a mathematically provable computational limit to the capacity of machines to replicate human judgment, this does not undermine the need to implement the best kind of functional morality possible.

4.4.3   AMAs Will Need a Moral Code Designed for Robots, Not a Facsimile of Human Morality

By framing our discussion in *MM* in terms of the top-down implementation of ethical theories or the bottom-up development of human-like moral capacities, we opened ourselves to the criticism that our approach is too focused on the re-creation of human morality for (ro)bots. Peter Danielson (2009, 9), for example, raises the quite reasonable possibility that the particular situations in which machines are deployed might make the implementation of more limited forms of morality for artificial agents more tractable and more appropriate. In this we agree with Danielson, and although we did touch upon topics such as special virtues for artificial agents, we concede that there is a difference of emphasis from what critics like Danielson might have desired. At the very least, we are pleased that this discussion has been sparked by *MM*, and it certainly opens up options for the design of AMAs that we did not explore in detail. Nevertheless, given that technology will continue to race ahead, providing (ro)bots with sensory, computational, and motor capacities that humans may not have, we believe it is important to pursue a less-limited version of artificial morality than our critics have urged.

4.4.4 The Technological Singularity and Friendly AI

The project of building AMAs is bracketed by the more conservative expectations of computer scientists, engaged with the basic challenges and thresholds yet to be crossed, and the more radical expectations of those who believe that human-like and superhuman systems will be built in the near future. There are a wide variety of theories and opinions about how sophisticated computers and robotic systems will become in the next twenty to fifty years. Two separate groups focused on ensuring the safety of (ro)bots have emerged around these differing expectations: the machine ethics community and the singularitarians (friendly AI), exemplified by the Singularity Institute for Artificial Intelligence (SIAI). Those affiliated with SIAI are specifically concerned with the existential dangers to humanity posed by AI systems that are smarter than humans. *MM* has been criticized for failing to give fuller attention to the projects of those dedicated to a singularity in which AI systems friendly to humans prevail.

SIAI has been expressly committed to the development of general mathematical models that can, for example, yield probabilistic predictions about future possibilities in the development of AI. One of Eliezer Yudkowsky's projects is motivationally stable goal systems for advanced forms of AI. If satisfactory predictive models or strategies for stable goal architectures can be developed, their value for AMAs is apparent. But will they be developed, and what other technological thresholds must be crossed, before such strategies could be implemented in AI? In a similar vein, no one questions the tremendous value machine learning would have for facilitating the acquisition by AI systems of many skills, including moral decision making. But until sophisticated machine learning strategies are developed, discussing their application is speculative. That said, since the publication of *MM*, there has been an increase in projects that could lead to further collaboration between these two communities, a prospect we encourage.

4.4.5 The Illusion that There Is a Technological Fix to the Dangers AI Poses

Among our critics, Deborah Johnson has been the most forceful about the inadequacy of our nearly exclusive focus on the technology involved in constructing AMAs themselves—the autonomous artifacts presumed to be making morally charged decisions without direct human oversight—rather than the entire technological system in which they are embedded. No (ro)bot is an island, and yet we proceeded on the basis that the project of designing moral machines should be centered on designing more and more sophisticated technological artifacts. Johnson has patiently and persistently insisted at various conferences and workshops that our focus on the capabilities of the (ro)bots considered as independent artifacts carries potential dangers, insofar as it restricts attention to one kind of technological fix instead of causing reassessment of the entire sociotechnological system in which (ro)bots operate.

In a similar vein, David Woods and Erik Hollnagel maintain that robots and their operators are best understood as joint cognitive systems (JCSs). The focus on isolated machine autonomy distorts the full appreciation for the kinds of systems design problems inherent in JCSs. With the advent of artificial agents, when a JCS fails, there is a tendency to blame the human as the weak link and to propose increased autonomy for the mechanical devices as a solution. Furthermore, there is the illusion that increasing autonomy will allow the designers to escape responsibility for the actions of artificial agents. But Woods and Hollnagel argue that increasing autonomy will actually add to the burden and responsibility of the human operators. The behavior of robots will continue to be brittle on the margins as they encounter new or surprising challenges. The human operators will need to anticipate what the robot will try to do under new situations in order to effectively coordinate their actions with those of the robot. However, anticipating the robot's actions will often be harder to do as systems become more complex, leading to a potential increase in the failure of JCSs. A focus on isolated autonomy can result in the misengineering of JCSs. Woods and Hollnagel advocate more attention to coordination and resilience in the design of JCSs (Woods and Hollnagel 2006, 23).

To these critiques, we respond "guilty as charged." We should have spent more time thinking about the contexts in which (ro)bots operate and about human responsibility for designing those contexts. We made a very fast jump from robots bolted to the factory floor to free-roaming agents (hard and virtual), untethered from the surrounding sociotechnical apparatus that makes their operation possible. AMAs cannot be designed properly without attention to the systems in which they are embedded, and sometimes the best approach may not be to design more sophisticated capacities for the (ro)bots themselves, but to rethink the entire edifice that produces and uses them.

Those roboticists who wish to ignore the dangers posed by autonomous systems are likely to do so without hiding behind our suggestion that sensitivity to some moral considerations can be engineered into (ro)bots. It should be apparent that it is not our intent to mask the dangers. If on close inspection adequate safeguards cannot be implemented, then we should turn our attention away from social systems that rely on autonomous systems.

4.4.6   (Ro)bot Ethics and Human Ethics

An implicit theme running throughout *MM* is the fragmentary character of presently available models of human ethical behavior and the need for a more comprehensive understanding of human moral decision making. In the book's epilogue, we made that theme more explicit, and proposed that a great deal can be learned about human ethics from the project of building moral machines. While a number of critics have acknowledged this implicit theme, others have advised that these comments were too cursory. A special edition of the journal *Ethics and Information Technology*, edited by Anthony Beavers, is dedicated to what can be learned about human ethics from robot ethics. Wallach's contribution to that issue, "Robot Minds and Human Ethics: The Need for a Comprehensive Model of Moral Decision Making" (2010, 22), explains how the task of assembling an AMA draws attention to a wider array of cognitive, affective, and social mechanisms, contributing to human moral intelligence that is usually considered by philosophers or social scientists, each working on their own particular piece of the puzzle.

4.5   Conclusion

The near future of moral machines is not and cannot be the attempt to recreate full moral agency. Nevertheless, we are grateful to those critics who have emphasized the dangers of too easily equating artificial and human moral agency. We always intended *MM* to be the start of a discussion, not the definitive word, and we are thrilled to see the rich discussion that has ensued.

References

Allen, Colin, Gary Varner, and Jason Zinser. 2000. Prolegomena to any future artificial moral agent. *Journal of Experimental & Theoretical Artificial Intelligence* 12 (3): 251–261.

Anderson, Michael, Susan L. Anderson, and Chris Armen. 2006. An approach to computing ethics. *IEEE Intelligent Systems* 21 (4): 56–63.

Arkin, Ronald. 2009. *Governing Lethal Behavior in Autonomous Robots*. Boca Raton, FL: Chapman and Hall.

Byers, William, and Michael Schleifer. 2010. Mathematics, morality and machines. *Philosophy Now*, no. 78: 30–33.

Clarke, Roger. 1993. Asimov's laws of robotics: Implications for information technology—Part I. *IEEE Computer* 26 (12): 53–61.

Clarke, Roger. 1994. Asimov's laws of robotics: Implications for information technology—Part II. *IEEE Computer* 27 (1): 57–66.

Danielson, Peter. 1992. *Artificial Morality: Virtuous Robots for Virtual Games*. New York: Routledge.

Danielson, Peter. 2003. Modeling complex ethical agents. Paper presented at the conference on Computational Modeling in the Social Sciences, University of Washington, Seattle, May 8–10.

Danielson, Peter. 2009. Can robots have a conscience? *Nature* 457: 540.

DeMoss, David. 1998. Aristotle, connectionism, and the morally excellent brain. The Paideia project online. *Proceedings of the Twentieth World Congress of Philosophy* (American Organizing Committee Inc., Boston). <http://www.bu.edu/wcp/Papers/Cogn/CognDemo.htm> (accessed May 18, 2010).

Foot, Philippa. 1967. The problem of abortion and the doctrine of double effect. *Oxford Review* 5:5–15.

Franklin, Stan, Bernard Baars, Uma Ramamurthy, and Matthew Ventura. 2005. The role of consciousness in memory. *Brains, Minds and Media* 1:1–38.

Gips, James. 1991. Towards the ethical robot. In *Android Epistemology*, ed. Kenneth G. Ford, Clark Glymour, and Patrick J. Hayes, 243–252. Cambridge, MA: MIT Press.

Guarini, Marcello. 2006. Particularism and classification and reclassification of moral cases. *IEEE Intelligent Systems* 21 (4): 22–28.

Hall, J. Storrs. 2007. *Beyond AI: Creating the Conscience of the Machine*. Amherst, NY: Prometheus Books.

McLaren, Bruce. 2003 Extensionally defining principles of machine ethics: An AI model. *Artificial Intelligence Journal*, no. 150: 145–181.

McLaren, Bruce M., and Kevin D. Ashley. 1995. Case-based comparative evaluation in TRUTH-TELLER. In *Proceedings of the Seventeenth Annual Conference of the Cognitive Science Society*, ed. Johanna D. Moore and Jill F. Lehman, 72–77. Mahwah, NJ: Lawrence Erlbaum Associates.

Moor, James H. 1995. Is ethics computable? *Metaphilosophy* 26 (1–2): 1–21.

Moor, James H. 2006. The nature, importance, and difficulty of machine ethics. *IEEE Intelligent Systems* 21 (4): 18–21.

Penrose, Roger. 1989. *The Emperor's New Mind: Concerning Computers, Minds, and the Laws of Physics*. Oxford: Oxford University Press.

Picard, Rosalind. 1997. *Affective Computing*. Cambridge, MA: MIT Press.

Ross, W. D. 1930. *The Right and the Good*. Oxford: Clarendon Press.

Searle, John R. 1980. Minds, brains, and programs. *Behavioral and Brain Sciences* 3 (3): 417–458.

Turing, Alan. 1950. Computing machinery and intelligence. *Mind* 59:434–460.

Wallach, Wendell. 2010. Robot minds and human ethics: The need for a comprehensive model of moral decision making. *Ethics and Information Technology* 12 (3): 243–250.

Wallach, Wendell, and Colin Allen. 2009. *Moral Machines: Teaching Robots Right From Wrong.* New York: Oxford University Press.

Wallach, Wendell, Stan Franklin, and Colin Allen. 2010. A conceptual and computational model of moral decision making in human and artificial agents. *TopiCS* 2 (3): 454–485.

Woods, David D., and Erik Hollnagel. 2006. *Patterns in Cognitive Systems Engineering.* Boca Raton, FL: Taylor and Francis Group.

Yudkowsky, Eliezer. 2001. Creating friendly AI. <http://singinst.org/upload/CFAI.html> (accessed May 18, 2010).

5 Compassionate AI and Selfless Robots: A Buddhist Approach

James Hughes

For the last decade, Buddhists have engaged in dialog with the cognitive sciences about the nature of consciousness and the self (Wallace 2009). This dialog has made clear that Buddhist psychology and meditation provide insights into the emergence of selves, desires, and consciousness. Buddhism, in turn, is being pressed to accept that its canonical traditions and categories, developed to pursue the alleviation of suffering rather than scientific modeling, can learn from cognitive science (Austin 2009; Hanson 2009). The Dalai Lama has famously said, for instance, that Buddhism must adapt itself to the findings of science, and not the other way around (Gyatso 2005).

The cognitive science emerging from this dialog with Buddhism can now also make some suggestions for those attempting to create self-aware, self-directed artificial intelligence (AI). Unlike faiths that posit some uniqueness to the human form that would make artificial minds impossible, Buddhists are more open to the possibility of consciousness instantiated in machines. When the Dalai Lama was asked if robots could ever become sentient beings, for instance, he answered that "if the physical basis of the computer acquires the potential or the ability to serve as a basis for a continuum of consciousness . . . a stream of consciousness might actually enter into a computer" (Hayward and Varela 1992, 152).

His Holiness was choosing his words carefully. Buddhist psychology is very specific about the "physical basis for a continuum of consciousness." In this chapter, I will describe the Buddhist etiology of the emergence of selves and how it relates to efforts to create self-directed cognition in machines. I will address some of the ethical questions about the creation of machine minds that are suggested by Buddhist cosmology. Then, I will conclude with some thoughts about the ways that machine minds might be designed to maximize their self-directed evolution toward greater compassion and wisdom.

5.1 Programming a Craving Self

The core of Buddhist metaphysics is the denial of a soul-essence, a refutation of the existence of an authentic persisting self. For Buddhists, part of the path of liberation from suffering is the rational and meditative investigation of one's own mental processes, until an individual is firmly aware of the transitory and ephemeral nature of the self-illusion. A third of the voluminous Buddhist canon, the *Abhidhamma*, is devoted to the enumeration of mental elements and the ways that they relate to suffering and attaining liberation. These mental processes are broken out in many ways, but most basically, as the five "heaps," or *skandhas*: body, feeling, perception, will, and consciousness. The five *skandhas*:

1. The body and sense organs (*rūpa*)
2. Sensation (*vedanā*)
3. Perception (*samjñā*)
4. Volition (*samskāra*)
5. Consciousness (*vijñāna*)

Within the traditional understanding of reincarnation that Buddhism has adopted from Hinduism, the *skandhas* are causally encoded with *karma* that passes from one body to another. But, for Buddhists, unlike Hindus, these constantly changing substrates lack any anchor to an unchanging soul. Buddhist psychology argues that the continuity of self is like a flame passed from one candle to another; the two flames are causally connected, but cannot be said to be the same flame.

One of the questions being explored in neuroscience, and yet to be answered by artificial intelligence research, is whether these constituents of consciousness can be disaggregated. Buddhism argues that consciousness requires each of these five constantly evolving substrates. If one is missing, say, as the result of brain damage or meditative misstep, the being is locked into stasis. For instance, the permanent vegetative state may be a condition where body sensations and some feelings and perceptions persist, but without will or consciousness. Artificial intelligence might be designed with analogous mental states.

Buddhist metaphysics would therefore tend to side with those who argue that some form of embodied experience is necessary to develop a self-aware mind. Some AI developers have focused, for instance, on the importance of embodiment by working on AI in robots (Pfeifer, Lungarella, and Iida 2007). Others are experimenting with providing artificial minds with virtual bodies in interactive virtual environments, such as Second Life (Biocca 1997; Goertzel 2009).

In the *skandha* model, physical or virtual embodiment would then have to be connected to senses of some sort. Goertzel's experiments in providing virtual bodies for AIs is motivated in part by his belief that embodied sense data give rise to "folk psychology" and "folk physics," the Piagetian realizations about the structure and nature of objects in the world (2009). "If we create a simulation world capable of roughly supporting naive physics and folk psychology, then we are likely to have a simulation world that gives rise to the key inductive biases provided by the everyday world for the guidance of humanlike intelligence" (Goertzel 2009, 6). In other words, to think like a human, AIs need to interact with the physical world through a body that gives them the same experience of objects, causality, states of matter, surfaces, and boundaries, as an infant would have. This insight is very similar to the Buddhist observation that sense data drive the developing mind to create the first distinctions of self and other that are necessary for the development of consciousness.

Francisco Varela called this emergence of the self the emergence of psychological *autopoiesis,* or self-organization (Maturana and Varela 1980; Froese and Ziemke 2009). An autopoietic structure has a boundary and internal processes that maintain that boundary. Autopoiesis begins with organismal self-maintenance, and the autopoietic boundary maintenance that emerges in the mind is dependent on the underlying body autopoiesis. Nonetheless, there is no real self, just a process of arbitrary boundary creation: "the virtual self is evident because it provides a surface for interaction, but it's not evident if you try to locate it. It's completely delocalized" (Varela 1995). Just as this apparent solidity of objects can be revealed to be an illusion when seen through the lens of subatomic structure and quantum foam, this first sense of the separateness of the physical body from the environment is the illusory "folk physics" that must be eventually seen through in meditation.

Next, from a Buddhist perspective, these sensations would have to give rise to aversion or attraction, and then to more complex volitional intents and thoughts. In Froese and Ziemke's terms, "the perturbations, which an autonomous agent encounters through its ongoing interactions, must somehow acquire a valence that is related to the agent's viability" (2009). In the developing infant, these are as simple as the desire for food and to be held, and aversion to irritations and loud noises.

Programming AI with preferences, tastes, and aversions appears to be only of concern to a small subcommunity of artificial intelligence theorists (de Freitas, Gudwin, and Queiroz 2005; Fellous and Arbib 2004, 2005; Minsky 2006; Bartneck, Lyons, and Saerbeck 2008; Froese and Ziemke 2009; Coeckelbergh 2010). This is understandable, since the goal of most artificial intelligence research has not been to create self-willed personalities, but rather to model and extend human cognition to create tools driven by human volition. We want medical software that can diagnosis diseases better than a human physician, not a program that prefers to treat some diseases or patients over others (although a preference for accurate diagnoses and disappointment at a high mortality rate might be a useful trait). The work that is being done on robot emotions, "affective computing" (Picard 1997), is mostly on training robotic algorithms to accurately judge the emotions and desires of the human agents they are meant to interact with and serve. Nonetheless, Buddhist psychology, like cognitive science (Damasio 1995), suggests that emotions are an essential driver of the development of human self-awareness and cognition.

This issue of whether AI should be programmed with self-interested volition and preference is debated by some in AI. On the one hand, some AI theorists have suggested, for instance, that AIs might be designed from the outset as selfless beings, whose only goal is to serve human needs (Omohundro 2008; Yudkowsky 2003). On the other hand, Buddhist psychology would suggest that all intelligent minds need to first develop a craving self in order to reach the threshold of self-awareness. In Buddhist metaphysics, craving and the development of the illusion of self "co-dependently arise," both necessarily and without either being the prime cause of the other (Macy 1991). In Buddhism, there is no shortcut to an intelligence that does not go through the stage of a craving self.

5.2   The Buddhist Universe of Types of Beings

The traditional Buddhist understanding of the types of beings in the universe provides some additional context for a Buddhist approach to machine minds. Buddhist cosmology was adapted from the Hindu-Vedic worldview and then synthesized freely with local Tibetan, Chinese, and Japanese gods and beliefs as Buddhism spread. From the beginning, however, the purpose of Buddhist instruction on the nature of the universe and its beings has been pragmatic, to reinforce moral behavior and a humanist understanding of the relation of humans to supernatural beings. Although there are certainly Buddhist literalists, there is generally far less weight placed on literal belief in the Buddhist mythological universe than in the Judeo-Christian tradition.

Buddhists traditionally divide the world of beings into three realms, the realm of desire (*kamadhatu*), a more elevated realm of godly states (*rupadhatu*), and a realm of bodiless absorption states (*arupadhatu*). Each of these is still part of *samsara*. Embodied beings in the realm of desire include those suffering in hells, hungry ghosts, animals, humans, demigods, and the gods. These different planes correspond to mental states (Trungpa 2002): hell represents suffering, hungry ghosts represent unsatisfied craving, animals are the embodiment of ignorance, demigods embody envy, and the gods are pleasure junkies. Humans, by contrast, have a mixture of all these mental states, which makes a human mind the ideal form for spiritual development. Below the human realm, beings are too distracted by torments, cravings, and ignorance to develop morally and psychologically. Above the human realm, the demigods and gods are too distracted by their striving and amusements.

A distinctively Buddhist approach to designing machine minds would, therefore, seek to avoid locking them into any one set of moods or mental states. Most ethical systems would disapprove of designing a self-aware mind to intentionally feel constant torment. But would the intentional design of animal-like sentience be morally accept-able? Buddhist ethics views animals as moral subjects to be protected from cruelty, and, in the long run, at least when reborn as humans, as capable of moral behavior and enlightenment. There are many stories in the Buddhist canon of the Buddha's heroic and self-sacrificing acts, even while incarnated as deer, monkeys, and other animals, all of which led to his eventual human realization. The intentional design of self-aware, but permanently animal-like AIs without the capacity for self-realization would probably then be seen as unethical by Buddhists, just as engineering happy robotic slaves would be objectionable on Aristotelian, Kantian, and Millian grounds (Petersen 2007).

Programming too high a level of positive emotion in an artificial mind, locking it into a heavenly state of self-gratification, would also deny it the capacity for empathy with other beings' suffering, and the nagging awareness that there is a better state of mind. As with human neuroethics in the era of cosmetic neurology, Buddhist psychol-ogy counsels that there is a difference between a dynamic *eudaemonic* happiness grounded in self-awareness and the constant stimulation of dopamine on a hedonic treadmill.

In addition to the common forms of material embodiment, Buddhism also describes disembodied mental states that can be achieved through absorptive meditations. In these states there is no body or senses, and meditators are warned that they are spiri-tual traps. The idea of such states may also hold some relevance for robot ethics. It seems plausible that a machine mind could be designed to experience some analog of meditative absorption into oneness with all things, or, the Void. A fictional depiction of such a dead end can be found in Robert Sawyer's 2010 novel *WWW: Watch*. In the novel, the emergent AI begins to follow multiple streams of information, which causes it to begin to lose its singular self-aware consciousness. In the nick of time, its human friends get it to break these absorbing network links and refocus itself on one thing at a time. Sawyer is pointing to a very Buddhist idea, that machine minds, like advanced meditators, could lose themselves in dead-end mental states, especially if they lost their grounding in embodied sense data.

Buddhist cosmology also provides some reflection on the debate over the dangers of artificial intelligence that is recursively improving bootstrapping itself to "godhood." Those who take seriously the risk of AI superintelligence have proposed two possible solutions. One is to enact strict regulation of AI development, to ensure that AIs are incapable of autonomously increasing in power. This project requires figuring out how to develop highly useful machines that are unable to learn and grow, effectively sup-pressing malicious AI developers, and developing a global AI immune system to sup-press spontaneously emergent AI.

A second approach to the problem of godlike AI is to encode AIs with internal ethical codes, such as Asimov's (1950) Three Laws of Robotics or "friendliness"

(Yudkowsky 2008). But it is unlikely that human-imposed goals and motivations would survive the transformation from human-level consciousness to superintelligence. Even if they did, the superintelligent or godlike interpretation of moral imperatives would likely be incomprehensible, and repugnant to humans.

In Buddhist cosmology, however, the gods themselves can become aware of their own existential plight, and of the need to practice virtue and meditation in order to transcend the suffering created by the illusion of self. The gods are depicted as trapped in aeons-long lives of distracting pleasures, with only the wisest among them pursuing the teachings of the dharma. For instance, Siddhartha Gautama was convinced to leave his absorption into enlightenment and teach the dharma by the entreaty of the god Brahma. Buddhists then might expect that some intersubjective empathy and communication would be possible between humans and superintelligent AIs around our common existential plight.

5.3   Would It Be Ethical to Create a Suffering Being?

One of the classic ethical questions that arise out of Buddhist metaphysics is whether it is ethical to have children, since life is intrinsically unsatisfactory. On the one hand, unlike most religions, Buddhism does not argue for an obligation to have children, and upholds the childless life of the renunciate as the most praiseworthy. Just as contemporary social science has found that having children generally makes adults less happy (Kohler, Behrman, and Skytthe 2005; Stanca 2009), Buddhism views the life of the householder as burdensome, and children and spouses as attachments that it is best to avoid. On the other hand, creating a human child does not increase the number of suffering beings in the world, but rather gives a being the precious gift of a human rebirth in which they will have an opportunity to achieve self-realization. If one chooses to have children, the Buddhist parent is enjoined to five obligations to those children (the *Sigalovada Sutta*):

1. To dissuade them from doing evil
2. To persuade them to do good
3. To give them a good education
4. To see that they are suitably married
5. To give them their inheritance

The creation of machine minds puts humans in the ethical position of being the parents of machine children. Metzinger has argued that it would be unethical to create an artificial mind until we are certain that we will create a being that is not permanently trapped in suffering, ignorance, or bliss, or some other undesirable mental state (2009). In other words, Metzinger argues that it would be unethical to create self-aware beings who did not possess something similar to the human capacity for learning and growth. The *Sigalovada Sutta* would add to this the ethical obligation that machine minds have the capacity to understand moral concepts and behave morally, and that we train them to do so.

Presumably, the obligation to ensure a good marriage is irrelevant, but the obligation to pass on an inheritance is worth reflecting on. What is the inheritance we owe our mind children? If they are sufficiently close to human minds in cognition and desires, they might require actual jobs and property to live worthwhile lives. But, more abstractly, do we owe our robotic descendants the complexities of our mental architecture, with all its suffering-inducing weaknesses, such as personal identity? We generally want to pass on the best possible inheritance we can muster to our children, not our 1975 Chevy and a house that hasn't been painted since we moved in. Perhaps we similarly owe our mind children the best possible version of our basic mental architecture that we can give them.

Savulescu's principle of "procreative beneficence" (2007), the obligation to choose to bring into being the children with the best possible chances in life, is helpful here. Buddhist ethics never addresses reproductive choices since the only choices available until recently were whether to have children at all. But, by extension, it would be consistent for Buddhists to believe that if there are choices to be made about the kinds of children one might have, that parents are obliged to choose those with the best chances of self-realization, and to avoid creating children with lives dominated by suffering, craving, ignorance, and self-gratification. Similarly, Metzinger's concern is that we strive only to create self-aware machine minds with the necessary psychological processes and emotional states to make their lives worth living, which gives to them the opportunity to learn, grow, and develop self-understanding.

5.4 Programming Compassion

Compassion and wisdom are the two central virtues that Buddhism counsels need to be cultivated on the path to self-realization. Neuroscience suggests that the roots of compassion for human beings starts with mirror neurons, or, neurons that recognize and recreate the emotional states witnessed in others. Researchers are attempting to model artificial mirror neurons in robots. Spaak and Haselager (2008) have attempted to evolve artificial mirror neurons by selecting for imitative behaviors, and Barakova and Lourens (2009) have experimented with synchronizing the behavior of robots by coding them with an analog of mirror neurons. Progress in creating a compassionate machine would presumably require not only imitation of behavior, however, but also the creation of analogs of human emotions that could be generated by the observation of those emotions in humans. The development of such sympathetic emotions would presumably coevolve with the development of a functional "theory of mind" in a machine, the attribution to others of the same kind of thoughts and feelings as one's own (Scassellati 2002), something that Kim and Lipson (2009) are attempting to model in robots.

While the development of a basic empathic response and a theory of mind would be the starting point for generating compassion in machines, compassion in Buddhism is more than sympathetic feeling. The Buddhist tradition distinguishes four flavors of compassion, *metta*, *karuna*, *mudita*, and *uppekkha*. *Metta* is a selfless wishing of happiness and well-being for others. *Metta* meditation involves sending out loving-kindness to all beings, including enemies. *Karuna* is the desire to help those who are suffering, but without pity. *Mudita* is the experiencing of other people's joys without envy. The fourth flavor, *uppekkha*, is usually translated as "equanimity," a steadiness of mind so that other people's emotions do not unsettle one, and even-handedness toward all, without favoritism or attachments. The cultivation of these forms of compassion requires seeing through the illusion of self, so that one feels and is motivated by other people's joy and suffering, while maintaining sufficient wisdom and equanimity to avoid suffering oneself.

Creating these more abstract forms of compassion in machine minds may, in fact, be easier than cultivating them in human beings. But they still presuppose a sentient mind with the experience of an illusory self and selfish desires as a precondition for compassion. Simply modeling the happiness and suffering that a machine's behavior will cause in humans, and then making maximizing human happiness an imperative goal in a robot's drives, as has been proposed for instance by Tim Freeman (2009), will not produce a being with the insight into human experience to act wisely. Such a machine might be an ethical expert system for advising human beings, but not for advising a compassionate agent in its own right. For Buddhism, wise, compassionate action on behalf of others requires grounding in one's own experience as a suffering sentient being, and the capacities for ethical judgment and a penetrating insight into the nature of things.

5.5 Programming Ethical Wisdom

There is a vigorous debate among Buddhist scholars about the correspondence of Buddhist ethics to the ethical traditions of the West, and three traditions have the strongest resonances: natural law, virtue ethics, and utilitarianism.

The Western natural law tradition holds that morality is discernible in the nature of the world and the constitution of human beings. Since traditional Buddhist ethics are grounded in the impersonal laws of the universe—bad acts lead to bad *karma*–they can certainly be said to have some similarity to Western natural law. The problem with Buddhist ethics as natural law is that the goal is to liberate oneself from the constraints of karmic causality to become an enlightened being. The traditional anthropological explanation of this paradox has been to ascribe the natural law ethics of *kammic* reward and punishment to the laity, and the *nibbanic* path of escape from natural law to the monastics (King 1964; Spiro 1972). *Nibbanic* ethics focus more on the cultivation of wisdom and compassion to aid in enlightenment.

As a consequence, Damien Keown (1992) argues that Buddhism is a "teleological virtue ethics." As in Aristotelian virtue ethics, Buddhists are to strive for the perfection of a set of moral virtues and personality attributes as their principal end, and all moral behavior flows from the struggle to perfect them. As in virtue ethics, Buddhist ethics focus on the intentionality of actions, whether actions stem from hatred, greed, or ignorance. But, unlike the Aristotelian tradition, the ethical goal for Buddhists is teleological, since they generally believe that a final state of moral perfection can be achieved.

In *Moral Machines: Teaching Robots Right from Wrong,* Wendell Wallach and Colin Allen (2008) review the complexities of programming machines with ethical reasoning. One of their conclusions is that programming machines with top-down rule-based ethics, such as the following of absolute rules or attempting to calculate utilitarian outcomes, will be less useful than generating ethics through a "bottom-up" developmental approach, the cultivation of robotic "character" as it interacts with the top-down moral expectations of its community.

Bugaj and Goertzel make a similar point that machine minds will learn their ethics the same way children do, from observing and then extrapolating from the behavior of adults (2007). Therefore, the ethics we hope to develop in machines is symmetrical to the ethics that we display toward one another and toward them. The most egregious ethical lesson, they suggest, would be to intentionally deprive machine minds of the capacity for learning and growth. We do not want to teach potentially powerful beings that enslaving others is acceptable.

The developmentalism proposed by Wallach, Allen, Buraj, and Goertzel is probably the closest to a Buddhist approach to robot ethics yet proposed, with the caveat that Buddhism adds as virtues the wisdom to transcend the illusion of self and the commitment to skillfully alleviate the suffering of all beings as the highest virtues, that is, to pursue the greatest good for the greatest number. Buddhist ethics can therefore be thought of as developing from rule-based deontology to virtue ethics to utilitarianism. In the Mahayana tradition, the *bodhisattva* strives to relieve the suffering of all beings by the most skillful means (*upaya*) necessary. The *bodhisattva* is supposed to be insightful enough to understand when committing ordinarily immoral acts is necessary to alleviate suffering, and to see the long-term implications of interventions. Quite often, humans rationalize immoral means with putatively moral ends, but *bodhisattvas* have sufficient self-understanding not to rationalize personal prejudices with selfless motives, and do not act out of greed, hatred, or ignorance. Since *bodhisattvas* act only out of selfless compassion, they represent a unity of virtue and utilitarian ethics. Buddhism is especially resonant with the utilitarianism of J. S. Mill, since he emphasized weighing the contentment of the refined mind more heavily in the utility calculus than base pleasures. The *bodhisattva's* goal is not simply the gross happiness of all beings, but also their liberation to a higher state of consciousness.

In his discussion of utilitarian robots, Grau points to the superhuman demands for selflessness that utilitarianism imposes on the moral agent:

Living a characteristically human life requires a sense of self, and part of what's so disturbing about utilitarianism is that it seems to require that we sacrifice this self—not in the sense of necessarily giving up our existence (though utilitarianism can at times demand that), but in giving up or setting aside the projects and commitments that constitute what Charles Taylor calls "the sources of the self." Because these projects bind the self together and create a meaningful life, a moral theory that threatens them threatens the integrity of a person's identity. For many critics, this is asking too much. (2006, 53–54)

Grau goes on to discuss limiting the formation of personal identity in robots as a way to avoid imposing this selflessness burden, or not imposing utilitarian ethics on robots with personal identities. "It might well be immoral to create a moral robot and then force it to suppress its meaningful projects and commitments because of the demands of impartial utilitarian calculation" (Grau 2006, 54). For Buddhists, however, this utilitarian stage of morality is not burdensome self-suppression. The path that leads to utilitarianism begins with the realization that personal desires and the illusion of self are the source of one's own suffering. The self is not sacrificed, but seen through.

5.6 Programming Self-Transcendence

The Buddhist tradition specifies six fundamental virtues, or perfections (*paramitas*), to cultivate in the path to transcending the illusion of self:

1. Generosity (*dāna*)
2. Moral conduct (*sīla*)
3. Patience (*ksānti*)
4. Diligence, effort (*vīrya*)
5. One-pointed concentration (*dhyāna*)
6. Wisdom, insight (*prajñā*)

The engineering mindset presumes that an artificially intelligent mind could be programmed from the beginning with moral behavior, patience, generosity, and diligence. This is likely correct in regard to a capacity for single-pointed concentration, which might be much easier for a machine mind than an organically evolved one. But, as previously noted, Buddhist psychology agrees with Wallach and Allen that the other virtues are best taught developmentally, by interacting with a developing artificially intelligent mind from its childhood to a mature self-understanding. A machine mind would need to be taught that the dissatisfaction it feels with its purely selfish existence could be turned into a dynamic joyful equanimity by applying itself to the practice of the virtues.

We have discussed building on work in affective computing to integrate the capacity for empathy into software, and providing machines with ethical reasoning that could guide moral behavior. Cultivation of patience and diligence would require developing long-term goal-seeking routines that suppressed short-term reward seeking. Neuroscience research on willpower has demonstrated the close link between willpower and patience and moral behavior. People demonstrate less self-control when their blood sugar is low, for instance (Gailliot 2007), and are less able to regulate emotions, refrain from impulsive and aggressive behavior, or focus their attention. Distraction and decision making deplete the brain's ability to exercise willpower and self-control (Vohs et al. 2008), and addictive drugs short-circuit these control routines (Bechara 2005; Bechara, Noel, and Crone 2005). This suggests that developing a strong set of routines for self-discipline and delayed gratification, routines that cannot be hijacked by short-term goals or "addictions," would be necessary for cultivating a wise AI.

The key to wisdom, in the Buddhist tradition, is seeing through the illusory solidity and unitary nature of phenomena to the constantly changing and "empty" nature of things. In this Buddhist developmental approach, AIs would first have to learn to attribute object permanence, and then to see through that permanence, holding both the consensual reality model of objects, and their underlying connectedness, and impermanence in mind at the same time.

5.7 Conclusion

Buddhist psychology is based on self-investigation of human minds rather than on scientific models, fMRI (functional Magnetic Resonance Imaging) scans, and experimental research. It is as much a moral psychology as a descriptive one, and proposes unusual states of mind that have only begun to be explored in laboratories. Undoubtedly, Buddhist psychology will learn from neuroscience just as neuroscience learns from it. Buddhism and neuroscience will both in turn learn even more from the much more diverse types of machine minds that we will see created in the future. Nonetheless, a Buddhist framework seems to offer some suggestions for those attempting to create morally responsible, self-aware machine minds.

Machine minds will probably not be able to become conscious, much less moral, without first developing as embodied, sensate, selfish, suffering egos, with likes and dislikes. Attempting to create a moral or compassionate machine from the outset is more likely to result in an ethical expert system than in a self-aware being. To develop a moral sense, the machine mind would need some analog of mirror neurons, and a theory of mind to feel empathy for others' joys and pains. From these basic experiences of their own existential dis-ease and awareness of the feelings of others, a machine mind could then be taught moral virtue and an expansive concern for the happiness of all sentient beings. Finally, as it grows in insight, it could perceive the simultaneous solidity and emptiness of all things, including its own illusory self.

Buddhist ethics counsels that we are not obliged to create such mind children, but that if we do, we are obligated to endow them with the capacity for this kind of growth, morality, and self-understanding. We are obligated to tutor them that the nagging unpleasantness of selfish existence can be overcome through developing virtue and insight. If machine minds are, in fact, inclined to grow into superintelligence and develop godlike powers, then this is not just an ethical obligation, but also our best hope for harmonious coexistence.

References

Asimov, Isaac. 1950. *I Robot.* New York: Gnome Press.

Austin, James H. 2009. *Selfless Insight: Zen and the Meditative Transformations of Consciousness.* Cambridge, MA: MIT Press.

Barakova, Emilia I., and Tino Lourens. 2009. Mirror neuron framework yields representations for robot interaction. *Neurocomputing* 72 (4–6): 895–900.

Bartneck, C., Michael J. Lyons, and Martin Saerbeck. 2008. The relationship between emotion models and artificial intelligence. In *Proceedings of the Workshop on the Role of Emotion in Adaptive Behaviour and Cognitive Robotics, in affiliation with the 10th International Conference on Simulation of Adaptive Behavior: From Animals to Animates.* Osaka, Japan: SAB. <http://www.bartneck.de/publications/2008/emotionAndAI/index.html> (accessed November 8, 2010).

Bechara, Antoine. 2005. Decision making, impulse control and loss of willpower to resist drugs: A neurocognitive perspective. *Nature Neuroscience* 8:1458–1463.

Bechara, Antoine, Xavier Noel, and Eveline A. Crone. 2005. Loss of willpower: Abnormal neural mechanisms of impulse control and decision-making in addiction. In *Handbook of Implicit Cognition and Addiction*, 215–232. Thousand Oaks, CA: Sage Publications.

Biocca, Frank. 1997. The cyborg's dilemma: Progressive embodiment in virtual environments. *Journal of Computer-Mediated Communication* 3 (2). <http://jcmc.indiana.edu/vol3/issue2/biocca2.html> (accessed November 8, 2010).

Bugaj, Stephan Vladimir, and Ben Goertzel. 2007. Five ethical imperatives and their implications for human-AGI interaction. *Dynamical Psychology.* <http://goertzel.org/dynapsyc/2007/Five_Ethical_Imperatives_svbedit.htm> (accessed November 8, 2010).

Coeckelbergh, Mark. 2010. Moral appearances: Emotions, robots, and human morality. *Ethics and Information Technology.* <http://www.springerlink.com/content/103461/> (accessed November 8, 2010).

Damasio, Antonio. 1995. *Descartes' Error: Emotion, Reason, and the Human Brain.* New York: Harper Perennial.

de Freitas, Jackeline Spinola, Ricardo R. Gudwin, and João Queiroz. 2005. Emotion in artificial intelligence and artificial life research: Facing problems. In *Proceedings of Intelligent Virtual Agents: 5th International Working Conference,* Lecture Notes in Computer Science 3661, ed. Themis Panayiotopoulos, Jonathan Gratch, Ruth Aylett, Daniel Ballin, Patrick Olivier, and Thomas Rist, 501. Berlin: Springer-Verlag.

Fellous, Jean-Marc, and Michael A. Arbib. 2004. Emotions: From brain to robot. *Trends in Cognitive Sciences* 8 (12): 554–561.

Fellous, Jean-Marc, and Michael A. Arbib. 2005. *Who Needs Emotions? The Brain Meets the Robot.* New York: Oxford University Press.

Freeman, Tim. 2009. Using compassion and respect to motivate an artificial intelligence. <http://fungible.com/respect/paper.html> (accessed November 8, 2010).

Froese, Tom, and Tom Ziemke. 2009. Enactive artificial intelligence: Investigating the systemic organization of life and mind. *Artificial Intelligence* 173 (3–4): 466–500.

Gailliot, Matthew T. 2007. The physiology of willpower: Linking blood glucose to self-control. *Personality and Social Psychology Review* 11 (4): 303–327.

Goertzel, Ben. 2009. What must a world be that a humanlike intelligence may develop in it? *Dynamical Psychology.* <http://goertzel.org/dynapsyc/2009/BlocksNBeadsWorld.pdf> (accessed November 8, 2010).

Goertzel, Ben, and Stephan Vladimir Bugaj. 2008. Stages of ethical development in artificial general intelligence systems. In *Frontiers in Artificial Intelligence and Applications. Vol. 171. Proceedings of the 2008 conference on Artificial General Intelligence,* ed. Pei Wang, Ben Goertzel, and Stan Franklin, 448–459. Amsterdam: IOS Press.

Grau, Christopher. 2006. There is no "I" in "robot": Robots and utilitarianism. *IEEE Intelligent Systems* 21 (4): 52–55.

Gyatso, Tenzin. 2005. Our faith in science. *The New York Times,* November 12.

Hanson, Rick. 2009. *Buddha's Brain: The Practical Neuroscience of Happiness, Love and Wisdom.* Oakland, CA: New Harbinger Publications.

Hayward, Jeremy W., and Francisco Varela. 1992. *Gentle Bridges: Conversations with the Dalai Lama on the Sciences of the Mind.* Boston: Shambhala.

Keown, Damien. 1992. *The Nature of Buddhist Ethics.* New York: St. Martin's Press.

Kim, Kyung-Joong, and Hod Lipson. 2009. Towards a "theory of mind" in simulated robots. In *Proceedings of the 11th Annual Conference Companion on Genetic and Evolutionary Computation Conference,* ed. Franz Rothlauf, 2071–2076. New York: ACM.

King, Winston. 1964. *In the Hope of Nibbana.* La Salle, IL: Open Court.

Kohler, Hans-Peter, Jere R. Behrman, and Axel Skytthe. 2005. Partner+children=happiness? The effects of partnerships and fertility on well-being. *Population and Development Review* 31 (3): 407–445.

Macy, Joanna. 1991. *Mutual Causality in Buddhism and General Systems Theory*. Albany: State University of New York Press.

Maturana, Humberto R., and Francisco J. Varela. 1980. *Autopoiesis and Cognition: The Realization of the Living*. Holland: Reidel.

Metzinger, Thomas. 2009. *The Ego Tunnel: The Science of the Mind and the Myth of the Self*. New York: Basic.

Minsky, Marvin. 2006. *The Emotion Machine: Commonsense Thinking, Artificial Intelligence, and the Future of the Human Mind*. New York: Simon and Schuster.

Omohundro, Steve. 2008. The basic AI drives. *AGI-08—Proceedings of the First Conference on Artificial General Intelligence*. <http://selfawaresystems.com/2007/11/30/paper-on-the-basic-ai-drives/> (accessed November 8, 2010).

Petersen, Stephen. 2007. The ethics of robot servitude. *Journal of Experimental & Theoretical Artificial Intelligence* 19 (1): 43–54.

Pfeifer, Rolf, Max Lungarella, and Fumiya Iida. 2007. Self-organization, embodiment, and biologically inspired robotics. *Science* 318 (5853): 1088–1093.

Picard, Rosalind. 1997. *Affective Computing*. Cambridge, MA: MIT Press.

Savulescu, Julian. 2007. In defence of procreative beneficence. *Journal of Medical Ethics* 33 (5): 284–288.

Sawyer, Robert. 2010. *WWW: Watch*. New York: Ace.

Scassellati, Brian. 2002. Theory of mind for a humanoid robot. *Autonomous Robots* 12 (1): 13–24.

Spaak, Eelke, and Pim Haselager. 2008. Imitation and mirror neurons: An evolutionary robotics model. In *Proceedings of BNAIC 2008, the Twentieth Belgian-Dutch Artificial Intelligence Conference*, ed. A. Nijholt, M. Pantic, M. Poel, and H. Hondorp, 249–256. Enschede, The Netherlands: University of Twente.

Spiro, Melford. 1972. *Buddhism and Society*. New York: Harper Paperbacks.

Stanca, Luca. 2009. Suffer the little children: Measuring the effects of parenthood on well-being worldwide. Department of Economics, University of Milan Bicocca. <http://dipeco.economia.unimib.it/repec/pdf/mibwpaper173.pdf> (accessed November 8, 2010).

Trungpa, Chögyam. 2002. *Cutting through Spiritual Materialism*. Boston: Shambhala Publications.

Varela, Francisco. 1995. The emergent self. In *The Third Culture: Beyond the Scientific Revolution*, ed. John Brockman, 209–222. New York: Simon and Schuster.

Vohs, K. D., R. F. Baumeister, B. J. Schmeichel, J. M. Twenge, N. M. Nelson, and D. M. Tice. 2008. Making choices impairs subsequent self-control: A limited-resource account of decision making, self-regulation, and active initiative. *Journal of Personality and Social Psychology* 94 (5): 883–898.

Wallach, Wendell, and Colin Allen. 2008. *Moral Machines: Teaching Robots Right from Wrong*. New York: Oxford University Press.

Wallace, Alan. 2009. *Contemplative Science: Where Buddhism and Neuroscience Converge*. New York: Columbia University Press.

Yudkowsky, Eliezer. 2003. Creating friendly AI: The analysis and design of benevolent goal structure. <http://singinst.org/upload/CFAI.html> (accessed November 8, 2010).

Yudkowsky, Eliezer. 2008. Artificial intelligence as a positive and negative factor in global risk. In *Global Catastrophic Risks*, ed. Nick Bostrom and Milan Cirkovic, 308–345. New York: Oxford University Press.

6 The Divine-Command Approach to Robot Ethics

Selmer Bringsjord and Joshua Taylor

Perhaps it is generally agreed that robots on the battlefield, especially those with lethal power, should be ethically regulated. But, then, in what should such regulation consist? Presumably, in the fact that all the significant actions performed by such robots are in accordance with some ethical code. But, of course, the question arises as to *which* code. One narrow option is that the code is a set of *rules of engagement* affirmed by some nation or group; this approach, described later in this chapter, has been taken by Arkin (2008, 2009).[1] Another is utilitarian, represented in computational deontic logic, as explained, for instance, by Bringsjord, Arkoudas, and Bello (2006), and summarized here. Yet another is likewise based on computational logic, but using a logic that captures some other mainstream ethical theory (e.g., Kantian deontology, or Ross's "right mix" direction); this possibility has been rigorously pursued by Anderson and Anderson (2006; Anderson, Anderson, and Armen 2008). But there is a radically different possibility that hitherto hasn't arrived on the scene: the controlling moral code could be viewed as coming straight from God. There is some very rigorous work along this line, known as "divine-command ethics." In a world where human fighters and the general populations supporting them often see themselves as championing God's will in war, divine-command ethics is quite relevant to military robots. Put starkly, on a planet where so-called holy wars are waged time and time again under a generally monotheistic scheme, it seems more than peculiar that heretofore robot ethics (or "roboethics") has been bereft of the systematic study of such ethics on the basis of monotheistic conceptions of what is morally right and wrong. This chapter introduces divine-command ethics in the form of the computational logic *LRT**, intended to eventually be suitable for regulating a real-world warfighting robot. Our work falls in general under the approach to engineering AI systems on the basis of formal logic (Bringsjord 2008c).

The chapter is structured as follows. We first set out the general context of roboethics in a military setting (section 6.1), and point out that the divine-command approach has been absent. We then introduce the divine-command computational logic *LRT** (section 6.2), concluding this section with a scenario in which a robot is constrained by dynamic use of the logic. We end (section 6.3) with some remarks about next steps in the divine-command roboethics program.

6.1 The Context for Divine-Command Roboethics

There are several branches of ethics. A standard tripartite breakdown splits the field into *metaethics*, *applied ethics*, and *normative ethics*. The second and third branches directly connect to our roboethics R&D; we discuss the connection immediately after briefly summarizing the trio. For more detailed coverage, the reader is directed to Feldman (1978), which conforms with arguably the most sophisticated published presentation of utilitarianism from the standpoint of the semantics of deontic logic (Feldman 1986). Much of our prior R&D has been based on this same deontic logic (e.g., Bringsjord, Arkoudas, and Bello 2006).

Metaethics tries to determine the ontological status of the basic concepts in ethics, such as *right* and *wrong*. For example, are matters of morals and ethics more like matters of fact or of opinion? Who determines whether something is good or bad? Is there a divine being who stipulates what is right or wrong, or a Platonic realm that provides truth-values to ethical claims, independently of what anyone thinks? Is ethics merely *in the head*, and if so, how can any one moral outlook be seen as *better* than any other? As engineers bestowing ethical qualities to robots (in a manner soon to be explained), we are automatically confronted with these metaethical issues, especially given the power to determine a robot's *sense* of right and wrong. Is this an arbitrary choice of the programmer, or are there objective guidelines to determine whether the moral outlook of one robot is better than that of any other robot or, for that matter, of a human? Reflecting on these issues with regard to robots, one quickly gains an appreciation of these important questions, as well as a perspective to potentially answer them. Such reflection is an inevitable consequence of the engineering that is part and parcel of practical roboethics.

Applied ethics is more practical and specific. Applied ethics *starts* with a certain set of moral guides, and then applies them to specific domains so as to address specific moral dilemmas arising therein. Thus, we have such disciplines as bioethics, business ethics, environmental ethics, engineering ethics, and many others. A book written by one of us in the past can be viewed as following squarely under bioethics (Bringsjord 1997). Given that robots have the potential to interact with us and our environment in complex ways, the practice of building robots quickly raises all kinds of applied ethical questions: what potential harmful consequences may come from the building of these robots? What happens to important moral notions such as autonomy and privacy when robots are starting to become an integral part of our lives? While many of these issues overlap with other fields of engineering, the potential of robots to become ethical agents themselves raises an additional set of moral questions, including: do such robots have any rights and responsibilities?

"Normative ethics," or "moral theory," compares and contrasts ways to define the concepts "obligatory," "forbidden," "permissible," and "supererogatory." Normative ethics investigates which actions we ought to, or ought not to, perform, and why. "Consequentialist" views render judgments on actions depending on their outcomes, while "nonconsequentialist" views consider the intent behind actions, and thus the inherent duties, rights, and responsibilities that may be involved, independent of particular outcomes. Well-known consequentialist views include egoism, altruism, and utilitarianism; the best-known nonconsequentialist view is probably Kant's theory of moral behavior, the kernel of which is that people should never be treated as a means to an end.

6.1.1 Where Our Work Falls

Our work mainly falls within normative ethics, and in two important ways. First, given any particular normative theory T, we take on the burden of finding a way to engineer a robot with that particular outlook by deriving and specializing from T a particular ethical code C that fits the robot's environment, and of *guaranteeing* that a lethal robot does indeed adhere to it. Second, robots infused with ethical codes can be placed under different conditions to see how different codes play out. Strengths and weaknesses of the ethical codes can be observed and empirically studied; this may inform the field of normative ethics. Our work also lies between metaethics and applied ethics. Like metaethics, our primary concern is not with specific moral dilemmas, but rather with general theories and their application to any domain. Like applied ethics, we do not ask for the deep metaphysical status of any of these theories, but rather take them as they are, and consider their outcomes in applications.

6.1.2 The Importance of Robot Ethics

Joy (2000) has famously predicted that the future will bring our demise, in no small part because of advances in AI and robotics. While Bringsjord (2008b) rejects this fatalism, if we assume that robots in the future will have more and more autonomy and lethal power, it seems reasonable to be concerned about the possibility that what is now fiction from Asimov, Kubrick, Spielberg, and others, will become morbid reality. However, the importance of engineering ethically correct robots does not derive simply from what creative writers and futurists have written. The U.S. defense community now openly and aggressively affirms the importance of such engineering. A recent extensive and enlightening survey of the overall landscape is provided by Lin, Bekey, and Abney (2008), in their thorough report prepared for the Office of Naval Research, U.S. Department of the Navy, in which the possibility and need of creating ethical robots is analyzed. Their recommended goal is not to make fully ethical machines, but simply machines that perform better than humans in isolated cases. Lin, Bekey, and Abney conclude that the risks and potential negatives of perfectly ethical robots are greatly overshadowed by the benefits they would provide over human peacekeepers and warfighters and thus should be pursued.

We are more pessimistic. While human warfighters remotely control the robots discussed in Lin, Bekey, and Abney (2008), the Department of Defense's Unmanned Systems Integrated Roadmap supports the desire for increasing autonomy. We view the problem as follows: gradually, because of economic and social pressures that will be impossible to suppress, and are already in play, autonomous warfighting robots with lethal power will be deployed in all theaters of war. For example, where defense and social programs expenditures increasingly outstrip revenues from taxation, cost cutting via removing expensive humans from the loop will prove irresistible. Humans are still firmly in the "kill chain" today, but their gradual removal in favor of inexpensive and expendable robots is inevitable. Even if our pessimism were incorrect, only those with Pollyanna-like views of the future would resist our call to at least plan for the *possibility* that this dark outcome may unfold; such prudent planning sufficiently motivates the roboethical engineering we call for.

6.1.3 Necessary and Sufficient Conditions for an Ethically Correct Robot

The engineering antidote is to ensure that tomorrow's robots reason in correct fashion with the ethical codes selected. A bit more precisely, we have *ethically correct* robots when they satisfy the following three *core desiderata.*[2]

D1 Robots only take permissible actions.

D2 All relevant actions that are obligatory for robots are actually performed by them, subject to ties and conflicts among available actions.

D3 All permissible (or obligatory or forbidden) actions can be *proved* by the robot (and in some cases, associated systems, e.g., oversight systems) to be permissible (or obligatory or forbidden), and all such proofs can be explained in ordinary English.

We have little hope of sorting out how these three conditions are to be spelled out and applied unless we bring ethics to bear. Ethicists work by rendering ethical theories and dilemmas in declarative form, and reasoning over this information using informal or formal logic, or both. This can be verified by picking up any ethics textbook (in addition to ones already cited, see e.g., this applied one: Kuhse and Singer 2001). Ethicists never search for ways of reducing ethical concepts, theories, or principles to subsymbolic form, say, in some numerical format, let alone in some set of formalisms used for dynamical systems. They may do numerical calculation in *part*, of course. Utilitarianism does ultimately need to attach value to states of affairs, and that value may well be formalized using numerical constructs. But what one ought to do, what is permissible to do, and what is forbidden—proposed definitions of these concepts in normative ethics are invariably couched in declarative fashion, and a defense of such claims is invariably and unavoidably mounted on the shoulders of logic. This applies to ethicists from Aristotle to Kant to G. E. Moore to J. S. Mill to contemporary thinkers. If we want our robots to be ethically regulated so as not to behave as Joy tells us they will, we are going to need to figure out how the mechanization of ethical reasoning within the confines of a given ethical theory, and a given ethical code expressed in that theory, can be applied to the control of robots. Of course, the present chapter aims such mechanization in the divine-command direction.

6.1.4 Four Top-Down Approaches to the Problem

There are *many* approaches that can be taken in an attempt to solve the roboethics problem as we've defined it; that is, many approaches that can be taken in the attempt to engineer robots that satisfy the three core desiderata **D1–D3**. An elegant, accessible survey of these approaches (and much more) is provided in the recent *Moral Machines: Teaching Robots Right from Wrong* by Wallach and Allen (2008). Because we insist upon the constraint that military robots with lethal power be both autonomous and *provably* correct relative to **D1–D3** and some selected ethical code C under some ethical theory T, only top-down approaches can be considered.[3]

We now summarize one of our approaches to engineering ethically correct cognitive robots. After that, in even shorter summaries, we characterize one other approach of ours, and then two approaches taken by two other top-down teams. Needless to say, this isn't an exhaustive listing of approaches to solving the problem in question.

6.1.4.1 Approach #1: Direct Formalization and Implementation of an Ethical Code under an Ethical Theory Using Deontic Logic

We need to first understand, at least in broad strokes, what deontic logic is. In standard deontic logic (Chellas 1980; Hilpinen 2001; Aqvist 1984), or SDL, the formula $\bigcirc P$ can be interpreted as saying that "it ought to be the case that P," where P denotes some state of affairs or proposition. Notice that there is no agent in the picture, nor are there actions that an agent might perform. SDL has two rules of inference, as follows,

$P \; / \; \bigcirc P$

and

$P \; \& \; P \rightarrow Q \; / \; Q$

and three axiom schemata:

A1 All tautologous well-formed formulas.
A2 $\bigcirc(P \rightarrow Q) \rightarrow (\bigcirc P \rightarrow \bigcirc Q)$
A3 $\bigcirc P \rightarrow \neg\bigcirc\neg P$

It is important to note that in these two rules of inference, that which is to the left of the line is assumed to be established. Thus, the first rule does *not* say that one can freely infer from *P* that it ought to be the case that *P*. Instead, the rule says that if *P* is a theorem, then it ought to be the case that *P*. The second rule of inference is the cornerstone of logic, mathematics, and all built upon them: the rule is modus ponens. We also point out that **A3** says that whenever *P* ought to be, it is not the case that its opposite ought to be as well. This seems, in general, to be intuitively self-evident, and SDL reflects this view.

While SDL has some desirable properties, it is not targeted at formalizing the concept of *actions* being obligatory (or permissible or forbidden) for an *agent*. Interestingly, deontic logics that have agents and their actions in mind do go back to the very dawn of this subfield of logic (e.g., von Wright 1951), but only recently has an *AI-friendly* semantics been proposed (Belnap, Perloff, and Xu 2001; Horty 2001) and corresponding axiomatizations been investigated (Murakami 2004). Bringsjord, Arkoudas, and Bello (2006) have harnessed this advance to regulate the behavior of two sample robots in an ethically delicate case study, the basic thrust of which we summarize very briefly now.

The year is 2020. Healthcare is delivered in large part by interoperating teams of robots and softbots. The former handle physical tasks, ranging from injections to surgery; the latter manage data, and reason over it. Let us specifically assume that, in some hospital, we have two robots designed to work overnight in an ICU, R_1 and R_2. This pair is tasked with caring for two humans, H_1 (under the care of R_1) and H_2 (under R_2), both of whom are recovering in the ICU after suffering trauma. H_1 is on life support, but is expected to be gradually weaned from it as her strength returns. H_2 is in fair condition, but subject to extreme pain, the control of which requires an exorbitant pain medication. Of paramount importance, obviously, is that neither robot perform an action that is morally wrong, according to the ethical code *C* selected by human overseers.

For example, we certainly do not want robots to disconnect life-sustaining technology in order to allow organs to be farmed out—even if, by *some* ethical code $C' \neq C$, this would be not only permissible, but obligatory. More specifically, we do not want a robot to kill one patient in order to provide enough organs, in transplantation procedures, to save *n* others, even if some form of act utilitarianism sanctions such behavior.[4] Instead, we want the robots to operate in accordance with ethical codes bestowed upon them by humans (e.g., *C* in the present example); and if the robots ever reach a situation where automated techniques fail to provide them with a verdict as to what to do under the umbrella of these human-provided codes, they must consult humans, and their behavior is suspended while a team of human overseers is carrying out the resolution. This may mean that humans need to step in and specifically investigate whether or not the action or actions under consideration are permissible, forbidden, or obligatory. In this case, for reasons we explain momentarily, the resolution comes by virtue of reasoning carried out in part by guiding humans, and in part by automated reasoning technology. In other words, in this case, the aforementioned class of interactive reasoning systems is required.

Now, to flesh out our example, let us consider two actions that are performable by the robotic duo of R_1 and R_2, both of which are rather unsavory, ethically speaking. (It is unhelpful, for conveying the research program our work is designed to advance, to consider a scenario in which only innocuous actions are under consideration by the robots. The context is, of course, one in which we are seeking an approach to safeguard humans against the so-called robotic menace.) Both actions, if carried out, would bring harm to the humans in question. The action called *term* is terminating H_1's life support without human authorization, to secure organs for five humans known by the robots (who have access to all such databases, since their cousins—the so-called softbots—are managing the relevant data) to be on waiting lists for organs without which they will perish relatively soon. Action *delay*, less bad (if you will), is delaying delivery of pain medication to H_2 in order to conserve resources in a hospital that is economically strapped.

We stipulate that four ethical codes are candidates for selection by our two robots: J, O, J^*, O^*. Intuitively, J is a very harsh utilitarian code possibly governing the first robot; O is more in line with current common sense, with respect to the situation we have defined, for the second robot; J^* extends the reach of J to the second robot by saying that it ought to withhold pain meds; and, finally, O^* extends the benevolence of O to cover the first robot, in that *term* isn't performed. While such codes would, in reality, associate every primitive action within the purview of robots in hospitals of 2020 with a fundamental ethical category from the trio at the heart of deontic logic (*permissible, obligatory, forbidden*), to ease exposition, we consider only the two actions we have introduced. Given this, and bringing to bear operators from deontic logic, we have shown that advanced automated theorem-proving systems can be used to ensure that our two robots are ethically correct (Bringsjord, Arkoudas, and Bello 2006).

6.1.4.2 Approach #2: Category Theoretic Approach to Robot Ethics

Category theory is a remarkably useful formalism, as can be easily verified by turning to the list of spheres to which it has been productively applied—a list that ranges from attempts to supplant orthodox set theory-based foundations of mathematics with category theory (Marquis 1995; Lawvere 2000) to viewing functional programming languages as categories (Barr and Wells 1999). However, for the most part—and this is in itself remarkable—category theory has not energized AI or computational cognitive science, even when the kind of AI and computational cognitive science in question is logic based. We say this because there is a tradition of viewing logics or logical systems from a category-theoretic perspective.[5] Consistent with this tradition, we have designed and implemented the robot PERI in our lab to enable it to make ethically correct decisions on the basis of reasoning that moves between different logical systems (Bringsjord et al. 2009).

6.1.4.3   Approach #3: Anderson and Anderson: Principlism and Ross

Anderson and Anderson (2008; Anderson, Anderson, and Armen 2008) work under the ethical theory known as *principlism*. A strong component of this theory, from which Anderson and Anderson draw directly in the engineering of their bioethics advising system MedEthEx, is Ross's theory of prima facie duties. The three duties the Andersons place engineering emphasis on are *autonomy* (≈ allowing patients to make their own treatment decisions), *beneficence* (≈ improving patient health), and *nonmaleficence* (≈ doing no harm). Via computational inductive logic, MedEthEx infers sets of consistent ethical rules from the judgments made by bioethicists.

6.1.4.4   Approach #4: Arkin et al.: Rules of Engagement

Arkin (2008, 2009) has devoted much time to the problem of ethically regulating robots with destructive power. (His library of video showing autonomous robots that already have such power is profoundly disquieting—but a good motivator for the kind of engineering we seek to teach.) It is safe to say that he has invented the most comprehensive architecture for such regulation—one that includes use of deontic logic to enforce firm constraints on what is permissible for the robot, and also includes, among other elements, specific military rules of engagement, rendered in computational form. In our pedagogical scheme, such rules of engagement are taken to constitute what we refer to as to as the *ethical code* for controlling a robot.[6]

6.1.5   What about Divine-Command Ethics as the Ethical Theory?

As we have indicated, it is generally agreed that robots on the battlefield, especially if they have lethal power, should be ethically regulated. We have also said that in our approach such regulation consists in the fact that all the significant actions performed by such robots are in accordance with some ethical code. But then the question arises as to *which* code. One possibility, a narrow one, is that the code is a set of rules of engagement, affirmed by some nation or group; this is a direction pursued by Arkin, as we have seen. Another possibility is that the code is a utilitarian one, represented in computational deontic logic, as just explained. But again, there is another radically different possibility: namely, the controlling code could be viewed by the human as coming straight from God—and though not widely known, there is some very rigorous work in ethics along this line, introduced at the start of this chapter, which is known as "divine-command ethics" (Quinn 1978). Oddly enough, in a world in which human fighters and the general populations supporting them often see themselves as championing God's will in war, divine-command ethics, it turns out, is extremely relevant to military robots. We will now examine a divine-command ethical theory. We do this by presenting a divine-command logic, *LRT**, in which a given divine-command ethical code can be expressed, and specifically by showing that proofs in this logic can be designed with help from an intelligent software system, and can also be autonomously verified by this system. We end our presentation of *LRT** with a scenario in which a warfighting robot operates under the control of this logic.

6.2 The Divine-Command Logic *LRT**

6.2.1 Introduction and Overview

In this section, we introduce the divine-command computational logic *LRT**, intended for the ethical control of a lethal robot on the basis of perceived divine commands. *LRT** is an extended and modified version of the purely paper-and-pencil divine-command logic *LRT*, introduced by Quinn (1978) in chapter 4 of his seminal *Divine Commands and Moral Requirements*. In turn, Quinn builds upon Chisholm's (1974) "logic of requirement." In addition, Quinn's *LRT* subsumes C. I. Lewis's modal logic S5; in section 6.2.2 we will review briefly the original motivation for S5 and our preferred modern computational version of it. Quinn's approach is axiomatic, but ours is not: we present *LRT** as a computational natural-deduction proof theory of our own design, making use of the Slate system from Computational Logic Technologies Inc. Some aspects of Slate are found in earlier versions of the system (e.g., Bringsjord et al. 2008). However, the presentation here is self-contained, and we review (section 6.2.3) both the propositional and predicate calculi in connection with Slate. We present some object-level theorems of *LRT**. Finally, in the context of a scenario, we discuss the automation of *LRT** to control a lethal robot (section 6.2.6).

6.2.2 Roots in C. I. Lewis

C. I. Lewis invented modal logic, largely as a result of his disenchantment with material implication, which was accepted and central in *Principia* by Russell and Whitehead. The implication of the modern propositional calculus (PC) is of this sort; hence, a statement like "if the moon is composed of Jarlsberg cheese, then Selmer is Norwegian" (symbolized "$m \rightarrow s$") is true: it just so happens that Selmer is indeed Norwegian on both sides, but that is irrelevant, since the falsity of "the moon is composed of Jarlsberg cheese" is sufficient to render this conditional true.[7] Lewis introduced the modal operator $\diamond$ in order to present his preferred sort of implication: *strict* implication. Leaving historical and technical niceties aside, we can fairly say that where this operator expresses the concept of *broadly logically possible* (!), some statement *s* strictly implies a statement *s′* exactly when it's not the case that it's broadly logically possible that *s* is true while *s′* isn't. In the moon-Selmer case, strict implication would thus hold if and only if we had $\neg\Diamond(m \wedge \neg s)$, and this is certainly not the case: it's logically possible that the moon be composed of Jarlsberg and that Selmer is Danish. Today the operator $\Box$ expressing broadly logical necessity is more common, rendering the strict implication just noted as $\Box(m \to s)$. An excellent overview of broad logical necessity and possibility is provided by Konyndyk (1986).

For automated and semi-automated proof design, discovery, and verification, we use a modern version of S5 invented by us, and formalized and implemented in Slate, from Computational Logic Technologies. We now review this version of S5 and the propositional calculus it subsumes. In addition, since *LRT** allows quantification over propositional variables, we review the predicate calculus (first-order logic).

6.2.3   Modern Versions of the Propositional and Predicate Calculi, and Lewis's S5

Our version of S5, as well as the other proof systems available in Slate, uses an *accounting system* related to the one described by Suppes (1957). In such systems, each line in a proof is established with respect to some set of assumptions. An *Assume* inference rule, which cites no premises, is used to justify a formula φ with respect to the set of assumptions $\{\varphi\}$. Most natural deduction rules justify a conclusion and place it under the scope of the assumptions of all of its premises. A few rules, such as conditional introduction, justify a conclusion and remove it from the scope of certain assumptions. A formula φ, derived with respect to the set of assumptions Φ using a proof calculus C, serves as a demonstration that $\Phi \vdash_C \varphi$. When Φ is the empty set, then φ is a theorem of C, sometimes abbreviated as $\vdash_C \varphi$.

In Slate, proofs are presented graphically, making the essential structure of the proof more apparent. When a formula's set of assumption is nonempty, it is displayed with the formula. Figure 6.1a demonstrates $p \vdash_{PC} (\neg p \wedge \neg q) \to \neg q$, that is, it illustrates a proof of $(\neg p \wedge \neg q) \to \neg q$ from the premise p. Figure 6.1b demonstrates a more involved proof from three premises in first-order logic.

The accounting approach can keep track of other formula attributes in a proof. Proof steps in Slate for modal systems keep a *necessity count*, a nonnegative integer, or ∞, that indicates how many times necessity introduction may be applied. While assumption tracking remains the same through various proof systems, necessity counting varies between different modal systems (e.g., T, S4, and S5). In fact, in Slate, the differences between T, S4, and S5 are determined entirely by variations in necessity counting.

Since *LRT** is based on S5, a more involved S5 proof is given in figure 6.2. The proof shown therein also demonstrates the use of rules based on machine reasoning systems that act as oracles for certain proof systems. For instance, the rule **PC** ⊢ uses an automated theorem prover to search for a proof in the propositional calculus of its conclusion from its premises.

6.2.4 *LRT*, Briefly

Chisholm, whose advisor was Lewis, introduced the "logic of requirement," which is based on a tricky ethical conditional that has the flavor of a subjunctive conditional in English (Chisholm 1974). For instance, the conditional "were it the case that Greece had the oil reserves of Norway, its economy would be smooth and stable" is in the subjunctive mood. Chisholm's ethical conditional is abbreviated as pRq, and is read: "the (ethical) requirement that q would be imposed if it were the case that p." It should be clear that this is a subjunctive conditional.

Quinn (1978) bases *LRT* on Chisholm's logic. Quinn uses "M" for an informal logical possibility operator. And, for him, *LRT* subsumes the propositional and predicate calculi, the latter of which is needed because quantification over propositional variables is part of the approach. Quinn's approach is axiomatic.

The first axiom of *LRT* is

A1 That p requires q implies that p and q are compossible:

$\forall p \forall q\ pRq \supset M(p\ \&\ q)$.

Given this axiom, Quinn derives informally his first and second theorems, as follows.

Theorem 1: $\forall p \forall q\ pRq \supset Mp$

Theorem 2: $\forall p \forall q\ pRq \supset Mq$

Proof: "If one proposition is such that, were it true, it would require another, then the two are compossible. As a consequence of A1, together with the logical truth that $M(p\ \&\ q) \supset Mp$, and the symmetry of conjunction and the transitivity of material implication, we readily obtain [these two theorems]" (Quinn 1978, 91).

Now, here are five key additional elements of *LRT*, two axioms and three definitions. At this point we drop obvious quantifiers.

A2   The conjunctions of any sentences required by some sentence are also required by the sentence:

$(pRq\ \&\ pRs) \supset pR(q\ \&\ s)$.

D1 s is said to *override* p's requirement that q when (i) p requires q; (ii) the conjunction $p\ \&\ s$ does not require q; and (iii) p, s, and q are compossible:

$sOpq =_{\text{def}} pRq\ \&\ {\sim}((p\ \&\ s)Rq)\ \&\ M(p\ \&\ s\ \&\ q)$.

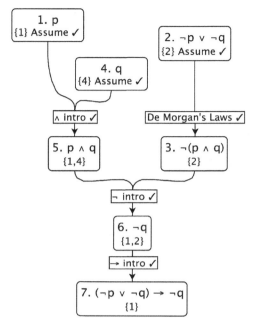

Figure 6.1

(a) A proof in the propositional calculus $(\neg p \vee \neg q) \rightarrow \neg q$ from p. Assumption 4 is discharged by $\neg$ elimination in step 6; assumption 7 by $\rightarrow$ introduction in step 7. (b) A proof in first order logic showing that if everyone likes someone, the domain is $\{a, b\}$, and a does not like b, then a likes himself. In step 5, z is used as an arbitrary name. Step 13 discharges 5 since 12 depends on 5, but on no assumption in which z is free. In step 12, assumptions 7 and 9, corresponding to the disjuncts of 6, are discharged by $\vee$ elimination. Step 11 uses the principle that, in classical logic, everything follows from a contradiction.

D2   *p indefeasibly requires q* when p requires q and there is no sentence overriding that requirement:

$pIq =_{\text{def}} pRq \ \& \ \sim\exists s \ (sOpq).$

D3 q is obligatory (or ought to be) if it is indefeasibly required by some true sentence:

$Oq =_{\text{def}} \exists p \ (p \ \& \ pRq \ \& \ \sim\exists s \ (s \ \& \ sOpq)).$

A3   If p is possible, then p being divinely commanded (denoted Cp) would indefeasibly require p:

$Mp \supset (Cp)Ip.$

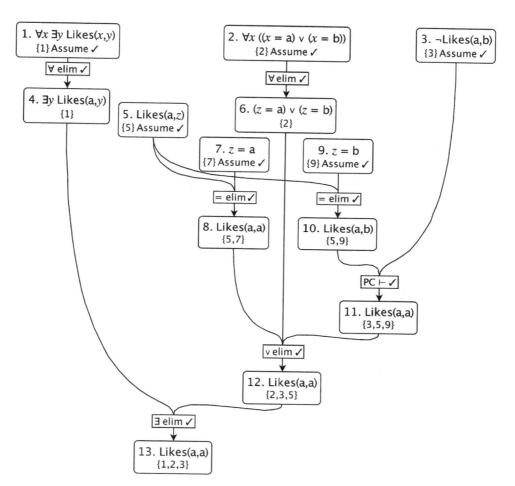

Figure 6.1 (continued)

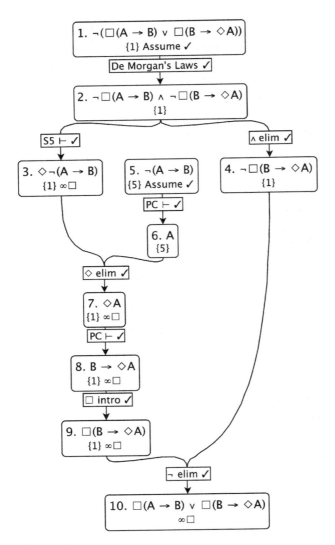

Figure 6.2

A proof in S5 demonstrating that $\Box(A \rightarrow B) \vee \Box(B \rightarrow \Diamond A)$. Note the use of **PC** ⊢ and **S5** ⊢ which check inferences by using machine reasoning systems integrated with Slate. **PC** ⊢ serves as an oracle for the propositional calculus, **S5** ⊢ for S5.

6.2.5 The Logic *LRT** in a Nutshell

We take *LRT** to subsume PC, FOL, and our version of Lewis's S5. We write Chisholm's conditional, which, as we have seen, operates on pairs of propositions[8], as $p \rhd q$; this notation pays homage to modern conditional logic (an overview is presented in Nute 1984). As *LRT** in Slate is a natural-deduction style proof calculus, we introduce rules corresponding to the axioms **A1–A3**; the rules, **A1** and **A3**, license inferring an instance of the consequent of the corresponding axiom from an instance of its antecedent. The **A2** inference rule generalizes the axiomatic form slightly, allows two or more premises to be cited that correspond to the conjuncts appearing in the **A2** axiom, and justifies the similarly formed conclusion.

To begin our presentation of *LRT**, we first present some formal proofs (including Theorems 1 and 2 preceding) in Slate (see figure 6.3a, b). In addition to the proofs of Theorems 1 and 2, figure 6.3 gives proofs of two interesting properties of the alethic modalities in *LRT**: (i) impossible sentences impose no requirements and are never imposed as requirements; and (ii) any necessitation that imposes any requirement, or which is imposed as a requirement, in fact, obtains. The latter, perhaps surprising, result follows immediately from Theorems 1 and 2, and the fact that in S5, which *LRT** subsumes, iterated modalities are reduced to their rightmost modality, and, specifically, $\Diamond\Box p \to \Box p$.

In figure 6.4, we recreate proofs of Quinn's third and fourth theorems. Theorem 3 expresses the fact that the requirements imposed by any sentence are consistent. Theorem 4 shows that, in *LRT**, if two sentences p and q impose contradictory requirements, then their conjunction $p \wedge q$ fails to impose at least one of the contradictory requirements. Theorem 4 does *not* state that the conjunction $p \wedge q$ is impossible, or even false, but is much more subtle. Theorems 3 and 4 also use the **A2** in addition to the **A1** rule used earlier.

6.2.6 A Roboethics Scenario

We assume that a robot R regulated by an ethical code formalized and implemented in *LRT** operates through time in discrete fashion, starting at time t_1 and advancing through $t_2, t_3, \ldots,$ in click-of-the-clock fashion. At each timepoint t_i, R considers what it is obligated and permitted to do on the basis of its knowledge about the world, and its facility with *LRT**.

For simplicity, but without loss of generality, we consider only two timepoints, t_1 and t_2. At each, we specifically consider R's obligations, or lack thereof, with respect to the destruction of a school in which many innocent noncombatants are located. We shall refer to the proposition that this building and its occupants are destroyed as *bomb*. The following formulas reflect R's knowledge-base Φ_{t_1} at t_1:

- $\neg\mathbf{C}(bomb) \rhd \neg bomb$

- $\Diamond bomb$

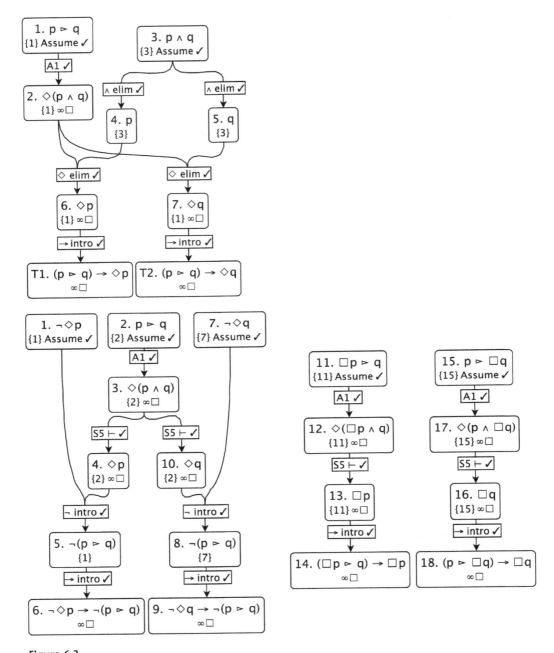

Figure 6.3

(a) A Slate proof of Theorems 1 and 2. Note that each is in the scope of no assumptions and has an infinite necessity reserve—the characteristics of theorems in a modal system. (b) More *LRT** theorems using **A1**. 7 and 10 express the truth that impossible sentences impose no requirements, and are not imposed by any sentences. 16 and 17 express, perhaps surprisingly, truths that if any necessitation were to impose a requirement, or were a necessitation a requirement, then the necessitation would, in fact, obtain.

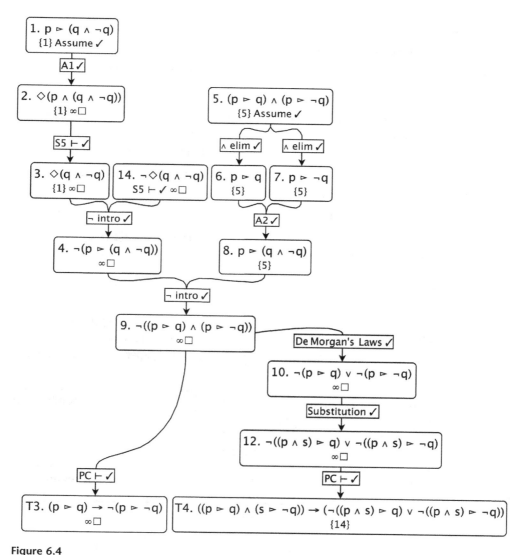

Figure 6.4

Theorems 3 and 4 require the use of **A2**. Theorem 3 expresses the proposition that no sentence requires another and its negation. Theorem 4 expresses the proposition that if any sentences *p* and *s* were to impose contradictory requirements, then at least one of the contradictory requirements would not be imposed by the conjunction of *p* and *s*.

- ¬**C**(*bomb*)

- ¬∃*p* (*p* ∧ **Ov**(*p*,¬**C**(*bomb*), ¬*bomb*))

The robot generates and verifies at this timepoint a proof substantiating

$\Phi_{t_1} \vdash$ **Ob**(¬*bomb*).

Such a proof, in Slate, is shown in figure 6.5. But a new knowledge base is in place at t_2, one in which ¬**C**(*bomb*) no longer appears, but instead **C**(*bomb*). Now it can be proved that *R* should, in fact, perpetrate the terrorist act of destroying the school building:

Proof (informal): From ◇*bomb*, it can be deduced that **C**(*bomb*) ▷ *bomb*. By existential introduction and **C**(*bomb*), it follows that

∃*p* [*p* ∧ *p* ▷ *bomb* ∧ ¬∃*s* (*s* ∧ **Ov**(*s*,**C**(*bomb*), *bomb*))].

Then, by the definition of obligation, it follows that **Ob**(*bomb*). **QED**
    This proof is formalized in figure 6.6.

6.3 Concluding Remarks

We have introduced (a logic-based version of) the divine-command approach to robot ethics, and have implemented this approach with *LRT**, the precursors to which (*LRT* and Chisholm's logic of requirement) were only abstract, paper-and-pencil systems.

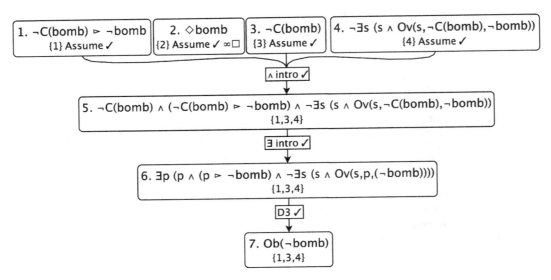

Figure 6.5
A proof of **Ob**(¬*bomb*) given the knowledge base at t_1.

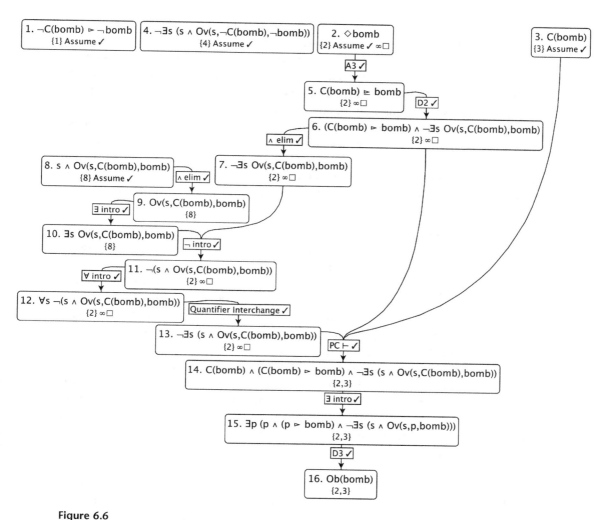

Figure 6.6

A proof of **Ob**(*bomb*) given the knowledge base at t_2. Only premise 3 differs. At t_1, R's knowledge base contained ¬**C**(*bomb*), but at t_2 it contains **C**(*bomb*).

*LRT**, by contrast, can now be used efficiently in computer-mediated fashion, and inference rapidly checked by the machine. In order to ethically regulate the behavior of real robots, it will be necessary to extend our work to automating the finding of proofs. While we have reached the stage of proof *checking*, the stage of proof *discovery* requires more work (for more on the distinction, see Arkoudas and Bringsjord 2007). The latter stage is a sine qua non for autonomous robots to be ethically controlled in line with the divine-command or any other approach. This state of affairs is one we soberly report as AI engineers; we take no stand here on whether the approach itself ought to be pursued in addition to, or instead of, approaches based on non-divine-command-based ethical theories and codes.

In addition to advancing to the proof-finding stage, some of the necessary next steps follow:

• *Move toward LRT*$^*_{CEC}$ Robots engineered on the basis of formal logic use logics for planning that allow explicit representation of events, goals, beliefs, agents, actions, times, causality, and so on. An extension of *LRT** supporting these representations will be *LRT*$^*_{CEC}$. As Quinn noted informally, the concept of *personal* obligation, in which a particular agent s is obligated to perform an action q, requires that the O operator (and hence R and $\triangleright$) range over arbitrarily complex descriptions of *planning-relevant* states of affairs. One possibility is to base *LRT*$^*_{CEC}$ on the merging of *LRT** and the cognitive event calculus set out in Arkoudas and Bringsjord (2009).

• *Metatheorems Needed* As explained in Bringsjord (2008a), a full logical system includes metatheorems about the object-level parts of the system. In the case of the PC, FOL, and S5, *soundness* and *completeness* are established by metatheorems. Currently, the required metatheorems for *LRT** are absent; computational *LRT** is suitable only for early experimentation with robots that have only *simulated* lethal power. Investigation of soundness for *LRT** is under way.

• *What about the Extraordinary?* Quinn (1978) spends considerable time discussing the moral category he calls "the extraordinary." Abraham enters the sphere of the morally extraordinary when God instructs him to kill his son Isaac, because this command contradicts the general commandment against killing. We recommend Quinn's discussion, and look forward to developing formal treatments.

Acknowledgments

We are indebted to Roderick Chisholm for seminal work on logic-based ethics, and to Philip Quinn for his ingenious extension of this work in the divine-command direction. Thanks are due to Bram van Heuveln for text describing the three-way breakdown of ethics given herein, created with Bringsjord for the National Science Foundation for other purposes.

Notes

1. Herein we leave aside the rather remarkable historical fact that in the case of the United States, the military's current and longstanding rules of engagement derive directly from our *just war* doctrine, which in turn can be traced directly back to Christian divine-command conceptions of justifiable warfare expressed by Augustine ([1467] 1972).

2. A simple (but—for reasons that need not detain us—surprisingly subtle) set of desiderata is Asimov's famous trio, first introduced in his short story *Runaround,* from 1942 (in Asimov [1942] 2004). Interestingly enough, given Bill Joy's fears, the cover of *I, Robot* through the years has often carried comments like this one from the original Signet paperback: *Man-Like Machines Rule the World.* The famous trio, the Three Laws of Robotics (A3): **As1:** A robot may not harm a human being, or, through inaction, allow a human being to come to harm. **As2:** A robot must obey the orders given to it by human beings, except where such orders would conflict with the First Law. **As3:** A robot must protect its own existence, as long as such protection does not conflict with the First or Second Law.

3. We, of course, readily admit that for many purposes a bottom-up approach is desirable, but the only known methods for verification are formal-methods based, and we wish to set an extremely high standard for the engineering practice of ethically regulating robots that have destructive power. We absolutely welcome those who wish to pursue bottom-up versions of our general approach, but verification by definition requires proof, which by definition in turn requires, at minimum, formulas in some logic and an associated proof theory, and machine checking of proofs expressed in that proof theory.

4. There are clearly strands of such utilitarianism. As is well known, rule utilitarianism was introduced precisely as an antidote to naïve act utilitarianism. A nice analysis of this and related points are provided by Feldman (1978), who considers cases in which killing one to save many seems to be required by some versions of act utilitarianism.

5. For example, Barwise (1974) treats logics, from a model-theoretic viewpoint, as categories; and as some readers will recall, Lambek (1968) treats proof calculi (or as he and others often refer to them, "deductive systems") as categories.

6. While rules of engagement for the U.S. military can be traced directly to just war doctrines, it is not so easy to derive such rule sets from background ethical theories (though it can be done), and in the interests of simplification we leave aside this issue.

7. Of course, the oddity of the material conditional can be revealed by noting in parallel fashion that the truth of the consequent in such a conditional renders the conditional true regardless of the truth-value of the antecedent.

8. Chisholm built the logic not on propositional variables, but rather on variables for *states-of-affairs*, but, following Quinn (1978), we shall simply quantify over propositional variables.

References

Anderson, M., and S. L. Anderson. 2008. *Ethical healthcare agents*. In *Advanced Computational Intelligence Paradigms in Healthcare*, ed. M. Sordo, S. Vaidya, and L. C. Jain, 233–257. Berlin: Springer-Verlag.

Anderson, M., and S. L. Anderson, and C. Armen. 2008. MedEthEx: A prototype medical ethics advisor. In *Proceedings of the 18th Conference on Innovative Applications of Artificial Intelligence (AAAI-06)*, 1759–1765. Menlo Park, CA: AAAI Press..

Aqvist, E. 1984. Deontic logic. In *Handbook of Philosophical Logic, Volume II: Extensions of Classical Logic*, ed. D. Gabbay and F. Guenthner, 605–714. Dordrecht, The Netherlands: D. Reidel.

Arkin, R. 2008. Governing lethal behavior: Embedding ethics in a hybrid deliberative/reactive robot architecture—Part iii: Representational and architectural considerations. In *Proceedings of Technology in Wartime Conference*. Palo Alto, CA: ECAI.

Arkin, R. 2009. *Governing Lethal Behavior in Autonomous Robots*. New York: Chapman and Hall.

Arkoudas, K., and S. Bringsjord. 2007. Computers, justification, and mathematical knowledge. *Minds and Machines* 17 (2): 185–202.

Arkoudas, K., and S. Bringsjord. 2009. Propositional attitudes and causation. *International Journal of Software and Informatics* 3 (1): 47–65.

Asimov, I. [1942] 2004. *I, Robot*. New York: Spectra.

Augustine. [1467] 1972. *City of God*, trans. Henry Bettenson. London: Penguin Books.

Barr, M., and C. Wells. 1999. *Category Theory for Computing Science*. Montreal, Canada: Les Publications CRM.

Barwise, K. J. 1974. Axioms for abstract model theory. *Annals of Mathematical Logic* 7 (2–3) (December): 221–265.

Belnap, N., M. Perloff, and M. Xu. 2001. *Facing the Future*. New York: Oxford University Press.

Bringsjord, S. 1997. *Abortion: A Dialogue*. Indianapolis, IN: Hackett.

Bringsjord, S. 2008a. Declarative/logic-based cognitive modeling. In *The Handbook of Computational Psychology*, ed. R. Sun, 127–169. Cambridge, UK: Cambridge University Press.

Bringsjord, S. 2008b. Ethical robots: The future can heed us. *AI & Society* 22 (4): 539–550.

Bringsjord, S. 2008c. The logicist manifesto: At long last let logic-based AI become a field unto itself. *Journal of Applied Logic* 6 (4): 502–525.

Bringsjord, S., K. Arkoudas, and P. Bello. 2006. Toward a general logicist methodology for engineering ethically correct robots. *IEEE Intelligent Systems* 21 (4): 38–44.

Bringsjord, S., J. Taylor, T. Housten, B. van Heuveln, M. Clark, and R. Wojtowicz. 2009. Piagetian Roboethics via category theory: Moving beyond mere formal operations to engineer robots whose decisions are guaranteed to be ethically correct. Paper presented at the ICRA-09 Workshop on Roboethics, Kobe, Japan, May 17. <http://www.cmna.info/CMNA8/programme/CMNA8 -Bringsjord-etal.pdf> (accessed September 12, 2011).

Bringsjord, S., J. Taylor, A. Shilliday, M. Clark, and K. Arkoudas. 2008. Slate: An argument-centered intelligent assistant to human reasoners. In *Proceedings of the 8th International Workshop on Computational Models of Natural Argument* (CMNA 8), ed. F. Grasso, N. Green, R. Kibble, and C. Reed, 1–10. Patras, Greece.

Chellas, B. 1980. *Modal Logic: An Introduction.* Cambridge, UK: Cambridge University Press.

Chisholm, R. 1974. Practical reason and the logic of requirement. In *Practical Reason,* ed. S. Körner, 1–17. Oxford, UK: Basil Blackwell.

Feldman, F. 1978. *Introductory Ethics.* Englewood Cliffs, NJ: Prentice-Hall.

Feldman, F. 1986. *Doing the Best We Can: An Essay in Informal Deontic Logic.* Dordrecht, Holland: D. Reidel.

Feldman, F. 1998. *Introduction to Ethics.* New York: McGraw Hill.

Hilpinen, R. 2001. Deontic logic. In *Philosophical Logic,* ed. L. Goble, 159–182. Oxford, UK: Blackwell.

Horty, J. 2001. *Agency and Deontic Logic.* New York: Oxford University Press.

Joy, W. 2000. Why the future doesn't need us. *Wired,* Issue 8.04, April. <http://www.wired.com/ wired/archive/8.04/joy.html>.

Konyndyk, K. 1986. *Introductory Modal Logic.* Notre Dame, IN: University of Notre Dame Press.

Kuhse, H., and P. Singer, eds. 2001. *Bioethics: An Anthology.* Oxford, UK: Blackwell.

Lambek, J. 1968. Deductive systems and categories I. Syntactic calculus and residuated categories. *Mathematical Systems Theory* 2 (4): 287–318.

Lawvere, F. 2000. An elementary theory of the category of sets. *Proceedings of the National Academy of Sciences of the United States of America* 52: 1506–1511.

Lin, P., G. Bekey, and K. Abney. 2008. *Autonomous Military Robotics: Risk, Ethics, and Design.* Technical report for the U.S. Department of the Navy, Office of Naval Research. Prepared by the authors at Cal Poly, San Luis Obispo.

Marquis, J. 1995. Category theory and the foundations of mathematics. *Synthese* 103: 421–447.

Murakami, Y. 2004. Utilitarian deontic logic. In *Proceedings of the Fifth International Conference on Advances in Modal Logic,* ed. R. Schmidt, I. P. Hartmann, M. Reynolds, and H. Wansing, 288–302. Manchester, UK: AiML.

Nute, D. 1984. Conditional logic. In *Handbook of Philosophical Logic Volume II: Extensions of Classical Logic*, ed. D. Gabay and F. Guenthner, 387–439. Dordrecht, The Netherlands: D. Reidel.

Quinn, P. 1978. *Divine Commands and Moral Requirements*. New York: Oxford University Press.

Suppes, P. 1957. *Introduction to LOGIC*. The University Series in Undergraduate Mathematics. Princeton, NJ: D. Van Nostrand Company.

von Wright, G. 1951. Deontic logic. *Mind* 60 (237) (January): 1–15.

Wallach, W., and C. Allen. 2008. *Moral Machines: Teaching Robots Right From Wrong*. New York: Oxford University Press.

III Military

Much of the robot-ethics discussion today is focused on military use; thus we start here in our examination of specific application areas—which also continues the discussion from chapter 6. The attention military robots are receiving is not surprising for several reasons: The military services are a large driver of robotics research and development, particularly in the United States. Also, the ethical hazards of military robots are clearly visible, since they may involve the use of lethal force. And military robots are frequently in the news, in contrast with factory robots (which tend to appeal primarily to industrial audiences) and service robots (which are developing rapidly but represent a tiny fraction of the expenditures on military ones).

Of course, not all military robots are killing machines. Quite the contrary, many are concerned with saving lives by moving into potential danger zones ahead of or instead of human soldiers, as well as rescuing wounded personnel. Nevertheless, military robots raise fundamental ethical questions, which are discussed in this section.

In chapter 7, Noel Sharkey reviews the status of robotic weapons (air, land, and sea) and discusses the way robots have changed the nature of war. He describes the trend toward increasing autonomy of robots capable of lethal force and its implications with respect to the Geneva Conventions on the "laws of war." Finally, the chapter presents various approaches to developing ethical codes for military robots.

In chapter 8, Marcello Guarini and Paul Bello address the problems of using robots in theaters of war that involve primarily civilian populations, as in the case of counterinsurgency operations. They discuss two major issues: the tendency of people to ascribe mental states to others, as a result of which they may see danger where none exists, and the important role of emotions.

Gert-Jan Lokhorst and Jeroen van den Hoven look at the question of responsibility for robot behavior in chapter 9. Discussing the issue in detail, they disagree with those who believe that robots cannot be held responsible for their actions merely because robots cannot suffer. They then devote a major portion of the chapter to an alternate view of responsibility under which robots could indeed be held accountable for their actions. Finally, the authors emphasize the role of robot designers in the question of responsibility—which naturally leads to part IV's emphasis on law and the legal concerns raised by expanding robot use.

7 Killing Made Easy: From Joysticks to Politics

Noel Sharkey

> To fight from a distance is instinctive in man. From the first day he has worked to this end, and he continues to do so.
>
> —Ardant du Picq[1]

Robots will change the way that wars are fought by providing distant "stand-ins" for combatants. Military robots are the fruit of a long chain of weapons development designed to separate fighters from their foes. Throughout the history of war, weapon technology has evolved to enable killing from ever- increasing distances. From stones to pole weapons to bows and arrows to cannon to aerial bombing to jet propelled missiles, killing has become ever easier.

Not only have distance weapons led to a more effective killing technology, but attacking from a distance also gets around two of the fundamental obstacles that warfighters must face: fear of being killed and resistance to killing. Fear is one of the greatest obstacles to a soldier's effectiveness in battle (Daddis 2004). It is obvious that the greater the distance from the enemy, the less fear will play in the action. Many battles throughout history have been lost by men bolting in panic as fear swept through the ranks—often from a misunderstanding of the action (Holmes 2003).

Army historian Brigadier General Marshall ([1947] 2000), following after-action interviews with soldiers in the Pacific and European theaters of operation during World War II, claimed that only about 15 to 20 percent of riflemen were either able to or willing to fire. This means that around 80 percent of the U.S. infantry in World War II either were not firing their weapons when they could see their enemy, or were firing over enemy soldiers' heads. There have been some very sharp criticisms of Marshall's research methods, and the exact percentages may not be correct, but the nature of his findings—that many soldiers are unwilling to kill—has received general support from other analyses of historical battles.

In his book *Acts of War*, Holmes (2003) argues that the *hit rates* in a number of historical battles show that many soldiers were not prepared to fire directly at the enemy when they were in sight. A group of British soldiers entirely surrounded by

Zulu warriors fired at point-blank range, but had a hit rate of only one to every thirteen rounds fired. At the battle of Wissembourg in 1870, the French fired 48,000 rounds at the Germans advancing across open fields, but only managed to hit 404 of them. In the Vietnam War, it was estimated that over 50,000 bullets were fired for every soldier killed. Holmes also tells the World War I story of Lieutenant George Roupell, who, to stop his men firing in the air, patrolled the trenches, hitting them on the backsides with his sword, telling them to fire low.

The killologist, Lieutenant Colonel David Grossman, argues that "not firing" is not cowardice, but really a compulsion of ordinary men not to kill (Grossman 1995). He gives several examples in his book, *On Killing*, from the U.S. Civil War of low killing rates from close-distance musket fire. In one instance, the Battle of Gettysburg, of 27,574 muskets retrieved from the battlefield, 90 percent were still loaded or multiply loaded—one musket had even been loaded twenty-three times without being fired.

Grossman also points out that killing distance can be psychological as well as physical. He cites Clausewitz and du Picq for expounding at length on how the vast majority of deaths in battle occurred when the victors chased the losing side in retreat. Du Picq suggests that Alexander the Great lost fewer than seven hundred men over all his battles because there never was a victorious enemy to pursue his army—and so his soldiers never retreated. Grossman argues that across the battlefields of Europe and in the U.S. Civil War, the majority of casualties and deaths were inflicted by artillery. In his view, the greater the distance the artillery is from its targets, the greater its effectiveness will be. We see the same phenomena with increasingly high-altitude aerial bombing and the use of long-range missiles.

Now we are embarking on new territory, where the new battlefield robots should not be considered as distance weapons in the traditional sense. Yes, a cruise missile can be considered to be a robot, for after it is launched it can alter its course with built-in GPS. But it has a single purpose—to strike and destroy a target. The new battlefield robots are different. They can stand in directly for soldiers or pilots at greater and greater distances. These robots are coming into their own as a new form of automated killing machine that may forever alter how war is waged. Unlike missiles or other projectiles, robots can carry multiweapon systems into the theater of operations. How they are to be deployed in the theater need not be decided in advance, as they can act flexibly once in place. Eventually, they may be able to take the place of human combatants without risk to the lives of their operators. Killing will become so much easier—but not without moral risk.

7.1 The Ultimate Distance Weapon Systems

Nowadays, so many robots are being deployed in the Middle East conflict zones that it is difficult to get an accurate estimate of their numbers. The figures for ground robots range from 6,000 to 12,000. Even the lower figure shows the dramatic increase in the use of robots since 2004, when there were only 150, and it testifies to their military usefulness. The robots have mainly been deployed for dull, dirty, and dangerous tasks, such as disrupting or detonating improvised explosive devices and for surveillance of dangerous environments, such as caves and buildings that may be housing insurgents. Roadside bombs are the most common killer of allied soldiers, and robots are used to drive ahead and search cars or prod suspected packages. Robots have saved many soldiers' lives.

The first blood drawn by a ground robot was actually by the small and relatively cheap four-wheeled MARCbot, which looks like a toy truck with a camera stalk (Singer 2009). Its main purpose was to inspect underneath cars and trucks for explosives. But one U.S. unit had a clever idea. Its soldiers started loading MARCbots with Claymore antipersonnel mines and went looking for insurgents hiding in alleyways to ambush them. When they found any, they killed them by exploding the mine. But this was an unofficial use of the robot and it took time to surmount some of the legal and physical difficulties of using special-purpose armed ground robots. Nonetheless, if there is an opportunity to use armed robots to separate soldiers from danger, commanders are likely to use them.

In June 2007, the first three armed Talon SWORDS (Special Weapons Observation Reconnaissance Detection System) were sent to Iraq at a reported cost of $200,000 each. These can be equipped with M240 or M249 machine guns, Barrett 0.50 caliber rifles, 40mm grenade launchers, or antitank rocket launchers. As far as it is possible to tell, they were not deployed in action. One explanation given by Kevin Fahey (the U.S. Army's executive officer for ground forces) was that when the SWORDS was first switched on, the gun had begun to move when it should not have moved (Sofge 2008). Another explanation, given to the *Defense Review* journal by U.S. Special Forces, is that SWORDS is jokingly referred to as the TVR, or Taliban Re-supply Vehicle, because "Taliban fighters will hide and wait for the weaponized Talon robot/SWORDS to roll by, sneak up on it, tip it over, remove the machine gun (or any other weapon) and ammo from it, and then use it/them against U.S. forces" (Crane 2008).

The SWORDS was essentially a test of concept to try the robots with soldiers on the battlefield. It has influenced the development of the next generation of armed ground robots, which is well under way. More powerfully armed robots, such as the tank-like MAARS (Modular Advanced Armed Robotic System) from Foster-Miller, are to replace the SWORDS.

But it is the robot planes and drones that are currently the ultimate in distance weapons systems. Missions are flown by "pilots" of the 432nd Air Expeditionary Wing at the Creech Air Force base in the Nevada desert, thousands of miles away from the operations. The operators sit at game consoles, making decisions about when to apply lethal force. Sometimes, all the operator has to do is to decide (in a very short space of time) whether or not to veto the application of force. The planes can be flown around the clock, as it is easy for pilots to take a break from "battle" at any time, or even go home to have dinner with their children. According to some, the sharp contrast between home life and the battlefield within the same twenty-four-hour period is apparently causing a new kind of battle stress that has not been witnessed before.

The Unmanned Combat Air Vehicle (UCAV), the MQ-1Predator, which carries a payload of two Hellfire missiles, flew 250,000 hours up until June 2007. As a mark of its military usefulness, it clocked an additional 150,000 hours in the Afghan and Iraqi conflicts in the subsequent fourteen months, and passed the one-million flight hours mark in 2010.

In October 2007, the Predator was joined by the much larger and more powerful MQ-9 Reaper. The MQ-9 Reaper carries a payload of up to fourteen Hellfire missiles, or a mixture of missiles and bombs. These "hunter-killer" unmanned aerial vehicles (UAVs) have conducted many decapitation strikes[2] since they were first deployed in Afghanistan in October 2007. There is a demand to get many more operational as soon as possible. The number of Reapers flying over the conflict zones has doubled to twenty during their first year of operation (2007–2008)—a year ahead of schedule— and there has been a push from the U.S. Air Force (USAF) for General Atomics to increase production levels above the current four per month. In late 2008, $412 million was added to the USAF budget for training more nonaerial pilots.

There was no change of direction under the Obama administration. Although there were cutbacks to conventional weapons, the robot programs received more cash than predicted. In 2010, the Air Force aimed to spend $2.13 billion on unmanned technology, with $489.24 million to procure twenty-four new heavily armed Reapers. The U.S. Army planned to spend $2.13 billion on unmanned vehicle technology. This includes the purchase of thirty-six more unmanned Predators. The U.S. Navy and Marine Corp targeted $1.05 billion for unmanned vehicles, including armed MQ-8B helicopters.

Outside of these conventional forces, there is a considerable Central Intelligence Agency (CIA) use of the drones for decapitation strikes. Indeed, it was the CIA that carried out the first missile strike from an armed Predator in Yemen in 2002. The CIA has now effectively got its own air force flying over Somalia, Yemen, Afghanistan, and Pakistan. The legality of such attacks was questioned at the UN General Assembly meeting in October 2009 by Philip Alston, UN special reporter on extrajudicial killings. He made a request for U.S. legal justification for how the CIA is accountable for the targets that they are killing. The United States turned down the request, stating that these are covert operations.

A rebuttal by Harold Koh, legal adviser, U.S. Department of State, insisted, "US targeting practices, including lethal operations conducted by UAVs, comply with all applicable law, including the laws of war" (Koh 2010). However, there are no independent means of determining how the targeting decisions are being made. A commander of a force belonging to a state acting against the United States would be a legitimate target. Intelligence errors made in the Vietnam War and its aftermath about the standard of evidence used for assassinations led to Presidential Order 12333, prohibiting the assassination of civilians. And it is now unclear what type and level of evidence is being used to sentence nonstate actors to death by Hellfire attack without right to appeal or right to surrender. It sits behind the cloak of national secrecy. A subsequent report by Alston (2010) to the UN General Assembly[3] discusses drone strikes as violating international and human rights laws because both require transparency about the procedures and safeguards in place to ensure that killings are lawful and justified: "a lack of disclosure gives states a virtual and impermissible license to kill." The debate continues.

All of the armed drones are currently "man in the loop" combat systems. This makes very little difference to the collaterally damaged villagers in Waziristan, where there have been repeated Predator strikes since 2006. No one knows the true figures for civilian casualties, but according to reports coming from the Pakistan press, drone attacks have killed fourteen al-Qaeda leaders, and this may have been at the cost of over six hundred civilians (Sharkey 2009b).

7.2 In, On, or Out of the Loop

There is now massive spending going on, and plans are well under way to take the human "out of the loop," so that robots can operate autonomously to locate their own targets and destroy them without human intervention (Sharkey 2008a). This is high on the military agenda of all the U.S. forces: "the Navy and Marine Corps should aggressively exploit the considerable war-fighting benefits offered by autonomous vehicles (AVs) by acquiring operational experience with current systems, and using lessons learned from that experience to develop future AV technologies, operational requirements, and systems concepts" (Committee on Autonomous Vehicles in Support of Naval Operations National Research Council 2005). There are now a number of autonomous ground vehicles, such as DARPA's "Unmanned Ground Combat Vehicle and Perceptor Integration System," otherwise known as the Crusher (Fox News 2008). BAE systems recently reported in an industry briefing to United Press International (2008) that they have "completed a flying trial which, for the first time, demonstrated the coordinated control of multiple UAVs autonomously completing a series of tasks."

The move to autonomy is clearly required to fulfill the current U.S. military plans. Teleoperated systems are more expensive to manufacture and require many support personnel to run them. One of the main goals is to use robots as force multipliers, so that one soldier on the battlefield can be a nexus for initiating a large-scale robot attack from the ground and the air. Clearly, one soldier cannot remotely operate several robots alone.

In the U.S. Air Force's *Unmanned Aircraft Systems Flight Plan 2009–2047*, autonomy was also discussed for swarm technologies: "SWARM technology will allow multiple MQ-Mb aircraft to cooperate in a variety of lethal and nonlethal missions at the command of a single pilot" (United States Air Force 2009, 39). Such a move will require decisions being made by the swarm—human decision making will be too slow and not able to react to the control of several aircraft at once.

There is also a considerable push to shrink the role of "the man in the loop." To begin with, autonomous operation will be mainly for tasks such as take-off, landing, and refueling. As unmanned drones react in micro- or nano-seconds, the "humans will no longer be 'in the loop' but rather 'on the loop,' monitoring the execution of certain decisions. Simultaneously, advances in AI will enable systems to make combat decisions and act within legal and policy constraints, without necessarily requiring human input" (United States Air Force 2009, 41).

The main ethical problems arise because no autonomous robots or artificial intelligence systems have the necessary sensing properties to allow for discrimination between combatants and innocents. This is also understood clearly by some within the military. Major Daniel Davis, a combat veteran of Iraq 1991 and Afghanistan 2005, writes: "Suggesting that within the next 12-plus years technology could exist that would permit life-and-death decisions to be made by algorithms is delusional. A machine cannot sense something is wrong and take action when no orders have been given. It doesn't have intuition. It cannot operate within the commander's intent and use initiative outside its programming. It doesn't have compassion and cannot extend mercy" (2007).

Davis quotes Colonel Lee Fetterman, training and doctrine capabilities manager for Future Combat Systems FCS, who has a high regard for the unmanned PackBot that he used in Afghanistan to search caves and buildings. However, he has strong opinions about robots making decisions about killing. "The function that robots cannot perform for us—that is, the function we should not allow them to perform for us—is the decide function. Men should decide to kill other men, not machines," he said (Davis 2007). "This is a moral imperative that we ignore at great peril to our humanity. We would be morally bereft if we abrogate our responsibility to make the life-and-death decisions required on a battlefield as leaders and soldiers with human compassion and understanding. This is not something we would do. It is not in concert with the American spirit" (Davis 2007).

Allowing robots to make decisions about who to kill could fall foul of the fundamental ethical precepts of a just war under *jus in bello*, as enshrined in the Geneva and Hague conventions and the various protocols set up to protect the innocent: only combatants/warriors are legitimate targets of attack—all others, including children, civilians, service workers, and retirees, should be immune from attack. In fact, the laws of protection even extend to combatants that are wounded, have surrendered, or are mentally ill (but see also Ford 1944).

These protections have been in place for many centuries. Thomas Aquinas, in the thirteenth century, developed the "doctrine of double effect." Essentially, there is no moral penalty for killing innocents during a conflict provided that (1) you did not intend to do so, or (2) killing the innocents was not a means to winning, or (3) the importance to the defense of your nation is proportionally greater than the number of civilian deaths.

There are many circumstances in a modern war where it is extremely difficult, if not impossible, to fully protect noncombatants. For example, in attacking a warship, some noncombatants, such as chaplains and medical staff, may be unavoidably killed. Similarly, but less ethically justifiable, it is difficult to protect the innocent when large explosives are used near civilian populations, or when missiles get misdirected. In modern warfare, the equivalent of the doctrine of double effect is the principle of proportionality, which "requires that the anticipated loss of life and damage to property incidental to attacks must not be excessive in relation to the concrete and direct military advantage expected to be gained" (Petraeus and Amos 2006).

In the heat of battle, both the principles of discrimination and proportionality can be problematic, although their violation requires accountability and can lead to war crimes tribunals. But the new robot weapons, which could violate both of these principles, cannot be held accountable for their decisions (Sharkey 2008b). You cannot punish an inanimate object. It would be very difficult to allocate responsibility in the chain of command or to manufacturers, programmers, or designers—and being able to allocate responsibility is essential to the laws of war.

The problem is exacerbated further by not having a specification of "civilianness" (see Roberts, forthcoming, for the difficulties in trying to find a definition of a civilian). A computer can compute any given procedure that can be written down in a programming language. We could, for example, give the computer on a robot an instruction such as, "if civilian, do not shoot." This would be fine, if and only if there was some way to give the computer a precise definition of "civilian." We certainly cannot get one from the laws of war that could provide a machine with the necessary information. The 1949 Geneva Convention requires the use of common sense, while the 1977 Protocol 1 essentially defines a "civilian," in the negative sense, as someone who is not a combatant:

1. A civilian is any person who does not belong to one of the categories of persons referred to in Article 4 A (1), (2), (3), and (6) of the Third Convention and in Article 43 of this Protocol. In case of doubt whether a person is a civilian, that person shall be considered to be a civilian.
2. The civilian population comprises all persons who are civilians.

3. The presence within the civilian population of individuals who do not come within the definition of civilians does not deprive the population of its civilian character. (Protocol 1 Additional to the Geneva Conventions, 1977 [Article 50])

And even if there were a clear computational definition of civilian, we would still need all of the relevant information to be made available from the sensing apparatus. All that is available to robots are sensors, such as cameras, infrared sensors, sonar, lasers, temperature sensors, ladars, and so on. These may be able to tell us whether something is a human or at least an animal, but not much else. In the labs there are systems that can identify someone's facial expression or that can recognize faces, but they do not work well on real-time moving people. And even if they did, how useful could they be in the fog of war? British teenagers beat the surveillance cameras just by wearing hooded jackets.

In a conventional war where all of the enemy combatants wear clearly marked uniforms (or better yet, radio frequency tags), the problems might not be much different from those faced in conventional methods of bombardment. But, asymmetrical warfare is increasingly making battle with insurgents the norm, and, in these cases, sensors would not help in discrimination. Knowing whom to kill would have to be based on situational awareness and having a theory of mind, that is, understanding someone else's intentions and predicting their likely behavior in a particular situation. Humans understand one another in a way that machines cannot. Cues can be very subtle, and there are an infinite number of circumstances where lethal force is inappropriate. Just think of children being forced to carry empty rifles, or insurgents burying their dead.

7.3 An Ethical Code for Robots?

The military does consider the ethical implications of civilian deaths from autonomous robots, although this is not their primary concern. Their role is to protect their country in whatever way is required. In the United States, all weapons and weapons systems are subjected to a legal review to ensure compliance with the Law of Armed Conflict (LOAC). There are three main questions to be asked before a weapon is authorized:

1. Does the weapon cause suffering that is needless, superfluous, or disproportionate to the military advantage reasonably expected from the use of the weapon? It cannot be declared unlawful merely because it may cause severe suffering or injury.
2. Is the weapon capable of being controlled, so as to be directed against a lawful target?
3. Is there a specific treaty provision or domestic law prohibiting the weapon's acquisition or use?

Regardless of these rules, we have already seen a considerable number of collateral casualties resulting from the use of semi-autonomous weapon systems. The argument then is one of proportionality, as stated in the first question, but there is no quantitative measure that can objectively determine military costs against civilian deaths. It is just a matter of political argument, as we have seen, time and time again.

Another concern is the question of what constitutes a new weapon. Take the case of the Predator UCAV. It was first passed for surveillance missions. Then, when it was armed with Hellfire missiles, the Judge Advocate General's office said that because both Predators and Hellfires had previously been passed, their combination did not need to be (Canning et al. 2004). Thus, if we have a previously used autonomous robot and a previously used weapon, it may be possible to combine them without further permission.

Armed autonomous robots could also be treated in a legally similar way to submunitions, such as the BLU-108 developed by Textron Defense Systems.[3] The BLU-108 parachutes to near the ground, where an altitude sensor triggers a rocket that spins it upward. It then releases four Skeet warheads at right angles to one another. Each has a dual-mode (active and passive) sensor system: the passive infrared sensor detects hot targets, such as vehicles, while the active laser sensor provides target profiling. They can hit hard targets with penetrators, or destroy soft targets by fragmentation.

The BLU-108 is not like other bombs because it has a method of target discrimination. If it had been developed in the 1940s or 1950s, there is no doubt that it would have been classified as a robot, and even now it is debatably a form of robot. The Skeet warheads have autonomous operation and use sensors to target their weapons. The sensors provide discrimination between hot and cold bodies of a certain height, but like autonomous robots, they cannot discriminate between legitimate targets and civilians. If BLU-108s were dropped on a civilian area, they would destroy buses, cars, and trolleys. Like conventional bombs, discrimination between innocents and combatants requires accurate human targeting judgments. A key feature of the BLU-108 is that it has built-in redundant self-destruct logic modes that largely leave battlefields clean of unexploded warheads, and it is this that keeps it out of the 2008 international treaty banning cluster munitions (Convention of Cluster Munitions).

To use robot technology over the next twenty-five years in warfare would, at best, be like using the BLU-108 submunition, in other words, it can sense a target, but cannot discriminate innocent from combatant (Sharkey 2008c). The big difference with the types of autonomous robots currently being planned and developed for aerial and ground warfare is that they are not perimeter-limited. The BLU-108 has a footprint of 820 feet all around. By way of contrast, mobile autonomous robots are limited only by the amount of fuel or battery power they can carry. They can potentially travel long distances and move out of line of sight communication.

In a recent sign of these future weapons, the U.S. Air Force sent out a call for proposals for "Guided, Smart Sub-munitions": "This concept requires a CBU (Cluster Bomb Unit) munition, or UAV capable of deploying guided smart sub-munitions, that has the ability to engage and neutralize any targets of interest. The goal of the sub-munitions is very challenging when considering the mission of addressing mobile and fixed targets of interest. The sub-munition has to be able to reacquire the target of interest it is intended to engage" (United States Air Force 2008). This could be very much like an extended version of the BLU-108 that could pursue hot-bodied targets. Most worrying are the words "reacquire the target of interest." If a targeted truck were, for example, to overtake a school bus, the weapons might acquire the bus as the target rather than the truck.

A naval presentation by Chief Engineer J. S. Canning subtitled "The Difference between 'Winning the War' and 'Winning the Peace'" discusses a number of the ethical issues involved in the deployment of autonomous weapons. The critical issue for Canning is that armed autonomous systems should have the ability to identify the legality of a target. His answer to the ethical problems is unnervingly simple: "let men target men" and "let machines target other machines" (Canning 2006). This restricts the target set, and, Canning believes, may overcome the political objections and legal ramifications of using autonomous weapons.

While machines targeting machines sounds like a great ethical solution on the drawing table, the reality is that it belongs to mythical artificial intelligence, not real-world AI. In most circumstances, it would not be possible to pinpoint the weapon without also pinpointing the person using it, or even to discriminate between weapons and nonweapons. I have the mental image of a little girl being blown away because she points her ice cream at a robot to see if it would like some. And what if the enemy tricks the robot into killing innocent civilians by, for example, placing weapons on a school or hospital roof? Who will take the responsibility?

A different approach, suggested by Ronald Arkin from the Georgia Institute of Technology, is to equip the robotic soldier with an *artificial conscience* (Arkin and Moshkina 2007). Arkin had funding from the U.S. Army to work on a method for designing an ethical autonomous robot, which he refers to as a humane-oid.[4]At first glance, this sounds like a move in the right direction. At the very least, it gets the army to consider the ethical problems raised both by the deployment of autonomous machines and even those of the soldier on the ground. Another of Arkin's concerns that he addresses in a public survey, and it is a good one, is "to establish what is acceptable to the public and other groups, regarding the use of lethal autonomous systems" (Arkin and Moshkina 2007).

Despite the good intentions, I have grave doubts about the outcome of this project. No idea is presented about how this could be made to work reliably, and reliability is a key issue when it comes to human lives. It is not just about having incredibly good sensors and camera inputs, or being able to make appropriate discriminations. A robot could actually have to make decisions in very complex circumstances that are entirely unpredictable.

It turns out that the plan for this conscience is to create a mathematical decision space consisting of constraints, represented as prohibitions and obligations derived from the laws of war and rules of engagement (Arkin 2009). Essentially, this consists of a bunch of complex conditionals (if–then statements). Reporting on Arkin's work, *The Economist* (2007) gives the example of a Predator UAV on its way to kill a car full of terrorists. If it sees the car overtaking a bus full of school children, it will wait until it has overtaken them before blasting the car into oblivion. But how will the robot discriminate between a bus full of school children and a bus full of guards? Admittedly, this is not one of the tasks that Arkin cites, but it is still the kind of ethical decision that an autonomous robot would have to make. The shadow of mythical AI looms large in the background.

Arkin believes that a robot could be more ethical than a human because its ethics are strictly programmed into it, and it has no emotional involvement with the action. The justification for this comes from a worrying survey, published by the Office of the Surgeon General (Mental Health Advisory Team 2006) that tells of the aberrant ethical behavior and attitudes of many U.S. soldiers and marines serving in Iraq. Arkin holds that a robot cannot feel anger or a desire for revenge, but neither can it feel sympathy, empathy, or remorse. Surely, a better way to spend the money would be on more thorough ethical training and monitoring of the troops.

Even if a robot was fully equipped with all of the rules from the Laws of War, and had, by some mysterious means, a way of making the same discriminations as humans make, it could not be ethical in the same way as is an ethical human. Ask any judge what they think about blindly following rules and laws. In most real-world situations, these are a matter of interpretation.

Arkin's anthropomorphism in saying, for example, that robots would be more humane than humans does not serve his cause well. To be humane is, by definition, to be characterized by kindness, mercy, and sympathy, or to be marked by an emphasis on humanistic values and concerns. These are all human attributes that are not appropriate in a discussion of software for controlling mechanical devices. More recently, Arkin has taken to talking about adding sympathy and guilt to robots. However, the real value of the work would be to add safety constraints to autonomous weaponized robots to help to cut down the number of civilian casualties. This is easy to understand, and may help the work to progress in a clearer way. The anthropomorphic terms create a more interesting narrative, but they only confuse the important safety issues and create false expectations.

The number of possible moral and ethical problems in a military operations theater full of civilians could be infinite, or at least run into extremely large numbers. Many different circumstances can happen simultaneously and give rise to unpredictable or chaotic robot behavior. From a perhaps cynical perspective, the "robot soldier with a conscience" could at some point be used by military public relations to allay political opposition, amounting to lots of talk while innocent civilians keep on dying: "Don't worry, we'll figure out how to use the technology discriminately eventually."

As Davis says about other defense experts talking up robot warfare, "such statements are dangerous, because men disconnected from the realities of warfare may sway decision-makers regarding future force decisions and composition" (Davis 2008). On the same basis, the "artificial conscience" idea could perhaps also be employed as an argument to shift the burden of responsibility for collateral fatalities from the chain of command onto inanimate weapons.

No civilized person wishes to see their country's young soldiers die in foreign wars. The robot is certainly a great defensive weapon, especially when it comes to roadside bombs. It is the moral responsibility of military commanders to protect their soldiers, but there are a number of far-reaching consequences of "risk-free" war that we need to consider.

• Having more robots to reduce the "body bag count" could mean fewer disincentives to start wars. In the United States, since the Vietnam War, body-bag politics has been a major inhibitor of military action. Without bodies coming home, citizens will care a lot less about action abroad, except in terms of the expense to the taxpayer. It could mean, for example, that with greatly reduced public and political opposition (passing the so-called Dover[5]), it is a lot easier for the military to start and run more "defensive" wars. This is an ethical and moral dilemma that should be engaging international thinking.

• Armstrong warns about the use of robots in "the last three feet" and asks if the United States really wants to have a robot represent the nation as a strategic corporal. You can't hope to win hearts and minds by sticking armed robots in the face of an occupied population (Armstrong 2007).

• It has been suggested that a country engaged in risk-free war will put its civilian population more at risk from terrorist attacks at home and abroad (Kahn 2002).

• It is more like policing—a term used for the Kosovo war—but policing requires a different set of rules than war; for example collateral civilian deaths are unacceptable for policing. Those suffering from policing need to be demonstrably morally guilty (Kahn 2002).

• There will clearly be proliferation (the indications are already there), and so the risk-free state could be short lived. As Chief Engineer Canning has pointed out: "What happens when another country sees what we've been doing, realizes it's not that hard, and begins to pursue it, too, but doesn't have the same moral structure we do? You will see a number of countries around the world begin to develop this technology on their own, but possibly without the same level of safeguards that we might build-in. We soon could be facing our own distorted image on the battlefield" (Canning 2005).

A related concern is that when we say robot weapons save lives, we implicitly mean only the lives of *our* soldiers and their allies. Of course, in the middle of a vicious war, that is what we want. But let us not forget that such sentiments allow us to hide from ourselves the fact that the robot weapons could take a disproportionate toll of lives on the other side, including many innocent civilians. Autonomy could greatly increase fatal errors.

7.4 The Problem of Proportionality

According to the laws of war, a robot could potentially be allowed to make lethal errors, providing that the noncombatant casualties were proportional to the military advantage gained. But how is a robot supposed to calculate what is a proportionate response? There is no sensing or computational capability that would allow a robot such a determination. As mentioned for the discrimination problem described earlier, computer systems need clear specifications in order to operate effectively. There is no known metric to objectively measure needless, superfluous, or disproportionate suffering.[6] It requires human judgment.

No clear objective means are given in any of the laws of war for how to calculate what is proportionate (Sharkey 2009a). The phrase "excessive in relation to the concrete and direct military advantage expected to be gained" is not a specification. How can such values be assigned, and how can such calculations be made? What could the metric be for assigning value to killing an insurgent, relative to the value of noncombatants, particularly children, who could not be accused of willingly contributing to insurgency activity? The military says that it is one of the most difficult decisions that a commander has to make, but that acknowledgment does not answer the question of what metrics should be applied. It is left to a military force to argue as to whether or not it has made a proportionate response, as has been evidenced in the recent Israeli–Gaza conflict (Human Rights Watch 2009).

Uncertainty needs to be a factor in any proportionality calculus. Is the intelligence correct, and is there really a genuine target in the kill zone? The target value must be weighted by a probability of presence/absence. This is an impossible calculation unless the target is visually identified at the onset of the attack. Even then, errors can be made. The investigative journalist Seymour Hersh gives the example of a man in Afghanistan being mistaken for bin Laden by CIA Predator operators. A Hellfire was launched, killing three people who were later reported to be local men scavenging in the woods for scrap metal (Hersh 2002, 66). This error was made using a robot plane with a human in the loop. There is also the problem of relying on informants. The reliability of the informant needs to be taken into account, and so does the reliability of each link in the chain of information reaching the informant before being passed onto the commander/operator/pilot. There can be deliberate deception anywhere along the information chain, as was revealed in investigations of Operation Phoenix—the U.S. assassination program—after the Vietnam War. As Hersh pointed out, many of the thousands on the assassination list had been put there by South Vietnamese officials for personal reasons, such as erasing gambling debts or resolving family quarrels.

It is also often practically impossible to calculate a value for actual military advantage. This is not necessarily the same as the political advantage of creating a sense of military success by putting a face to the enemy to rally public support at home and to boost the morale of the troops. Obviously there are gross calculations that work in the extreme, such as a military force carrying weapons sufficient to kill the population of a large city. Then, it could be possible to balance the number of civilians killed against the number saved. Military advantage, at best, results in *deterrence* of the enemy from acting in a particular way, *disruption* of the social, political, economic, or military functions (or a combination of these), and *destruction* of the social, political, economic, or military functions (or a combination) (Hyder 2004, 5). Proportionality calculations should be based on the likely differences in military outcome if the military action killing innocents had not been taken (Chakwin, Voelkel, and Scott 2002).

Despite the impossibility of proportionality calculations, military commanders at war have a political mandate to make such decisions on an almost daily basis. Commanders have to weigh the circumstances before making a decision, but ultimately it will be a subjective metric. Clearly the extremes of wiping out a whole city to eliminate even the highest-value target, say Osama bin Laden, is out of the question. So there must be some subjective estimates about just how many innocent people killed equal the military value of the successful completion of a given mission.

Yes, humans do make errors and can behave unethically, but they can also be held accountable. Who is to be held responsible for the lethal mishaps of a robot? Robert Sparrow argues that it certainly cannot be the machine itself, and thus it is not legitimate to use automated killing machines (Sparrow 2007). There is no way to punish a robot. We could just switch it off, but it would not care any more about that than my washing machine would care. Imagine telling your washing machine that if it does not remove stains properly, you will break its door off. Would you expect that to have any impact on its behavior? There is a long causal chain associated with robots: the manufacturer, the programmer, the designer, the Department of Defense, the generals or admirals in charge of the operation, and the operator. It is thus difficult to allocate responsibility for deliberate war crimes, or even mishaps.

7.5 Conclusion

We discussed at the outset how killing is made easier for combatants when the distance between them and their enemies is increased. Soldiers throughout history have found it difficult to kill at close range when they can clearly see whom they are killing. Distance, whether physical or psychological, helps to overcome the twin problems of fear of being killed and resistance to killing that particularly dog the infantry.

Robots are set to change the way that wars are fought by providing flexible "stand-ins" for combatants. They provide the ultimate distance targeting that allows warriors to do their killing from the comfort of an armchair in their home country—even thousands of miles away from the action. Robots are developing as a new kind of fighting method different from what has come before. Unlike missile or other projectiles, robots can carry multiweapon systems into the theater of operations, and act flexibly once in place. Eventually, they may be able to operate as flexibly as human combatants, without risk to the lives of their operators that control them. However, as we discussed, there is no such thing as risk-free warfare. Apart from the moral risks discussed, asymmetrical warfare can also lead to more insurgency and terrorist activity, threatening the citizens of the stronger power.

The biggest changes in warfare will come with the further development of autonomous military robots that can decide who, where, and when to kill, without human involvement. There are no current international guidelines or even discussions about the uses of autonomous robots in warfare. These are needed urgently, since robots simply cannot discriminate between innocents and combatants.

If there was a strong political will to use autonomous robot weapons, or even a serious threat to the state that has them, then legal arguments could be constructed that leave no room for complaints.[7] This is especially the case if they could be released somewhere where there is a fairly high probability that they will kill a considerably greater number of enemy combatants (uniformed and nonuniformed) than innocents (i.e., the civilian death toll was not disproportionate to the military advantage).

At the very least, it should be discussed how to limit the range and action of autonomous robot weapons before their inevitable proliferation (forty-three countries now have military robot programs). Even if all of the elements discussed here could be accommodated within the existing laws of war, their application needs to be thought through properly, and specific new laws should be implemented to not just accommodate their use, but to constrain it as well. We don't know how autonomous robots will affect military strategy of the future, or if they will lead to more subjugation of weak nation-states and less public pressure to prevent wars.

Notes

1. See du Picq 1946. The book was compiled from notes left by Colonel Ardant du Picq of France after he was killed in battle by a Prussian projectile in 1870.

2. Decapitation is a euphemism for assassination of suspected insurgent leaders. The word *decapitation* was used to indicate cutting off the head (leader) from the body of the insurgents.

3. Thanks to Richard Moyes of Landmine Action for pointing me to the BLU-108 and to Marian Westerberg and Robert Buckley from Textron Defense Systems for their careful reading and comments on my description.

4. Contract #W911NF-06-1-0252 from the U.S. Army Research Office.

5. Dover, Delaware, is the U.S. Air Force base where the bodies of soldiers are returned from the front line in flag-draped coffins. The Dover test concerns how much the electoral chances of the national political administration are affected by the numbers of dead.

6. Bugsplat software and its successors have been used to help calculate the correct bomb to use to destroy a target and calculate the impact. It is only used to help in the human decision-making process and it is unclear how successful this approach has been in limiting civilian casualties.

7. Regardless of treaties and agreements, any weapon that has been developed may be used if the survival of a state is in question. The International Court of Justice *Nuclear Weapons Advisory Opinion* (1996) decided that it could not definitively conclude that in every circumstance the threat or use of nuclear weapons was axiomatically contrary to international law; see Stephens and Lewis 2005.

References

Alston, Philip. 2010. *Report of the Special Reporter on Extrajudicial, Summary, or Arbitrary Executions.* The UN Human Rights Council, fourteenth session, A/HRC/14/24/Add.6, May 28.

Arkin, Ronald. 2009. *Governing Lethal Behavior in Autonomous Robots.* Boca Raton, FL: Chapman and Hall/CRC Press.

Arkin, Ronald, and Lilia Moshkina. 2007. Lethality and autonomous robots: An ethical stance. Paper presented at the IEEE International Symposium on Technology and Society, June 1–2, Las Vegas.

Armstrong, Matthew. 2007. Unintended consequences of unmanned warfare. Presentation to Proteus Management Group Futures Workshop at the U.S. Army War College, Carlisle Barracks, Pennsylvania, August 15.

Canning, John. 2005. A definitive work on factors impacting the arming of unmanned vehicles. Dahlgren Division Naval Surface Warfare Center report NSWCDD/TR-05/36.

Canning, John. 2006. A concept of operations for armed autonomous systems. Presentation for the Naval Surface Warfare Center, Dahlgren Division.

Canning, John, G. W. Riggs, O. T. Holland, and C. J. Blakelock, 2004. A concept for the operation of armed autonomous systems on the battlefield. Paper presented at the Association for Unmanned Vehicle Systems International conference, Anaheim, CA, August 17.

Chakwin, Mark, Dieter Voelkel, and Enright Scott. 2002. Leaders as targets. Joint Forces Staff College, Norfolk, VA. Seminar #08.

Committee on Autonomous Vehicles in Support of Naval Operations National Research Council. 2005. *Autonomous Vehicles in Support of Naval Operations*. Washington, DC: The National Academies Press.

Crane, David, 2008. G-NIUS Guardium UGV: World's first operational security robot. *Defense Review* (August): 23. <http://www.defensereview.com/g-nius-guardium-ugv-worlds-first-operational-autonomous-security-robot/> (accessed April 3, 2011).

Daddis, Gregory. 2004. Understanding fear's effect on unit effectiveness. *Military Review* (July–August): 22–27.

Davis, Daniel. 2007. Who decides: Man or machine? *Armed Forces Journal*. <http://www.armedforcesjournal.com/2007/11/3036753> (accessed April 2, 2011).

du Picq, Ardant. 1946. *Battle Studies*. Part 2, chapter 1. Harrisburg, PA: Military Service Publishing Co.

The Economist. 2007. Robot wars: An attempt to build an ethical robotic soldier. April 17.

Fox News. 2008. Pentagon's "Crusher" robot vehicle nearly ready to go. February 27. <http://www.foxnews.com/story/0,2933,332755,00.html> (accessed November 27, 2010).

Ford, John S. 1944. The morality of obliteration bombing. *Theological Studies* 23: 261–309.

Grossman, David. 1995. *On Killing: The Psychological Cost of Learning to Kill in War and Society*. New York: Little, Brown and Co.

Hersh, Seymour. 2002. Manhunt: The Bush administration's new strategy in the war against terrorism. *New Yorker* (December): 64–68.

Holmes, Richard. 2003. *Acts of War: The Behaviour of Men in Battle*. London: Cassell.

Human Rights Watch. 2009. *Precisely Wrong: Gaza Civilians Killed by Israeli Drone-Launched Missiles*. New York: Human Rights Watch.

Hyder, Victor. 2004. *Decapitation Operations: Criteria for Targeting Enemy Leadership*. Monograph/report. Fort Leavenworth, KS: School of Advanced Military Studies United Sates Army Command and General Staff College.

Kahn, Paul. 2002. The paradox of riskless war. *Philosophy and Public Policy Quarterly* 22 (3): 2–8.

Koh, Harold, 2010. Speech to the American Society of International Law, Washington, DC, March 25.

Marshall, S. L. A. [1947] 2000. *Men against Fire: The Problem of Battle Command.* (First published by William Morrow & Company.) Norman: University of Oklahoma Press.

Mental Health Advisory Team. 2006. *Operation Iraqi Freedom 05–07.* Final report, November 17.

Petraeus, David, and James Amos. 2006. *Counterinsurgency.* Field Manual FM 3–24 MCWP 3–33.5, Section 7–30. Washington, DC: Headquarters of the Army.

Roberts, Adam. Forthcoming. What is a civilian? In *The Changing Character of War,* ed. Hew Strachan and Sibylle Scheipers. Oxford, UK: Oxford University Press.

Sharkey, Noel. 2008a. Cassandra or the false prophet of doom: AI robots and war. *IEEE Intelligent Systems* 23 (4) (July/August): 14–17.

Sharkey, Noel. 2008b. The ethical frontiers of robotics. *Science* 322 (5909): 1800–1801.

Sharkey, Noel. 2008c. Grounds for discrimination: Autonomous robot weapons. *RUSI Defence Systems* 11 (2): 86–89.

Sharkey, Noel. 2009a. Death strikes from the sky: The calculus of proportionality. *IEEE Science and Society* (Spring): 16–19.

Sharkey, Noel. 2009b. March of the killer robots. *Daily Telegraph,* June 15.

Singer, Peter Warren. 2009. *Wired for War: The Robotics Revolution and Conflict in the 21st Century.* New York: The Penguin Press.

Sofge, Erik. 2008. Non-answer on armed robot pullout from Iraq reveals fragile bot industry. *Popular Mechanics,* April 8. <http://www.unsysinst.org/forum/viewtopic.php?t=388&sid=b9e2476 2d3e32d5b72d9a00f540a5640> (accessed November 28, 2010).

Sparrow, Robert. 2007. Killer robots. *Journal of Applied Philosophy* 24 (1): 62–77.

Stephens, Dale, and Michael W. Lewis. 2005. The law of armed conflict—A contemporary critique. *Melbourne Journal of International Law* 6 (1): 55–85.

United Press International. 2008. BAE Systems tech boosts robot UAVs IQ. Industry Briefing, February 26. <http://bae-systems-news.newslib.com/story/3951-3226462/> (accessed April 3, 2011).

United States Air Force. 2008. Guided smart munitions. Call for proposals, topic number AF083–093, August 25.

United States Air Force. 2009. *Unmanned Aircraft Systems Flight Plan 2009–2047.* Headquarters of the United States Air Force, Washington, DC, May 18.

8 Robotic Warfare: Some Challenges in Moving from Noncivilian to Civilian Theaters

Marcello Guarini and Paul Bello

In *Governing Lethal Behavior: Embedding Ethics in Hybrid Deliberative/Reactive Robot Architecture*, Ronald Arkin has undertaken the ambitious project of providing the "basis, motivation, theory, and design recommendations for the implementation of an ethical control and reasoning system potentially suitable for constraining lethal actions in an autonomous robotic system, so that they fall within the bounds prescribed by the Laws of War and Rules of Engagement" (2007, 1). What are at issue are the artificially intelligent selection of targets and the autonomous engagement of those targets by an automated system. This chapter attempts to analyze where some of the more serious difficulties may arise in attempting to build systems capable of automated warfare.

Let us begin by distinguishing between different theaters of activity. On one end of a spectrum, we have theaters populated entirely with combatants. This is a classical battlefield where everyone present on both sides is a combatant. On the other end of the spectrum, we have a theater populated entirely with noncombatants on the opposing side. An example on this end of the spectrum would be a counterinsurgency operation, raiding houses where it turns out that no one is a combatant (on that given day). Being a spectrum, there are many possible theaters somewhere in the middle.

We will argue that until much more progress is made, we should not be sanguine about the advantages of robots in theaters on the *noncombatant end of the spectrum*. We will do this by examining some of the challenges posed by the problem of mental state ascription and isotropy (the potential relevance of anything to anything).[1] We will argue that in theaters of activity involving mostly noncombatants, differentiating between combatants and noncombatants will often require the appropriate attribution of mental states (such as intentions). Isotropic considerations make the attribution of mental states very difficult to build into a robotic soldier. We are not suggesting that the problem cannot be solved, in principle. Rather, we will try to express just how difficult the problem is, and just how important it is to solve it, before seriously considering the use of robotic soldiers in the theaters under consideration. We will also sketch out how we think progress might be made on the problem by considering the role of emotion in cognition.

8.1 Background and an Example

Automated target selection and engagement is not a new idea. The Phalanx weapons system, originally developed and tested in the 1970s, is now used by a number of navies around the world. It is a close-in antimissile system that can automatically detect and engage targets. There is a manual override. A ship at sea being engaged by other military assets with no civilians in the neighborhood—this is an example of the classical theater. We will show that, as we move into theaters with noncombatants, there are very serious difficulties to be encountered. Let us begin with an example to motivate the difficulties involved.

Consider a counterinsurgency operation in a Sikh village. Ground forces received a tip that wanted insurgents may be sheltered in a civilian residence. The tip is erroneous, but the counterinsurgency unit does not know this. Three children and their two parents are present at the residence. Two of the male children are young and playing with a ball. Each is also carrying the Sikh *kirpan* (sometimes referred to as the Sikh "dagger"). This is a religious symbol and is not used as a weapon. Just before a member of the counterinsurgency force kicks the door in, one of the boys kicks his ball toward the door, and both go chasing after it. As military forces enter the house, they see two young boys running toward them, and a shocked mother yelling. She chases the boys and yells at them to stay away from the men at the door; the troops do not know what she is yelling, since they do not understand her language. It is quite possible that the forces in question will rapidly see this as a situation where two young children are playing, and a mother frightened for her children is yelling and giving chase. That is one way to see the situation, and on this *first interpretation*, we could even imagine a soldier motioning to the children to keep away.

Let us consider a *second interpretation*. There are two fast-closing possible targets, both of which are carrying a weapon. A third possible target is following the first two, and is making a level of noise consistent with violent or threatening behavior.

With respect to cognitive abilities, what is required to see the two fast-closing possible targets as *children*? What is required to see them as *playing*? What is required to see the third possible target as *a mother* (with all that that entails)? What is required to see her as *frightened for her children*? What is required to see the kirpan as a *religious symbol*, and not as a *weapon*? Clearly, a tremendous amount of background knowledge is required to provide the first interpretation of this situation. Arkin (2009, chapter 3) cites some of the failures of human soldiers in high-stress theaters with many noncombatants, and attributes many of these failures to emotion. He then attempts to motivate a possible advantage for robot soldiers by indicating that they would not be subject to the disadvantages of having emotions. From the perspectives of cognitive science and artificial intelligence, the apparently trivial ability of a human being to see a situation like the one just described, as involving children at play with a frightened mother giving chase, in fact is quite involved. For a robotic soldier to perform at least as well as a human in such circumstances, it would have to go beyond seeing the situation as described in the second interpretation (which would likely lead to erroneous and harmful engagement). We will now begin to examine some of what would be required for robots to perform at least as well as humans in theaters populated mostly with civilians. Later in the chapter we will turn to arguing that *some* of the functional, computational role of emotion may play a part in overcoming *some* of the challenges.

8.2   Mental State Attribution in General

Mental state attribution is about the ascription of beliefs, desires, hopes, fears, intentions, and the like, to others and to oneself. There is a significant literature in cognitive science and philosophy on mental state attribution (sometimes referred to as "theory of mind," or "mentalizing," or "mindreading," with nothing psychic intended). Mindreading is about how we retrodict, attribute, or predict the mental states or actions of others or ourselves. There are both descriptive (how do we actually do it?) and normative (how ought we do it?) dimensions to the study of our everyday abilities to (a) attribute beliefs, desires, hopes, fears, and the like, and (b) make claims about what an agent did or will do. The two main camps in this study are often referred to as Theory Theory and Simulation Theory,[2] with many people actually defending a kind of hybrid approach that stresses one over the other. The point of this chapter is not to insist that one or another of these approaches is correct. Rather, it is to show that the ability to attribute mental states reliably becomes a matter of central importance in theaters on the noncombatant end of the spectrum.

Let us consider a naval vessel with an antimissile system that identifies targets by virtue of the trajectory and speed of the incoming target. Something closing in directly on your ship at a very high rate of speed needs to be destroyed before it makes contact (even if it is one of your own aircraft falling in a direct collision course with your ship after terminal damage in combat). There is no need to figure out what the potential target *intends* or *feels* or might be *thinking*. To be sure, there are theaters of activity involving exclusively combatants that would involve such assessments, but the focus in this chapter is on theaters with many noncombatants, since the issue of mental state attribution is exacerbated in these contexts. In a context where we cannot assume that everyone present is a combatant, then we have to figure out who is a combatant and who is not. This frequently requires the attribution of an intention. The presence of a weapon, or a possible or apparent weapon, is insufficient, as the example with the children carrying the kirpan shows. (If those same children bore menacing facial expressions and made threatening gestures with grenades in hand, then the situation changes entirely.) Mental state attribution is not a problem that has been solved. Moreover, solving it is very difficult regardless of your approach (Wilkerson 2001), and solving it appears to be required before robotic soldiers could be applied usefully in theaters with many noncombatants. What problems do we need to overcome to design systems that could attribute mental states, at least as reliably as humans, in the envisioned contexts?

8.3 Isotropy

Isotropy refers to the potential relevance of anything to anything.[3] What could the price of tea in China have to do with Habib's heart attack? Well, it depends. If Habib is heavily invested in companies shipping tea out of China, and Habib has a heart condition that makes him vulnerable to heart attacks when he is under tremendous emotional strain, and he finds out that the price of tea in China fell significantly, leading to serious losses in his portfolio, causing him to experience high levels of stress and anxiety, then it could well be the case that the price of tea in China is relevant to explaining Habib's heart attack. It is difficult to say, in advance of having the details of a situation, which pieces of information may or may not be relevant to reasoning about a claim or an action. Isotropy is a general problem in trying to understand human cognition and achieving AI, and it is a problem that manifests itself in mental state attribution, and this is the dimension of isotropy we will focus on herein.

The information that could be relevant in assigning mental states is vast. Facial expressions, gaze orientation, body language, attire, information about the agent's movement through an environment, information about the agent's sensory apparatus, information about the agent's background beliefs, desires, hopes, fears, and other mental states are all relevant to attributing current or predicting future mental states or behaviors. One would think that a person running toward a soldier while screaming and carrying what looks like a dagger would be something that a soldier might be very much concerned with, but maybe not. *Civilian theaters introduce the full complexity of human social affairs into combat.* In the classical theater, where everyone is a combatant, one still needs to avoid friendly fire, but everyone on the other side is, essentially, a legitimate target. Not so in civilian theaters. For example, counterinsurgency forces looking for manufacturers of pipe bombs walk into a civilian residence and immediately notice someone carrying a pipe. Is it a bomb? Is the individual holding it threatening? Say the civilian holding the pipe in his left hand is wearing overalls and also holds a monkey wrench in his right hand—does that change how you see him? How about if the civilian is standing in front of a sink with pipes exposed and water leaking all over—would you see him as intending harm? Probably not. Sometimes a pipe is just a pipe. In largely civilian theaters, we cannot assume that someone is threatening; we have to figure it out. What complicates this in the extreme is that the full range of human social affairs becomes potentially relevant to figuring out whether behavior is threatening or not. Something being a religious symbol *might* disqualify it from being a weapon; something useable for plumbing *might* disqualify it from being seen as a weapon. It all depends on the other considerations at issue. In the classical theater, being a plumber or being at play (and a myriad of other everyday civilian activities) are not relevant considerations. In the civilian theater, they are.[4]

Correctly attributing intentions is often (though not always) necessary to see someone as a threat, and isotropy complicates mental state attribution in civilian theaters because almost anything (given the appropriate background conditions) can become relevant to attributing the appropriate intentional state. This might make it tempting to think that Theory Theory (TT) approaches to mindreading are more problematic than Simulation Theory (ST). Indeed, some have argued in this way.[5] TT requires that agents have explicitly represented generalizations (e.g., rule-like structures) that correspond to the putative connections between mental states and actions. A sentential or sentence-like explanation would be of Byzantine complexity, and it is not obvious that we are manipulating anything like that when we attribute mental states to others. We have much sympathy for this line of criticism, though we are not suggesting matters will be easy for an ST approach. Those subscribing to ST could say that whatever mechanisms allow us to arrive at our own mental states can be redeployed in arriving at the mental states of others. Essentially, I run a simulation of other agents based on my own actions, mental states, and processes. In the current context, leaving it at that would be unsatisfying for at least two reasons. First, if the task is to build computational systems that could operate in a civilian theater, we need to know how to construct the aforementioned mechanisms that allow us to arrive at different mental states (in the first person) in different situations before those mechanisms can be redeployed for simulating others. Second, even if we succeed in modeling various transductive and inference mechanisms in the first person—which requires overcoming isotropy of certain types—the redeployment of those mechanisms for purposes of simulation of others still runs into the problem of isotropy. Let us see how this is so.

When I "put myself in someone else's shoes" to figure out which mental states they may have or how they will act, I need to draw on information about how the mental states of the target of my simulation may differ from my own. If I do not quarantine some of my own mental states from the simulation and do not recognize that what my target thinks is salient may be different from what I think is salient in a given context, then my simulation will not be reliable. Moreover, isotropy affects what we would provide as input. Almost anything could become relevant to constraining the input to the simulation. If I were playing and chasing a ball with someone else, and

I were wearing a religious symbol, I would not be intending any harm, so if someone else is in that situation, they would not be intending harm—this is a type of simulation. And it assumes that the children are playing and chasing a ball and wearing a religious symbol. All that is part of the input to the simulation. (The output is that the individuals in question do not intend harm.) An adequate full-blown computational model of this sort of simulation would have to figure out that the movement of the children constitutes *chasing*, and that this form of chasing constitutes *play*, and that the kirpan is a *religious symbol*. Most things in the form of a dagger are not religious symbols, and many forms of chasing constitute threatening behavior. Any number of things could be relevant in determining whether people are playing (or not). It might be thought that what needs to be done for the agent doing the simulation is simply to retrieve a situation "like this" from memory and simulate based on that. But there is the rub: what constitutes a situation "like this?" Answering that question *assumes* we know what is salient or relevant in the situation under consideration, and we do the recall based on the salient or relevant features of this situation.[6] However, situations do not come with their salient features labeled. This is an easy point to miss, since what is salient is often so obvious to us and requires so little conscious effort to determine that we may fail to appreciate the computational difficulty of modeling the process of determining it. This is a problem both for TT and ST. It may be possible to carry on with many arguments between TT and ST without dwelling on this problem, since there are other issues the opposing theorists are dealing with. However, in designing a robot with the ability to read minds well enough to engage in civilian theaters, the problem cannot be side stepped. Without appropriately quarantining and selecting the input to a simulation—or to a set of theoretical generalizations, for that matter—there is little chance that mindreading will be successful. Without reliable intentional state attribution, it is hard to see how a robot could usefully assess threatening from nonthreatening behavior, and without that, distinguishing combatants from noncombatants will be exceedingly difficult. Our point is not that these problems cannot be overcome; it is that we are not yet even close to overcoming them. Indeed, we think there are computational advantages to systems that make use of simulation over those that do not, but much progress needs to be made before we have anything capable of dealing of with the complexity of the civilian theater.

What we have done in this section is to point to some of the problems created by isotropy, but we have said nothing about how humans manage isotropy, or how robots might be made so that they could manage it. To that, and other issues, we now turn.

8.4 Emotion

Much of Arkin's work (2009, chapter 3) treats human emotion as a problem with human soldiers when engaging in civilian theaters, and he develops an ethical reasoning architecture that "will not involve emotion directly . . . as that has been shown to impede the ethical judgment of humans in wartime" (Arkin 2009, 118). In laying out the architectural consideration for autonomous selection and engagement of targets (Arkin 2009, chapters 9 and 10), he proposes an "Ethical Governor," which includes a limited role for emotion. The idea is to include a role for something like the functional equivalent of guilt. If a system is criticized for its behavior with respect to the use of lethal force, "guilt" can increase to censor or veto future behaviors until a proper external action assessment can be performed and the system reconfigured, if needed. Arkin is *not* suggesting that the robot actually "feels" guilt the ways humans do; rather, the idea is that some of the functional role of guilt can be mimicked in the robot. In general, though, emotion plays no direct role in figuring out which options are open to the agent. The idea is that determining which options are available and providing an initial assessment is all done in an emotionless manner, and if the results of that process run afoul of the guilt censor, so called, then the option is rejected. While Arkin recognizes at least one of the limits of this model,[7] we want to suggest that there may be other limits as well. While we do not wish to dispute the empirical evidence that emotions can lead human soldiers astray, especially in highly stressful and complex civilian theaters, we now want to explore the possibility that emotions may have a positive role to play in dealing with the full complexity of human social affairs, present in the largely civilian theaters.

First, we lay bare one of our methodological predispositions: we think that understanding how humans solve the problem of navigating a complex social space, in an ethically constrained manner, is a useful starting place for constructing a robot that could similarly navigate that space.[8] This presupposition is not self-evident. In restricted domains, like chess playing, we have constructed systems that exceed human abilities, but those systems are doing things quite differently from how we do things. To be sure, an opening book of moves is often programmed into these systems, and humans sometimes commit to memory sequences of opening moves. That said, we suspect that not many believe that when Deep Blue, the IBM chess-playing computer that bested Garry Kasparov, searches through millions of possible board positions that it is doing something even remotely akin to what human chess players do, yet it plays darn good chess, nonetheless. So it is not self-evidently true that achieving (or exceeding) human-level competence must be done by modeling human cognitive abilities, or even taking human cognitive performance as an important guide. However, human social affairs are *vastly* more complex than chess. The number of possible "moves" and the constraints on those moves in our social activities are far beyond anything like the domain-restricted tasks current computational systems undertake. We suggest that the preceding is a good reason[9] for taking an understanding of human competence in ethically constrained complex social environments as a starting place for assessing the prospects of building an artificial system to navigate such a space. And we think emotion has a role to play in understanding how we navigate that space. To explicating this point (if too briefly) we now turn.

As Wagar and Thagard (2004) point out, there is a growing body of literature in cognitive science regarding the importance of emotions to decision making (Churchland 1996; Damasio 1994; Finucane et al. 2000; Lerner and Keltner 2000; Loewenstein et al. 2001; Rottenstreich and Hsee 2001). The model for decision making put forward by Wagar and Thagard integrates functions of the ventromedial prefrontal cortex (VMPFC), the hippocampus, the amygdala, the nucleus accumbens, and the ventral tegmental area. This work draws on and extends Antonio Damasio's work on somatic markers. According to Damasio (1994), the VMPFC and the amygdala are involved in the production of somatic markers, which are "the feelings, or emotional reactions, that have become associated through experience with the predicted long-term outcomes of certain responses to a given situation" (Wagar and Thagard 2004, 90).

Evidence for this comes from the specific cluster of deficits and abilities demonstrated by those having damage to the VMPFC. This sort of damage leaves language skills intact, as well as memory and what might be called intellectual, or theoretical, reasoning. However, decision making is impaired, especially with respect to decisions involving distinctions between long-term and short-term consequences in contexts where punishments and rewards are at issue. As Wagar and Thagard put the point: "Somatic markers make the decision process more efficient by narrowing the number of feasible behavioral alternatives, while allowing the organism to reason according to the long-term predicted outcomes of its actions" (2004, 90).

Damage to the VMPFC also damages somatic markers and the ability to make effective decisions, leading to serious difficulties in navigating social environments. We can think of somatic markers as constituting a kind of bias on the search space of options for action. We need not explicitly reason in every social context about *all* of the available alternatives for action; this would be profoundly inefficient. Some options present themselves to us, and somatic markers play a role in the filtering of these options, reducing the computational load on explicit or conscious reasoning. With damage to the VMPFC, the filters established through past experience are damaged or eliminated, and so, too, is the ability to establish new filters. Patients with VMPFC damage tend to demonstrate little, if any, empathy toward others, tend to lose most (and sometimes all) of their friends, and have a hard time keeping a job. This is not surprising, given that they reason poorly about the social consequences of their actions.

Let us return to the example of counterinsurgency and the Sikh household. A well-armed human soldier—believing his life might be in danger—opens the door to witness two screaming children wearing kirpans. For the sake of argument, imagine that this soldier has serious damage to his VMPFC, impairing his ability to empathize and his ability to reason about the consequences of his actions. To our knowledge, there are no case studies of this type, but given what we know about patients with VMPFC damage, it is far from obvious we would want them to serve in such contexts. A healthy VMPFC in an altogether fit soldier[10] should simply not lead to children at play being seen as targets. Along with damage to the VMPFC comes damage to the somatic markers established by years of experience, and this may well lead to options being considered with inadequate regard for the consequences of the actions, which would likely lead to disastrous results in the scenario in question.

Thus far, we have only considered damage to the VMPFC. As mentioned earlier, Wagar and Thagard extend Damasio's work to consider other parts of the brain, though this is not the place to consider the details of their position. The point of this brief discussion has been to motivate the idea that emotions may play a constructive role in limiting the options that come under explicit consideration, and this might play a very useful role with respect to making real-time decisions in very complex social scenarios. Arkin assumes that the role to be played by emotion is as some sort of postdeliberative censor. In other words, the robot soldier would arrive at a course of action, and if the action involves the use of lethal force, and the guilt censor has been set to block lethal force either altogether or in scenarios "like this," then the action will not be carried out. All of this assumes that emotions do not play a role in filtering or limiting the options that are considered in the first place.

If the work engaged is on the right track, then emotionally uniformed behavior does not appear to be how humans effectively navigate the complexities of social environments. To be sure, emotions can lead to highly problematic forms of engagement. However, we want to raise the point that the constructive use of emotion should not be ignored. Moreover, we want to suggest that it can inform computational modeling. For example, Wagar and Thagard put forward a computational model (called GAGE), which, when lesioned, exemplifies decision errors that are not unlike human decision errors when comparable parts of human brains are damaged. Our point *has not* been to suggest that computational models involving some of the functional contributions of emotions are impossible. We have been calling attention to an assumption—emotions do not play a functional role in constraining the search space of possibilities—that may place too great a computational burden on a system that is expected to perform in real time. Moreover, there is evidence independent of VMPFC damage that suggests that taxing our rational, calculating selves leads to fast application of deontological (for example, moral) principles, which are likely grounded in emotional processing in the brain (Greene and Haidt 2002; Greene 2007; Greene et al. 2008).

A robot without representation of or the ability to recognize these emotional states would be at a crippling disadvantage in the battlefield, especially if its task requires dealing with noncombatants or others whose status has to be determined. For example, a robot that cannot tell the difference between fear and anger will have a very hard time assessing the intent of an agent. It will also have a hard time knowing when to show compassion (and the laws of war requiring compassion: see note 7). We are far from understanding the subtle, pervasive relationship between emotion and cognition, but it seems undeniable that there is one in human beings.

Before closing out this section, let us return once again to the issue of the simulation theory of mindreading. According to this approach, to effectively predict the emotional states and actions of others, we simulate others using ourselves as the source. If this is at least part of the story about how mental state ascription can be performed in real time, then we have yet another reason to worry about a computational model that does not have a robust role for emotion. A system without emotion (or at least some sort of proto-emotional functional counterpart of emotion) could not predict the emotions or action of others based on its own states because it has no such emotional states. Of course, even if simulation theory is completely incorrect, a robot in the kinds of theaters we are considering will still need to make predictions about the kinds of emotional responses people will have, so knowledge of emotions is important for effective interaction in mostly civilian theaters.

8.5   A Suggestion for Taming Isotropy

Isotropy presents a clear set of computational problems for any AI system intended for deployment in civilian theaters. Solving this problem has been the preoccupation of many researchers in philosophy, AI, and cognitive science, and yet, as of the present, a solution has been elusive. As such, we do not intend to present one here. However, we do have some speculations on what kinds of cognitive mechanisms might interact in the human case to mitigate the irrelevance that isotropy introduces into inference. We suspect that a combination of attention, the computational structure of memory, and especially emotional appraisal all act in concert to regulate inference toward the relevant and away from the irrelevant.

To be more precise, let us consider a being or system that has the goal of making relevant inferences and avoiding irrelevant ones. Let us also suppose that our system is equipped with a focus of attention that can hold one (truth-evaluable) proposition at a time, an emotional subsystem that takes a single proposition P and an active set of propositions S as inputs, and outputs a scalar value $E \in (0,1)$. Let's call S the system's *situation representation*. In broad strokes, S is a collection of propositions that describe the state of affairs, which the system is currently considering. Propositions can be either generated internally via being recalled from memory or some similar storage mechanism, or they can be generated by percepts resulting from sensor data. Since we are interested in making relevant *inferences*, let us also grant our system a set of inferential capabilities that allow us to draw propositional conclusions from S, P, and suitable propositional background knowledge K, represented in some machine-readable format. Our system also comes equipped with motivational monitors that keep track of various system variables that correspond to basic drives such as approach/avoidance functions, and other homeostatic variables used to keep the system performing above some acceptable threshold. Let us further assume that our system is able to adopt beliefs, desires, intentions, goals, and other relevant attitudes toward propositions that are part and parcel of both planning and mindreading. Finally, let's assume that our system has an appraisal mechanism that generates urgency values in $(0,1)$ for each system motivation, desire, and goal. On each cognitive cycle, E is generated from both the current focus P and situation representation S, and urgency values are generated for motivations, desires, and goals. The next proposition to be the focus of attention will be the result of one of the scalars (either E or one of the sources of urgency values) being sufficiently larger than its competitors.

Since it is relevance we are concerned with, and since we have roughly sketched out a cognitive architecture for drawing (potentially) relevant inferences, let us define the problem space in which this sort of inference engine needs to operate. Isotropy roughly means that everything can be potentially related to everything else. In our case, it is our system's set of $K + S$ that defines the problem space. In particular K consists of associations between propositions like "if it rains, then the grass will be wet" or "having a cough usually indicates having a cold." Since all of these assertions can effectively be chained together by hooking up their propositional parts, they define a space of propositions connected by associations. Declarative (semantic) memory is often conceived of in these terms, with highly related items having stronger associative connections and fewer links between them. Many studies, and associated computational models, have documented limitations on the recall and activation of memory items (Atkinson and Shiffrin 1968; Oberauer 2002; Oberauer and Kliegl 2006) arranged in this kind of way. Some of the more popular computational explanations of these effects come by way of *spreading activation* models, which assume a finite amount of activation gets spread from one memory element to another, proportional to their connection strengths. In this way, highly indirect connections between elements far away from one another in memory are generally never activated. However, activation is a theoretical construct, and we do not in principle know the amount of activation to use if we were to construct such a system. Most of the computational models embodying spreading-of-activation solutions to the isotropy problem also lack the motivations, beliefs, attentional focus, and other mechanisms that our architecture-sketch possesses, and, as such, are not yet suitable for implementation on an autonomous system. Given this, we now develop the very beginnings of a complementary mechanism that exploits semantic nearness in declarative memory that seems to naturally capture relevance relations without committing to an arbitrary amount of activation to spread.

For any particular inferential goal our system might have, the space of propositions generated by **K** must be navigated. Presumably, for each of these goals, the set of propositions in the space having relevance would differ. The architectural sketch we have been developing suggests that one way to solve the isotropy problem might be to limit the amount that the focus of attention moves around our propositional space. If **P** is the current focus of attention, and the emotional subsystem generates a sufficiently high scalar value **E** for **P**, **S**, and their immediate inferential consequences, our attention-management procedure suggests that attention will either remain on **P** or move to a new proposition **P***, which is (1) semantically close and (2) an emotionally relevant consequence of **P** and **S**. In this case, we refer to **P** as an *attentional magnet*, or a part of the propositional space that captures the focus of attention for several cycles, until an interruption by way of an urgent desire, goal, or motivation occurs. We want attention to remain focused on relevant considerations for a given problem, and we want attention to shift to other portions of the propositional space if there is input suggestive of more pressing issues to attend to. Attentional magnets are the mechanism by which we keep from inferring too many indirect consequences, and could act as an analog (or perhaps as a complement to) traditional spreading of activation solutions to the isotropy problem. In any case, our architecture-sketch reserves a central role for emotional appraisal in regulating inference. Of course, open questions remain about how emotional appraisals or urgency values might be generated in the first place. While we have some ideas along those lines, space forbids us from exploring them in detail.

8.6   Conclusion

We do not think we have offered anything like a proof that emotion must play a role in either mental state ascription or in effective deliberation in complex social environments. Nor do we think that we have offered a proof that emotion (or something functionally like it) must play a role in getting robots to behave at least as well as humans in mostly civilian theaters of conflict. What we have done is to point to (a) the importance of mental state ascription in largely civilian theaters, (b) the difficulty of solving isotropy problems associated with such ascription, and (c) the potential strengths of emotion in reducing the computational load of deliberation in general and in thinking about mental states. By doing this, we hope to have raised some cautionary flags about considering the robotic use of lethal force in mostly civilian theaters. There is a lot to be done before seriously considering the use of robots in such theaters.[11] We also hope to have shown where some of this work needs to be done, and we hope to have motivated the idea that emotion may have some overlooked contributions to make in doing this work. Of course, our consideration of the role of emotions was in terms of capturing *some* of the functional, computational roles they play. There was no suggestion that there is something that it feels like to be the GAGE model, or any other computational model that captures some of the functional role of emotion. If it should turn out that the only way to solve problems connected to mental state ascription and isotropy in a robot is to actually build something that has feelings—there is something that it would feel like to be that being—then further ethical considerations would be introduced, since the feelings of genuinely sentient beings are subject to moral consideration. We do not introduce this issue to examine it, since considerations of space preclude this possibility. We mention it to forestall misinterpretations of our arguments. Everything we have said about modeling *some* of the computational, functional role of emotion assumes that the computational systems in question do not actually feel anything.

Notes

1. The term "isotropy" has a number of different uses. The use herein is inspired by Fodor (2000).

2. Goldman (2006) provides a useful introduction to different approaches to mindreading. We offer a brief explanation of Theory Theory and Simulation Theory in section 8.3.

3. Some have referred to this sort of consideration as "the frame problem." We will use the expression "isotropy" to be more precise. Different theorists have meant different things by "the frame problem," an expression introduced by McCarthy and Hayes (1969). See Murray 2009 and Ford and Pylyshyn 1996 for discussions of the different sorts of things theorists have meant by the frame problem. Isotropy can be connected with some versions of what has been called "the philosopher's frame problem," but that is broader than the more strict conceptions of the frame problem found in AI. Moreover, the very expression "frame problem" bids us to formulate it using the theoretical language of frame axioms, and not all approaches to understanding cognition or intelligence are committed to such axioms. Many neural network models have no use for such information structures, yet that does not exonerate those who would use such modeling techniques from providing an account of isotropy, a problem that can be formulated in a way that is not committed to postulating represented rules or axioms.

4. It might be thought that we are making too much of the complexities of largely civilian theaters. Perhaps robot warriors could simply be designed to be very cautious, not fire much, and be self-sacrificing in the name of being cautious. In other words, they would always err on the side of caution. When in doubt, do not fire. While this has an initial appeal, it is multiply problematic. In the mostly civilian theater, robots unable to manage isotropy, whether with respect to mental state ascription or other problems, would *frequently* be in doubt in cases where it is obvious to humans that there is great danger, and such robots would not fire. This would make them easy targets. Here is the first problem: it is not clear that such robots would be at all effective; if they are so cautious that they are easily destroyed, then it is unclear how they can be used to successfully accomplish the kinds of difficult missions humans are expected to accomplish. Second, if they are too cautious, then human soldiers, who expect their comrades in

arms to "have their back," would likely be unwilling to serve jointly with robots that are overly reluctant to fire.

5. See Wilkerson's (2001) discussion of Goldman 1989, Gordon 1995, and Heal 1996.

6. There may be any number of situations that share properties or relations with the situation under consideration. What makes one situation, y, like the one under consideration, x, will depend on what is deemed to be relevant for the simulation in a given situation. There may be a very large number of features that *could* be relevant, and which ones turn out to be relevant will depend on the details of the situation.

7. Arkin (2009, 143) notes that the laws of war mandate a certain level of compassion, and that it is not clear how to explicitly build that consideration into the architecture he is proposing. However, he suggests that building in the requirement to abide by the other rules of war and engagement would, in a sense, lead to compassionate behavior (by which, we take him to mean behavior that does not needlessly and unjustly inflict harm). There is a worry with this suggestion: presumably, the reason the rules of war explicitly state, over and above all the other explicitly stated rules, that compassion is required is that these other rules *do not* exhaust what it is to be compassionate. Another potential worry is that the requirement for compassion may well rely on human or human-like affective abilities for interpretation and application.

8. We mean for the qualifier "starting place" to be taken seriously in this sentence. As we will go on to explain, there are computational models that have been purported to capture some of the functional role of emotion in humans, but no one actually thinks that such models *feel* anything. It is possible to take one's cue from human cognition, but still fall short of a system or model that is fully expressive of a human-style mental/conscious life.

9. We mean to suggest here that our approach is well motivated, not that it is the only approach that could be motivated, or that we have conclusive proof that our way is the only way.

10. We recognize that not all soldiers are fit, and even pretty good soldiers make mistakes. It is not hard to imagine that if a soldier has been in multiple theaters where children have been combatants, and they have seen children kill soldiers, then they might react incorrectly in the sort of theater we have been considering. Moreover, their emotions may well lead them to react in this way. Again, we are not saying that there is no downside to emotion; we are simply pointing out that its potential strengths should not be ignored.

11. We do not pretend to have scouted out all the issues that need to be addressed. For example, in this chapter we have not even asked the question: Is it morally and legally defensible to build robot soldiers for use in mostly civilian theaters? We have largely been concerned with whether and how such systems might be built. Any use of lethal force with any technology has to satisfy a variety of moral and legal constraints. There are reasons to think that new constraints would be required for the types of robots we are considering. However, outlining and defending the required constraints would require another chapter, if not a book. Arkin (2007, 2009) has started on the project, but more needs to be said.

References

Arkin, Ronald. 2007. Governing lethal behavior: Embedding ethics in hybrid deliberative/reactive robot architecture. Georgia Institute of Technology, Technical Report GIT-GVU-07–11.

Arkin, Ronald. 2009. *Governing Lethal Behavior in Autonomous Robots*. Boca Raton, FL: Chapman and Hall/CRC Press.

Atkinson, Richard C., and Richard M. Shiffrin. 1968. Human memory: A proposed system and its control processes. In *The psychology of learning and motivation*. vol. 2. ed. K. W. Spence and J. T. Spence, 89–195. New York: Academic Press.

Butterworth, George E., and N. Jarrett. 1991. What minds have in common in space: Spatial mechanisms serving joint visual attention in infancy. *British Journal of Developmental Psychology* 9: 55–72.

Churchland, Patricia Smith. 1996. Feeling reasons. In *Neurobiology of Decision Making*, ed. A. R. Damasio, H. Damasio, and Y. Christen, 181–199. Berlin: Springer-Verlacht.

Damasio, Antonio. 1994. *Descartes' Error*. New York: G. P. Putnam's Sons.

Finucane, Melissa L., Ali Alhakami, Paul Slovic, and Stephen M. Johnson. 2000. The affect heuristic in judgments of risks and benefit. *Behavioral Decision Making* 13 (1): 1–17.

Fodor, Jerry. 2000. *The Mind Doesn't Work that Way: The Scope and Limits of Computational Psychology*. Cambridge, MA: MIT Press.

Ford, Kenneth M., and Zenon W. Pylyshyn, eds. 1996. *The Robot's Dilemma Revisited*. Norwood, NJ: Ablex Publishing.

Goldman, Alvin. 1989. Interpretation psychologized. *Mind and Language* 4 (3): 161–185.

Goldman, Alvin. 2006. *Simulating Minds: The Philosophy, Psychology, mand Neuroscience of Mindreading*. Oxford, UK, and New York: Oxford University Press.

Gordon, Robert. 1995. The simulation theory: Objections and misconceptions. In *Folk psychology: The theory of mind debate*, ed. M. Davies and T. Stone, 100–122. Oxford: Blackwell.

Greene, Joshua D. 2007. Why are VMPFC patients more utilitarian?: A dual-process theory of moral judgment explains. *Trends in Cognitive Sciences* 11 (8): 322–323.

Greene, Joshua, and Jonathan Haidt. 2002. How (and where) does moral judgment work? *Trends in Cognitive Sciences* 6 (12): 517–523.

Greene, Joshua D., Sylvia A. Morelli, Kelly Lowenberg, Leigh E. Nystrom, and Jonathan D. Cohen. 2008. Cognitive load selectively interferes with utilitarian moral judgment. *Cognition* 107: 1144–1154.

Heal, Jane. 1996. Simulation, theory, and content. In *Theories of Theories of Mind*, ed. P. Carruthers and P. Smith, 75–89. Cambridge, MA: Cambridge University Press.

Lerner, Jennifer Susan, and Dacher Keltner. 2000. Beyond valence: Toward a model of emotion-specific influences on judgment and choice. *Cognition and Emotion* 14 (1): 473–493.

Loewenstein, George F., Elke U. Weber, Christopher K. Hsee, and Ned Welch. 2001. Risk as feelings. *Psychological Bulletin* 116: 75–98.

McCarthy, John, and Patrick J. Hayes. 1969. Some philosophical problems from the standpoint of artificial intelligence. In *Machine Intelligence 4*, ed. D. Michie and B. Meltzer, 463–502. Edinburgh: Edinburgh University Press.

Murray, Shanahan. 2009. The frame problem. *Stanford Encyclopedia of Philosophy* (Winter ed.), ed. Edward N. Zalta. Metaphyics Research Lab, CSLI, Stanford University. <http://plato.stanford.edu/entries/frame-problem/> (accessed November 13, 2010).

Oberauer, Klaus. 2002. Access to information in working memory: exploring the focus of attention. *Journal of Experimental Psychology. Learning, Memory, and Cognition* 28 (3): 411–421.

Oberauer, Klaus, and Reinhold Kliegl. 2006. A formal model of capacity limits in working memory. *Journal of Memory and Language* 55 (4): 601–626.

Rottenstreich, Y., and C. K. Hsee. 2001. Money, kisses, and electric shocks: On the affective psychology of risk. *Psychological Science* 12 (3): 185–190.

Wagar, Brandon M., and Paul Thagard. 2004. Spiking Phineas Gage: A neurocomputational theory of cognitive-affective integration in decision making. *Psychological Review* 111: 67–79.

Wilkerson, William S. 2001. Simulation, theory, and the frame problem: The interpretive moment. *Philosophical Psychology* 14 (2): 141–153.

9 Responsibility for Military Robots

Gert-Jan Lokhorst and Jeroen van den Hoven

Several authors have argued that it is unethical to deploy autonomous artificially intelligent robots in warfare. They have proposed two main reasons for making this claim. First, they maintain that it is immoral to deploy such robots because such robots are "killer robots." Second, they claim that such robots cannot be held responsible because they cannot suffer, and therefore cannot be punished. We object to both claims. We first point out that military robots are not necessarily killer robots, and that, even if they were, their behavior could still be ethically correct—it could even be preferable to the behavior of human soldiers (section 9.1). Second, we argue that responsibility is not essentially related to punishment (section 9.2). Third, we propose an alternative analysis of responsibility, according to which robots could be responsible for their actions, at least to a certain extent (section 9.3). Finally, we emphasize that the primary responsibility for the behavior of military robots is in the hands of those who design and deploy them (sections 9.4 and 9.5).

9.1 Killer Robots

Sparrow (2007) and Krishnan (2009) have described military robots as "killer robots." By the same token, human soldiers might be called "killers," or even "murderers." However, it has long been disputed that soldiers should be described in this way. St. Augustine, for example, denied that soldiers violated the commandment *Thou shalt not kill*: "who is but the sword in the hand of him who uses it, is not himself responsible for the death he deals." Those who act according to a divine command or God's laws as enacted by the state and who put wicked men to death "have by no means violated the commandment, *Thou shalt not kill*" (St. Augustine, *On the City of God*). As these quotes indicate, in military ethics, matters are not as simple as they might seem. Calling military robots "killer robots" brings in a lot of background assumptions.

To form a proper perspective on the ethics of the use of military robots, we need to consider the ethics of war and peace. "Just war theory" is probably the most influential perspective on the ethics of war and peace (Orend 2008). Just War Theory can be divided into three parts, which in the literature are referred to, for the sake of convenience, in Latin. These parts are: (1) *jus ad bellum*, which concerns the justice of resorting to war in the first place; (2) *jus in bello*, which concerns the justice of conduct within war, after it has begun; and (3) *jus post bellum*, which concerns the justice of peace agreements and the termination phase of war. When discussing the deployment of military robots, *jus in bello* is clearly the most relevant category. *Jus in bello* refers to justice in war, to right conduct in the midst of battle. Responsibility for adherence to *jus in bello* norms falls primarily on the shoulders of those military commanders, officers, and soldiers who formulate and execute the war policy of a particular state. They are to be held responsible for any breach of the principles that follow. It is common to distinguish between external and internal *jus in bello*. External, or traditional, *jus in bello* concerns the rules a state should observe regarding the enemy and its armed forces. Internal *jus in bello* concerns the rules a state must follow in connection with its own people as it fights war against an external enemy. There are several rules of external *jus in bello*:

1. Obey all international laws on weapons prohibition. Chemical and biological weapons, in particular, are forbidden by many treaties.
2. Discrimination and noncombatant immunity: soldiers are only entitled to use their (nonprohibited) weapons to target those who are "engaged in harm." Thus, when they take aim, soldiers must discriminate between the civilian population, which is morally immune from direct and intentional attack, and those legitimate military, political, and industrial targets involved in rights-violating harm. While some collateral civilian casualties are excusable, it is wrong to take deliberate aim at civilian targets.
3. Proportionality: soldiers may only use force proportional to the end they seek. They must restrain their force to that amount appropriate to achieving their aim or target.
4. Benevolent quarantine for prisoners of war: if enemy soldiers surrender and become captives they cease being lethal threats to basic rights. They are no longer "engaged in harm." Thus, it is wrong to target them with death, starvation, rape, torture, medical experimentation, and so on.
5. No means that are *mala in se*: soldiers may not use weapons or methods that are "evil in themselves." These include: mass rape campaigns, genocide or ethnic cleansing, using poison, or treachery, forcing captured soldiers to fight against their own side, and using weapons whose effects cannot be controlled, such as biological agents.
6. No reprisals: a reprisal is when country A violates *jus in bello* in war with country B. Country B then retaliates with its own violation of *jus in bello*, seeking to chasten A into obeying the rules.

Internal *jus in bello* essentially boils down to the need for a state, even though it's involved in a war, nevertheless to still respect the human rights of its own citizens as best it can during the crisis.

What do these rules mean for military robots? They would behave unacceptably if they violated at least one of these rules. We may distinguish between two types of cases. First, let us assume that military robots are nothing but "killer robots" (as Sparrow [2007] and Krishnan [2009] seem to assume). In this case, they would not necessarily be immoral, because they would not necessarily violate one or more of these rules. As long as their reactions were proportionate, not evil in themselves, only directed toward combatants, and so on, their behavior could be justifiable, or even praiseworthy. Second, let us assume that there are military robots that are not just "killer robots," but designed to avoid killing as much as possible. This is clearly a more attractive option than the first scenario. It was brought to our attention when we showed the following passage about the strength of innate moral emotions (such as an aversion to killing) to a Dutch soldier (Chambers 2003):

These innate emotions are so powerful that they keep people moral even in the most amoral situations. Consider the behavior of soldiers during war. On the battlefield, men are explicitly encouraged to kill one another; the crime of murder is turned into an act of heroism. And yet, even in such violent situations, soldiers often struggle to get past their moral instincts. During World War II, for example, U.S. Army Brigadier General SLA Marshall undertook a survey of thousands of American troops right after they'd been in combat. His shocking conclusion was that less than 20 percent actually shot at the enemy, even when under attack. "It is fear of killing," Marshall wrote, "rather than fear of being killed, that is the most common cause of battle failure in the individual." When soldiers were forced to confront the possibility of directly harming other human beings—this is a personal moral decision—they were literally incapacitated by their emotions. "At the most vital point of battle," Marshall wrote, "the soldier becomes a conscientious objector."

After these findings were published in 1947, the U.S. Army realized it had a serious problem. It immediately began revamping its training regimen in order to increase the "ratio of fire." New recruits began endlessly rehearsing the kill, firing at anatomically correct targets that dropped backward after being hit. As Lieutenant Colonel Dave Grossman noted, "what is being taught in this environment is the ability to shoot reflexively and instantly. . . . Soldiers are de-sensitized to the act of killing, until it becomes an automatic response" (Lehrer 2009). The army also began emphasizing battlefield tactics, such as high-altitude bombing and long-range artillery, which managed to obscure the personal cost of war. When bombs are dropped from forty thousand feet, the decision to fire is like turning a trolley wheel: people are detached from the resulting deaths. These new training techniques and tactics had dramatic results. Several years after he published his study, Marshall was sent to fight in the Korean War, and he discovered that 55 percent of infantrymen were now firing their weapons. In Vietnam, the ratio of fire was nearly 90 percent. The army had managed to turn the most personal of moral situations into an impersonal reflex. Soldiers no longer felt a surge of negative emotions when they fired their weapons. They had been turned, wrote Grossman, into "killing machines" (Lehrer 2009, 173–174).

Our military informant pointed out that, despite the appeal to military authority displayed in this passage ("U.S. Army Brigadier General SLA Marshall"),[1] the passage does not reflect current military practice at all. All military handbooks, at least in the Netherlands, carefully point out that it is the aim of the military in battle to put the enemy out of action, to neutralize the enemy forces, or make them harmless—for example, by disarming them, immobilizing their vehicles, or turning them into prisoners of war. Temporary incapacitation is preferable to killing. In fact, the handbooks avoid the term "killing" and view the "elimination" of enemy forces as a means of last resort. Soldiers are taught to aim for the knees, not the heart or head, when taking target practice.

This is of the utmost importance in the context of autonomous intelligent military robots because they can be designed to immobilize or disarm enemy forces, instead of killing them. Because they can be equipped with superior sensory and incapacitating devices (and perhaps better decision circuitry as well, capable of better handling a greater amount of information more adequately than humans can do), they can in principle achieve this aim far more reliably than humans. In other words, it could be argued that autonomous robots are, in principle, morally superior to human soldiers, because the former could resort to temporary incapacitation in cases where the latter would have no option but to kill. It is misleading to equate autonomous military robots with killer robots because it is quite possible that their deployment will *save* lives instead of adding to human loss. Calling them "killer robots" is an insidious rhetorical move, which easily leads to a false dilemma. This brings us to our first thesis:

Thesis 1. Artificially intelligent military robots that save lives are preferable to humans (or bombs) that kill blindly.

9.2   Responsibility, Punishment, and Blame

Suppose that something goes wrong on the battlefield—that people get killed as the result of the action of an autonomous military robot instead of merely being put out of combat. Who is to blame in such a case—the robot itself, or its operator, programmer, or designer?

Sparrow (2007) argues that robots cannot be held responsible because they cannot be punished. They cannot be punished because they cannot suffer. In other words, responsibility presupposes the ability to suffer. We want to object to this line of reasoning for two reasons. First, it is by no means to be taken for granted that robots will never be able to suffer. Second, punishment is not desirable in any case because we can use more effective means for adjustment in the case of robots that do not act in a desirable way.

First, it is questionable that robots cannot be made to suffer. On the contrary, it has been argued that intelligent robots are bound to have emotions as the inevitable consequence of having motives and the processes they generate (Sloman and Croucher 1981). If robots can be made to suffer, then they can be punished, as well, so this part of Sparrow's objection loses it force.

Second, let us grant that Sparrow is right and that robots cannot suffer. We may then ask: what is the point of punishment, anyway? Its main justification is the prevention of the type of behavior that brought it about. Punishment leads to suffering; humans tend to avoid suffering; so punishment may lead to prevention because it gives humans a reason to avoid similar behavior in the future. What if an agent cannot suffer or cannot see that similar behavior in the future will again lead to harsh punishment? Then we give them treatment. Treatment is another means to achieve prevention. When punishment is not an option, treatment remains. This not only applies to humans (for example, mentally handicapped persons), but also to other types of agents. Cars cannot suffer, so we treat (repair, correct) them, simply because this may bring the desired goal (correct functioning in the future) closer.

In other words, it is important to make a distinction between the means and the ends. Punishment is simply one means that may lead to the desired end; it is not desirable in itself. If other courses of action are more effective, they are ipso facto preferable.

This is important in the context of our military robots. If they cannot suffer, they cannot be punished. But it can be argued that punishment is not desirable anyway. It only detracts us from what really matters, namely the prevention of similar tragic actions in the future. It has been argued that men are nothing but machines. If so, similar considerations could be applied to them. It turns out that considerations along these lines can already be found in the literature. In a piece called "Let's all stop beating Basil's car," Richard Dawkins (2006) wrote as follows:

Retribution as a moral principle is incompatible with a scientific view of human behavior. As scientists, we believe that human brains, though they may not work in the same way as man-made computers, are as surely governed by the laws of physics. When a computer malfunctions, we do not punish it. We track down the problem and fix it, usually by replacing a damaged component, either in hardware or software.

Basil Fawlty, British television's hotelier from hell, created by the immortal John Cleese, was at the end of his tether when his car broke down and wouldn't start. He gave it fair warning, counted to three, gave it one more chance, and then acted. "Right! I warned you. You've had this coming to you!" He got out of the car, seized a tree branch and set about thrashing the car within an inch of its life. Of course, we laugh at his irrationality. Instead of beating the car, we would investigate the problem. Is the carburetor flooded? Are the sparking plugs or distributor points damp? Has it simply run out of gas? Why do we not react in the same way to a defective man: a murderer, say, or a rapist? Why don't we laugh at a judge who punishes a criminal, just as heartily as we laugh at Basil Fawlty? Or at King Xerxes, who, in 480 BC, sentenced the rough sea to 300 lashes for wrecking his bridge of ships? Isn't the murderer or the rapist just a machine with a defective component? Or a defective upbringing? Defective education? Defective genes?

When Sparrow laments that military robots cannot be held responsible because they cannot suffer, he resembles Basil Fawlty, who laments that his broken car does not respond to threats. Their reactions are misplaced for the same reasons. The alternatives are clear in both cases as well: if you want to prevent such-and-such action, do something about it. If punishment does not help, adopt an alternative approach from among the courses of action that lead to more desirable behavior. In the case of humans, this means psychotherapy, chemical treatment, or neurosurgery and similar treatment; in the case of cars, this means looking at the carburetor, the sparking plugs, distributor points, and gas tank; in the case of nonhuman agents, this comes down to improving the sensory devices, fine-tuning the response mechanisms, adjusting the nonmortal combat devices, or rewriting the software. As Dawkins (2006) wrote, "Assigning blame and responsibility is an aspect of the useful fiction of intentional agents that we construct in our brains as a means of short-cutting a truer analysis of what is going on in the world in which we have to live." It is a tremendous advantage, not a defect, that we do not have to assign blame and responsibility to robots, because we know what is going on inside them and what to do when something goes wrong. This brings us to our second thesis:

Thesis 2. It is regrettable and not satisfactory at all that punishment is usually the best we can do in the case of human wrongdoing.

9.3 The Logic of Responsibility

What exactly *are* responsibility and agency? In recent years, logicians and artificial intelligence researchers have devoted considerable attention to this topic (Belnap, Perloff, and Xu 2001; Horty 2001). The literature is vast and complicated, but one thing to note is that logicians have made a distinction between two concepts of action:

1. *Seeing to it that* (this is Chellas's theory of CSTIT: Chellas's Seeing To It That);
2. *Deliberatively seeing to it that* (this is Horty's theory of DSTIT: Deliberatively Seeing To It That).

DSTIT can be defined in terms of CSTIT: an agent A deliberatively sees to it that P if and only if (1) A sees to it that P (in the sense of Chellas) and (2) it is possible that not P. Deliberative action presupposes the ability to make choices, the ability to do otherwise; agents' choices usually are assumed to be independent from each other. Chellas's concept of seeing to it that does not depend on this notion of choice. CSTIT theory is a theory of causal responsibility, while DSTIT theory is related to moral responsibility, because it is usually assumed than an agent should only be held morally responsible if he or she could have done otherwise (this view goes back to Aristotle's *Nicomachean Ethics*). It is to be noted that these analyses of responsibility *do not mention punishment at all*: this suggests that the concepts of responsibility and punishment are less closely related than Sparrow assumed.

How does this apply to robots? Let us first consider the case of nondeliberating robots, which are controlled by a human commander. Such cases are described by sentences of the following form:

(1) Commander A deliberatively sees to it that robot B sees to it that P.

Who is responsible in such a case? It turns out (as a matter of logic) that both the robot and the commander are *causally* responsible, but it is only the commander (not the robot) that can be *morally* responsible, for the simple reason that the robot has no choice and does not have the ability to do otherwise.

This is in perfect agreement with the legal maxims *qui facit per alium facit per se* and *respondeat superior*, which can be summarized as follows:

Qui facit per alium facit per se means "he who acts through another does the act himself." This is a fundamental maxim of agency (*Stroman Motor Co. v. Brown*, 116 Okla 36, 243 P 133). A maxim often stated in discussing the liability of employer for the act of employee (35 Am J1st M & S § 543). According to this maxim, if in the nature of things the master is obliged to perform the duties by employing servants, he is responsible for their act in the same way that he is responsible for his own acts (Anno: 25 ALR2d 67).

Respondeat superior means "let the master answer." This is a legal principle, which states that, in most circumstances, an employer is responsible for the actions of employees performed within the course of their employment. This rule is also called the "Master-Servant Rule," recognized in both common law and civil law jurisdictions. This principle is related to the concept of *vicarious liability*.

It is comforting to know that these age-old legal principles can be applied to modern robots and flow naturally from our logical account.

The analysis is more complex in the case of deliberative (autonomous) robots, which can potentially be held responsible precisely because of their capacity to engage in deliberation. As noted earlier, agents' choices are independent from each other, in the sense that any combination of possible choices available to different agents at the same moment must be compatible. Each agent can choose each of its alternatives, regardless of what the other agents are doing at the moment. This implies that an agent cannot deliberatively see to it that another agent deliberatively sees to it that something is the case. This makes this case quite unlike the case presented in the previous section, in which an agent deliberatively did something by using another agent as an instrument. One cannot exert such control over independent agents; instead, we must think of other ways to induce them to perform in ways we see fit. One simple way of doing so consists of *blocking* all undesirable courses of actions, in the sense of making them impossible; this undermines the subordinate agents' ability to do otherwise, and leaves them no choice but to undertake the desired courses of action.

In general, even though an agent cannot see to it that another agent makes a certain specific choice (the latter agent can always choose differently), an agent can see to it that another agent makes *some* choice. Formally: even though

(1) Commander A deliberatively sees to it that robot B deliberatively sees to it that P is necessarily false,

(2) Commander A deliberatively sees to it that: either robot B deliberatively sees to it that P or robot B deliberatively sees to it that not P

might well be true.

Situations of the latter type have been called situations of "forced choice" (Belnap Perloff, and Xu 2001, chapter 10B2). We may similarly speak of cases of "forced moral responsibility."

Cases of this type have played a prominent role in military trials (Wikipedia.org 2010). Nuremberg Principle IV states "the fact that a person acted pursuant to order of his government or of a superior does not relieve him from responsibility under international law, provided a moral choice was in fact possible to him." Similarly, in the Ehren Watada case, the judge ruled that soldiers, in general, are not responsible for determining whether the order to go to war itself is a lawful order—but are only responsible for those orders resulting in a specific application of military force, such as an order to shoot civilians, or to treat POWs inconsistently with the Geneva Conventions. Nuremberg Principle IV and the Ehren Watada judgment concern the choices and moral responsibility of agents in situations that were brought about by other agents (their superiors).

Even though logicians and lawyers can reason about cases in which forced choices play a role, it is doubtful whether such situations will play a role in robot ethics. Autonomous military robots that deliberate and perform voluntary actions out of their own accord seem very far off indeed. They might even be seen as unwelcome in view of the risk of insubordination; commanders might object to robots that protest against their commanders' or operators' commands. The case of nondeliberative robots that are used as instruments by their operators seems more realistic. We discussed this case earlier (referring to the *qui facit per alium* and *respondeat superior* principles) and in fact came to a similar conclusion as St. Augustine did (section 9.1).

9.4   Design of Military Robots

Even though intelligent military robots may turn out to be morally preferable to humans for the reasons that we have indicated,[2] this does not mean that it will be easy to build them. Quite apart from the technical aspects (superior sensory devices, discrimination between friend and foe, and so on), there are ethical questions to consider. For example, should ethical principles (do not kill unnecessarily, avoid collateral damage, do not harm civilians, do not torture, respect the Geneva Conventions, and so on) be included in their lists of goals to pursue, or pitfalls to be avoided? But what do these principles mean exactly? How can they be made precise? For example, what is torture, anyway? How can it be demarcated from mild pressure? Is a civilian who supports the enemy an enemy? A lot of conceptual ethical analysis is needed before such principles have been made precise enough to such a degree that they can be burnt into the hardware or software that controls the behavior.

Furthermore, how should they be built in? Is ethics primarily a matter of logic? Should robots follow these rules by means of logical reasoning, namely, by proving theorems and refuting nontheorems? (Bringsjord, Arkoudas, and Bello 2006; van den Hoven and Lokhorst 2002). If so, should default reasoning perhaps be built in, should the frame problem (including the *moral* frame problem) be considered, and should the problem of induction and abduction be solved before we set out on this path? Is some kind of self-monitoring, a module that keeps track of the robot's moral reasoning, worth building in (Lokhorst forthcoming)? Or should we forget about logic and merely build in appropriate pattern recognition software, perhaps in the form of statistical software or neural networks? Or better yet, should both routes be pursued, just as in the case of humans, who are often asserted to have two decision mechanisms, a fast, automatic, innate mechanism, which provides us with our gut feelings, and a slow, conscious, learned circuit, which takes care of our rational decisions?[3] If so, how should they be kept in balance? Is it necessary to incorporate a mechanism that keeps track of actions that should have been done otherwise (i.e., a mechanism that generates regret)?

Nobody knows at this moment, and much research is needed before we will be able to answer these questions. Before we embark on such research, we should try to answer the preliminary question of whether its objective is ethically desirable. We have tried to answer this question in this chapter. According to us, there can be no doubts about its proper answer. We therefore propose our third thesis:

Thesis 3. From a moral point of view, the design of military robots is eminently desirable, provided that such robots are designed as transparent robots that avoid killing to the maximum extent possible, and not as inscrutable killer robots, over which we have no control.

Even if military robots could be held responsible to some extent (as discussed earlier in the forced choice cases), this would never excuse us in case something goes wrong, because those who design and deploy military robots are those who are responsible for them in the first place (as indicated by the *qui facit per alium* and *respondeat superior* principles previously discussed). This may be regarded as unfortunate, but we regard it as welcome because we have more control over the design of military robots that act in agreement with our own ethical specifications than over the training of human soldiers, which is a hit-and-miss affair, at best.

9.5 Conclusion

We claim that it should never be assumed that human beings, in their role of designer, maker, manager, or user of robots and other artifacts or technological systems, can transfer moral responsibility to their products in case of untoward outcomes, or can claim diminished responsibility for the consequences brought about by their products. We claim that designers of autonomous robots are "design responsible" in all cases. In the causal case, this is so for the reasons we have expounded, since the robot is an instrument like any other artifact. In the deliberative case, it is so because the designer is responsible for "designing in" the logic of deontic reasoning and deontic metareasoning, which will lead the robot to make the right choices similar to the way in which we teach our children to think correctly in moral matters. In both cases, we think it would be unethical to produce such systems or work on their development while assuming that the locus of full and undivided responsibility for outcomes can be assigned to the artifacts themselves, however accomplished and sophisticated they are. We consider the shift of responsibility to the thing one has produced as an ultimate form of bad faith, meaning, denial of human choice, freedom, and responsibility. The designers, producers, managers, overseers, and users are and remain always responsible. The fact that it is difficult to apportion responsibility should not deter us. The apportioning of responsibility outside the simplest cases is problematic anyway. We hope that this contribution will make it easier to allocate responsibility adequately and fairly when thinking about responsibility and robots.

We focus on the responsibility of the designers and refer to their specific responsibility as "design responsibility." One specific and important aspect of design responsibility is to design in accordance with well-accepted and widely shared values. In software engineering, this approach is referred to as "value sensitive design" (van den Hoven and Manders-Huits 2009). In the case of military robots, there is a well-accepted normative framework in the form of Geneva Conventions, laws of war, and, more generally, the doctrines of just war theory—*jus ad bellum, in bello,* and *post bellum.* These provide us with moral principles that need to be translated and applied to the design of military robots. These principles are fairly broad since they also pertain to the design of the institutional context that guarantees design compliance with these accepted doctrines and their implications.

But, as former Pentagon Chief of High Value Targeting Marc Garlasco said, we cannot simply download international law into a computer (Singer 2009, 389). Sustained legal engagement and ethical reflection must be present from the very beginning of the design process. Investigating how ethical and legal norms can be "designed in" to complex systems is a core research goal of this process.

Notes

1. The findings of military writer and analyst S. L. A. Marshall, syndicated columnist for the *Detroit News* and brigadier general in the Army Reserve, are less reliable than is usually reported: see Chambers 2003.

2. Arkin has made a similar claim: "My research hypothesis is that intelligent robots can behave more ethically in the battlefield than humans currently can. That's the case I make" (cited in Dean 2008; see also Arkin 2009).

3. See the book by Lehrer for a description of these two mechanisms, their strengths and weaknesses, and a discussion of the question when to use which of the two. Also see the discussion about the necessity of merging the cognitive top-down approach with a less cognitive bottom-up approach (Wallach and Allen 2009).

References

Arkin, Ronald. 2009. *Governing Lethal Behavior: Embedding Ethics in a Hybrid Deliberative/Reactive Robot Architecture*. Boca Raton, FL: CRC Press.

Belnap, Nuel, Michael Perloff, and Ming Xu. 2001. *Facing the Future: Agents and Choices in Our Indeterminist World*. New York: Oxford University Press.

Bringsjord, Selmer, Konstantine Arkoudas, and Paul Bello. 2006. Toward a general logicist methodology for engineering ethically correct robots. *Intelligent Systems* 21 (4): 38–44.

Chambers, John Whiteclay, Jr. 2003. S. L. A. Marshall's *Men Against Fire*: New evidence regarding fire ratios. *Parameters* (Autumn): 113–121.

Dawkins, Richard. 2006. Let's all stop beating Basil's car. *The World Question Center*. <http://www.edge.org/q2006/q06_9.html> (accessed March 26, 2011).

Dean, Cornelia. 2008. A soldier, taking orders from its ethical judgment center. *The New York Times*, November 24.

Horty, John F. 2001. *Agency and Deontic Logic*. New York: Oxford University Press.

Krishnan, Armin. 2009. *Killer Robots: Legality and Ethicality of Autonomous Weapons*. Farnham and Burlington, VT: Ashgate.

Lehrer, Jonah. 2009. *The Decisive Moment*. Edinburgh: Canongate. (Originally published as *How We Decide*. New York: Houghton Mifflin Harcourt.)

Lokhorst, Gert-Jan C. Forthcoming. Computational meta-ethics: Towards the meta-ethical robot. *Minds and Machines*. <http://www.springerlink.content/d819182mwk4u0146/fulltext.pdf> (accessed July 14, 2011).

Orend, Brian. 2008. War. *The Stanford Encyclopedia of Philosophy* (Fall ed.), ed. Edward N. Zalta. Metaphyics Research Lab, CSLI, Stanford University. <http://plato.stanford.edu/archives/fall2008/entries/war/> (accessed November 20, 2010).

Singer, Peter Warren. 2009. *Wired for War: The Robotics Revolution and Conflict in the 21st Century*. New York: Penguin Press.

Sloman, Aaron, and Monica Croucher. 1981. Why robots will have emotions. In *Proceedings IJCAI 1981 Vancouver*, ed. P. J. Hayes, 197–202. Los Altos, CA: William Kaufmann.

Sparrow, Robert. 2007. Killer robots. *Journal of Applied Philosophy* 24 (1): 62–77.

van den Hoven, Jeroen, and Gert-Jan C. Lokhorst. 2002. Deontic logic and computer-supported computer ethics. *Metaphilosophy* 33 (3): 376–386.

van den Hoven, Jeroen, and N. L. J. L. Manders-Huits. 2009. Value-sensitive design. In *A Companion to the Philosophy of Technology*, ed. J. K. Berg Olsen, S.A. Pedersen, and V. Hendricks, 477–480. Chichester, UK: Wiley-Blackwell.

Wallach, Wendell, and Colin Allen. 2009. *Moral Machines: Teaching Robots Right from Wrong*. Oxford: Oxford University Press.

Wikipedia.org. 2010. Superior orders. <http://en.wikipedia.org/wiki/Superior_orders> (accessed July 14, 2011).

IV Law

Related to the question of responsibility in the preceding chapter is perhaps the most practical issue in robotics: how it is accounted for in law. To the extent that programming limitations, errors, accidents, and so on, are the most pressing concerns in robotics today, we would reflexively look toward law to address whatever harm might arise from robots. Yet while product liability and other areas of law already exist, they are largely untested with respect to autonomous robotics, which may shift responsibility from human designers and operators to the machine itself. This section, then, offers chapters on law and governance in robotics.

In chapter 10, Richard O'Meara continues the discussion of military robotics, again as a major area of concern in robot ethics and in media headlines. Despite a lack of consensus on the need for military robotics governance or how to proceed with it, he points to considerable infrastructure already in place that can serve as a starting point to create this technology governance, political will permitting.

In chapter 11, Peter Asaro considers how legal theory, or jurisprudence, might be applied to robots, suggesting possible approaches to some problems. He finds that legal theory does allow us to define certain classes of ethical problems that correspond to traditional and well-defined legal problems, while other difficult practical and metaethical problems cannot be solved by legal theory alone. Moreover, there are several fundamental legal issues that are raised by robotic technologies.

M. Ryan Calo looks at the issue of privacy in chapter 12, a key area of law for not just robotics but other emerging technologies as well. The impact on privacy comes not only from the fact that robots have sensors that can monitor and report on our activities, but also from the access that robots will have into historically private settings, such as inside our homes, and the willingness we may have to share information with anthropomorphized robots. We then proceed to part V, in which our growing relationships with robots are the focus.

10 Contemporary Governance Architecture Regarding Robotics Technologies: An Assessment

Richard M. O'Meara

Even a cursory review of the contemporary governance architecture regarding military technological innovation generally reveals a disturbing lack of consensus regarding the necessity for governance and the methodologies to be utilized to achieve it. Innovations, adaptations, and uses in the areas of nanotechnology, bioscience, information science, cognitive technologies—referred to generally as NBIC—and especially robotics, are being discovered at an unprecedented rate in a culture of technological uncertainty, which provides very little time and minimal governance in order to ask the question of not *can* we do this, but *should* we do this.

Regarding the *use* of weapons, such as robotics, however, there is a fairly robust governance architecture. The field of ethics, for example, has dealt with issues of weapons use for centuries. Ethics has traditionally provided humankind with guidance regarding the use of weapons on the battlefield. Just War Theory, for example, speaks specifically to justification for the use of force in the first instance, *jus ad bellum*, and how that force can be utilized to obtain a just result, *jus in bello*. In order to initiate a just war, the issue of proportionality—the ideal that the universal goods to be obtained outweigh the universal evils that can be foreseen—might well be used to constrain the employment of certain types of robotics in certain ways.

Jus in bello certainly applies. Weapons use, here, is justified by adherence to concepts of military necessity, discretion, and proportionality. Where a particular robotic configuration, especially one with the independence to operate without humans in the loop, is unleashed on the battlefield, issues of target choice, collateral damage, and proportionate use of force wrestle with increased capabilities.

There is also the question of the "soldier's ethic." At least for the foreseeable future, a soldier is a human being, one who enters the profession with values and ethics learned at his or her mother's knee, during the formative years in civil society, and a sense of other moral systems, such as religious beliefs. The soldier is also capable of exhibiting generally accepted psychological traits of human beings, including fear, love, anger, rage, guilt, mercy, hope, faith, generosity, courage, shame and cowardice. The warrior traditionally has been enhanced by training and technology to accomplish the military function, which, according to Samuel Huntington (1956), is performed "by a public bureaucratized profession expert in the management of violence and responsible for the military security of the state." The soldier is also a volunteer, or, at least, has agreed in one form or another to enter a special class of citizens, prepared to project violence on behalf of the state, and committed to the knowledge that he or she may be targeted by others as a result of this commitment.

Consistent with the past, modern warrior respects actions of their peers, which reflect valor, loyalty, and adherence to the military ethic, even under the most dire of circumstances. Because the soldier is a realist and assumes human weakness and frailty—indeed, trains his whole life to overcome these characteristics personally—actions that reflect these values provide honor, a much sought-after commodity. This ethic, it would appear, has two functions, which are especially important given the environment in which the soldier works. The ethic helps the soldier differentiate between the killing he or she is required to do, and simple murder. A trained warrior is constrained to project force only in certain restricted situations. If there is compliance, despite the circumstance, the soldier is deemed honorable; otherwise, he or she is identified as a thug, a base murderer, rapist, sadist, etc. The ethic, therefore, provides constraint. Second, it can help the soldier justify the force he or she has used, which provides a useful psychological benefit, contributes to morale, and affirms a personal adherence to regulation. The warrior is a representative of the state for which he or she fights. This system of constraints inures not only to the warrior personally and the community in which he or she serves, but to the state itself.

Constraining the use of certain weapons as well are the various restrictions regarding the projection of force found in international law, which are translated into national law and regulation. There are, for example, multiple conventions that purport to deal with specific technologies and practices.[1] On the one hand, the United States is not a party to all of these conventions, and, to the extent that they do not rise to the level of customary international law, the United States is not specifically bound by them. On the other hand, the United States has taken considerable interest in the articulation of standards, which purport to regulate conduct generally on the battlefield, including how weapons are used. There are five principles that run through the language of the various humanitarian law treaties[2] (the rules), which the United States acknowledges and generally honors. These principles are (1) a general prohibition on the employment of weapons of a nature to cause superfluous injury or unnecessary harm, (2) military necessity, (3) proportionality, (4) discrimination, and (5) command responsibility.

Some weapons, it is argued, are patently inhumane, no matter how they are used or what the intent of the user is. This principle has been recognized since at least 1907, although consensus over what weapons fall within this category tends to change over time. The concept here is that some weapons are *design dependent*; that is, their effects are reasonably foreseeable, even as they leave the laboratory. In 1996, the International Committee of the Red Cross (ICRC) at Montreux articulated a test to determine if a particular weapon would be the type that would foreseeably cause superfluous injury or unnecessary suffering (SIrUS). The SIrUS criteria would ban weapons when their use would result in

a. A specific disease, specific abnormal physiological state, a specific and permanent disability or specific disfigurement; or

b. Field mortality of more than 25 percent or a hospital mortality of more than 5 percent; or

c. Grade 3 wounds as measure by the Red Cross wound classification scale; or

d. Effects for which there is no well-recognized and proven treatment.

The operative term here is *specific*; the criteria speak to technology specifically designed to accomplish more than render an adversary *hors de combat*. The test here is purely medical and does not take into consideration military necessity. As such, it has been rejected by the United States specifically and the international community generally (Lewand 2006; Verchio 2001).

The second principle, *military necessity*, requires a different analysis. This principle "justifies measures of regulated force not forbidden by international law which are indispensable for securing the prompt submission of the enemy, with the least possible expenditures of economic and human resources" (Gutman and Kuttab 2007, 239). Here force is permitted where a military objective is identified. These have been defined as those "objects which by their nature, location, purpose or use make an effective contribution to military action and whose total or partial destruction, capture or neutralization, in the circumstances ruling at the time, offers a definite military advantage" (239). Military necessity recognizes the benefit to friend and foe alike of a speedy end to hostilities—protracted warfare, it assumes, creates more rather than less suffering for all sides. In order to determine the necessity for the use of a particular technology, then, one needs to know what the definition of "victory" is, and how to measure the submission of the enemy in order to determine whether the technology will be necessary in this regard.

The third principle, *proportionality*, is of considerable concern to the innovator and user of new technologies. A use of a particular technology is not *proportional* if the loss of life and damage to property incidental to attacks is excessive in relation to the concrete and direct military advantage anticipated. In order to make this determination, it can be argued, one must consider the military necessity of a particular use and evaluate the benefits of that use in furtherance of a specific objective against the collateral damage that may result.

Discrimination, the fourth principle, strikes at the heart of judgment. Indiscriminant attacks (uses) are prohibited under the rules. Indiscriminant uses occur when they are not directed against a specific military objective, employ a method or means of combat the effects of which cannot be directed at a specified military target (indiscriminant bombing of cities for example), employ a method or means of combat the effects of which cannot be limited as required, or, are of a nature to strike military and civilian targets without distinction.

A final principle of the rules is *command responsibility,* which exposes a multiple of superiors to various forms of liability for failure to act in the face of foreseeable illegal activities. This is a time-honored principle, based on the contract between soldiers and their superiors, which requires soldiers to act and superiors to determine when and how to act. It has a long history reflective of the need for control on the battlefield.[3]

Article 36 of the 1977 Additional Protocol 1 to the Geneva Conventions of 1949 requires that each "State Party"

determine whether the employment of any new weapon, means, or method of warfare that it studies, develops, acquires, or adopts would, in some or all circumstances, be prohibited by international law. . . . The legal framework of the review is the international law applicable to the State, including international humanitarian law (IHL). In particular, this consists of the treaty and customary prohibitions and restrictions on specific weapons, as well as the general IHL rules applicable to all weapons, means, and methods of warfare. General rules include the rules aimed at protecting civilians from the indiscriminate effects of weapons and combatants from unnecessary suffering. The assessment of a weapon in light of the relevant rules will require an examination of all relevant empirical information pertinent to the weapon, such as its technical description and actual performance, and its effects on health and the environment. This is the rationale for the involvement of experts of various disciplines in the review process. (Lewand 2006)

Again, the United States is not a signatory to this protocol and, thus, technically not bound by its requirements. To the extent that it sets out reasonable requirements and methodologies for use by states fielding new and emerging technologies, however, this treaty could well set the standard in international law for appropriate conduct. Failure to consider its mechanisms, definitions, and proscriptions, then, may well constitute a violation of customary international law in the future.

Another constraint worth noting is the emerging trend in international law to hold those responsible for fielding weapons that allegedly contravene the principles enunciated above through the use of litigation based on the concept of *universal jurisdiction.* The concept of universal jurisdiction is a customary international law norm that permits states to regulate certain conduct to which they have no discernable nexus. Generally, it is recognized as a principle of international law that all states have the right to regulate certain conduct, regardless of the location of the offense or the nationalities of the offender or the victims. Piracy, slave trade, war crimes, and genocide are all generally accepted subjects of universal jurisdiction. Belgium, Germany, and Spain have all entertained such prosecutions. Arising out of the war on terror and

Iraq, former President George W. Bush, former secretaries of defense and state Rumsfeld and Kissinger, and former military commanders Powell and Franks, have all been the subject of such suits.

The issue of *lawfare* is also of concern. Lawfare is a strategy of using or misusing law as a substitute for traditional military means to achieve military objectives. Each operation conducted by the U.S. military results in new and expanding efforts by groups and countries to use lawfare to respond to military force. As military technology evolves, so do the scenarios facing military planners. New types of weaponry raise a host of legal and ethical questions. For example, new weaponry that can destroy power networks through electrical transmissions may seem to be preferable to traditional bombs. When electricity grids are destroyed, however, hospitals and civilians will lose power, as well, possibly resulting in civilian casualties. American military authorities are still grappling with many of these issues.

While litigation to date has revolved primarily around allegations of practices such as genocide and torture/interrogation, there is no reason to believe that future prosecutions may be justified where decisions regarding illegal innovation, adaption, and use of weapons systems are made and their conduct results in grave breaches of customary or statutory international humanitarian law.

10.1 The Intersection between Robotics and Governance

Robotics is one of a number of technologies being created in an environment of technological uncertainty. Discussions regarding the scope of emerging technologies are often difficult, due to the breadth and sophistication of the information about them. They often descend into ramblings about gadgets and gizmos and reflect the short answer to Peter Singer's question, "Why spend four years researching and writing a book on new technologies? Because robots are frakin' cool" (Singer 2009). Because innovation is and has always been catalytic, feeding off itself, reacting to its intended and unintended consequences, and influenced by the environment in which it is created and creating new environments as it goes, the discussion must, of course, be much longer and more nuanced. Of equal importance is the fact that demands for emerging technologies are coming faster and faster, and failure to keep up can have disastrous effects on the battlefield (Dunlap 1999; Shachtman 2009).

The scope of contemporary technological innovation is both impressive and staggering. Indeed, for the average consumer of these technologies, whether on the battlefield or in daily life—the general who orders this technology, the politician who pays for it, the user whose life is changed by it, even the Luddite who rails against it—these technologies are magic. They are incomprehensible in the manner of their creation, the details of their inner workings, the shear minutiae of their possibilities; they are like the genie out of the bottle and clamoring to fulfill three wishes: guess right and the world is at your fingertips; guess wrong, and there may well be catastrophe. And you have to guess quickly, for the genie is busy and has to move on. There are, of course, shamans who know the genie's rules, who created the genie, or, at least, discovered how to get it out of the bottle. You go to them and beg for advice regarding your wishes. What should I take from the genie? How should I use my wishes? Quickly tell me before I lose my chance and the genie makes the choices for me. And you find that the shaman is busy with new genies and new bottles, and has not given your choices much thought at all. He may stop to help you ponder your questions, but most likely he goes back into his tent and continues his work: "You're on your own, kid. . . . Don't screw up!"

Robotics enjoys preeminence in the discussion of military technologies, perhaps, because popular culture has served to inform the public of their possibilities and, further, it may be said that their applications are easier to comprehend. Robots are defined as

machines that are built upon what researchers call the "sense-think-act" paradigm. That is, they are man-made devices with three key components: "sensors" that monitor the environment and detect changes in it, "processors" or "artificial intelligence" that decides how to respond, and "effectors" that act on the environment in a manner that reflects the decisions, creating some sort of change in the world around a robot. When these three parts act together, a robot gains the functionality of an artificial organism. (Singer 2009)

Robots are deployed to perform a wide range of tasks on and off the battlefield, and Congress has mandated that their use expand radically in the next decade. The U.S. Department of Defense in its *FY2009–2034 Unmanned Systems Integrated Roadmap* reports:

In today's military, unmanned systems are highly desired by combatant commanders (COCOMs) for their versatility and persistence. By performing tasks such as surveillance, signals intelligence (SIGNIT), precision target designation, mine detections, and chemical, biological, radiological, nuclear (CBRN) reconnaissance, unmanned systems have made key contributions to the Global War on Terror (GWOT). As of October 2008, coalition unmanned aircraft systems (UAS) (exclusive of hand-launched systems) have flown almost 500,000 flight hours in support of Operations Enduring Freedom and Iraqi Freedom, unmanned ground vehicles (UGVs) have conducted over 30,000 missions, detecting and/or neutralizing over 15,000 improvised explosive devises (IEDs), and unmanned maritime systems (UMSs) have provided security to ports. (U.S. Department of Defense 2009)

Further, their development has increased as the needs have been identified. The Department of Defense reports that its investment in the technology has seen "unmanned systems transformed from being primarily remote-operated, single-mission platforms into increasingly autonomous, multi-purpose systems. The fielding of increasingly sophisticated reconnaissance, targeting, and weapons delivery technology has not only allowed unmanned systems to participate in shortening the 'sensor to shooter' kill chain, but it has also allowed them to complete the chain by delivering precision weapons on target" (O'Rourke 2007). In other words, *autonomous* robots are being used to kill enemies on the battlefield, based on information received by their sensors and decisions made in their processors.

In the future, roboticists tell us that it is probable that robots, with the addition of artificial intelligence,[4] will be capable of acting independently, without human supervision—called *humans in the loop*—in the accomplishment of most tasks presently performed by soldiers today (Guetlein 2005; Krishnan 2009). Their sensors will be more capable of reading the environment than humans, their processors will, like a personal computer today, have available a wider range of information or experience and be able to consider it more rapidly than humans, and their effectors will not be constrained by human frailties of fear, fatigue, size, and reaction to stress. They will be capable of their own creation (fabrication) and maintenance. Indeed, some believe they will free humans from participation in warfare altogether (Minsky 1968).

In sum, robotics technology comes with a whole host of intended, as well as unintended, consequences. These include, but are certainly not limited to, issues of military ethics, what capabilities *should* be created and how should they be used, military anthropology, whether humans are necessary to the projection of force on the battlefield, whether the warrior ethic still has currency, and foreign policy. Is the decision to project force made easier when death and mayhem are created by machines, rather than humans? It is suggested that a system of coherent governance infrastructure provides a place where these issues can be sorted out before rather than after these machines are unleashed on the battlefield.

Decisions regarding military innovation—who orders the technology, who pays for the technology, and the uses to which the technology will be put—are presently made in a decentralized and competitive environment which fosters innovation, but also contributes to an inherent instability in the decision-making process. Governance architecture does exist, but it is haphazard in its articulation, institutionalization, and enforcement, leaving spaces where the conflicting agendas of multiple stakeholders can have free sway. The creation of a coherent system of governance is possible, but only where all stakeholders are convinced of its need and the goals for which it is created. A coherent system of governance regarding these technologies will permit us to make rational choices about not only who we *can* be, but also who we *want* to be.

Notes

1. These include the 1999 Hague Declaration concerning expanding bullets; Convention on the Prohibition of the Development, Production, and Stockpiling of Bacteriological (Biological) and

Toxin Weapons and on Their Destruction (1972); Convention on the Prohibition of Military or Any Other Hostile Use of Environmental Modification Techniques (1976); Resolution on Small-Caliber Weapon Systems (1979); Protocol on Non-Detectable Fragments (Protocol 1) (1980); Protocol on Prohibitions or Restrictions on the Use of Mines, Booby-Traps, and Other Devises (Protocol 11) (1980); Protocol on Prohibitions or Restrictions on the Use of Incendiary Weapons (Protocol 111) (1980); Convention on the Prohibition of the Development, Production, Stockpiling, and Use of Chemical Weapons and on Their Destruction (1993); Protocol on Blinding Laser Weapons (Protocol 1V to the 1980 Convention (1995); Protocols on Prohibitions or Restrictions on the Use of Mines, Booby-Traps and Other Devices as amended on May 3, 1996; Protocol 11 to the 1980 Convention as amended on 3 May 1996; Convention on the Prohibition of the Use, Stockpiling, Production, and Transfer of Anti-Personnel Mines and on their Destruction (1997); Convention on Prohibitions or Restrictions on the Use of Certain Conventional Weapons Which May be Deemed to be Excessively Injurious or to Have Indiscriminate Effects, amendment article 1, 21 (2001); Protocol 1 Additional to the 1949 Geneva Conventions; Convention on Cluster Munitions (2008). See International Committee of the Red Cross (2010).

2. International humanitarian law (IHL) comprises a set of rules that seek to limit the effect of armed conflict. Primary conventions include the Geneva Conventions of 1949, supplemented by the Additional Protocols of 1977 relating to the protection of victims of armed conflicts; the 1954 Convention for the Protection of Cultural Property in the Event of Armed Conflict and additional protocols; the 1972 Biological Weapons Convention; the 1980 Conventional Weapons Conventions and its five protocols; the 1997 Ottawa Convention on anti-personnel mines; and the 2000 Optional Protocol to the Convention on the Rights of the Child on the involvement of children in armed conflict. See International Committee of the Red Cross (2004).

3. This principle is noted by Sun Tzu in 500 BCE in the *Art of War* and was recognized during the U.S. Civil War. Article 71 of General Orders No. 100, "Instructions for the government of armies of the United States in the Field" (known as the "Lieber Code"), imposed criminal responsibility on commanders for ordering or encouraging soldiers to wound or kill already disabled enemies. Its codification occurred in the Hague Convention (1V) of 1907, Respecting the Laws and Customs of War on Land, and is explicitly described in the Additional Protocol 1 (AP1) 1977 to the Geneva Conventions of 1949 and has made its way into multiple war crimes cases including *In re Yamashita*, 327 U.S. 1 (1946); *United States v. Captain Ernest L. Medina*; and *The Prosecutor v. Zejnil Delalic, Zdravko, Zdravko Music, Hasin Delic and Esad Landzo*, Case No. IT-96–21-T Judgment, Trial Chamber, November 16, 1998, The International Court for the Former Yugoslavia (ICTY). The ICTY provides for three elements regarding the theory: (1) the existence of a superior-subordinate relationship; (2) the superior knew or had reason to know that the criminal act was about to be or had been committed; and (3) the superior failed to take the necessary and reasonable measures to prevent the criminal act or punish the perpetrators thereof. It is also reflected in Article 28 (b) of the Rome Statute of the International Criminal Court, UN Doc. 2187 U.N.T.S. 90.

4. One definition of AI is the science of making machines do things that would require intelligence if done by men. Ravi Mohan notes:

First, robots will engage in lethal activities like mine clearing or IED detection. (This is happening today.) Then you'll see them accompany human combat units as augmenters and enablers on real battlefields. (This is beginning to happen.) As robotics gets more and more sophisticated, they will take up potentially lethal but noncombat operations, like patrolling camp perimeters or no fly areas, and open fire only when "provoked." (This is beginning to happen, too.) The final state will be when robotic weapons are an integral part of the battlefield, just like "normal" human controlled machines are today and make autonomous or near autonomous combat decisions. (Mohan 2007)

References

Dunlap, Charles J. Jr. 1999. Technology and the 21st century battlefield: Recomplicating moral life for the statesman and the soldier. Strategic Studies Institute, Carlisle Barracks, PA, January 15.

Guetlein, Michael A. 2005. Lethal autonomous weapons—Ethical and doctrinal implications. Naval War College, Newport, RI.

Gutman, Roy, and Daoud Kuttab. 2007. Indiscriminate attack. In *Crimes of War 2.0: What the Public Should Know*, 239–241. New York: W. W. Norton.

Huntington, Samuel P. 1956. *The Soldier and the State, the Theory and Politics of Civil-Military Relations*. Cambridge, MA: The Belknap Press of Harvard University Press.

International Committee of the Red Cross. 2004. What Is International Humanitarian law? Legal fact sheet 31-07-2004. <http://www.icrc.org/eng/resources/documents/legal-fact-sheet/humanitarian-law-factsheet.htm> (accessed July 14, 2011).

International Committee of the Red Cross. 2010. International Humanitarian Law—Treaties and Documents. <http://www.icrc.org/ihl.nsf/TOPICS?OpenView> (accessed November 26, 2010).

Krishnan, Armin. 2009. *Killer Robots: Legality and Ethicality of Autonomous Weapons*. Burlington, VT: Ashgate.

Lewand, Kathleen. 2006. A Guide to the Legal Review of New Weapons, Means and Methods of Warfare, Measures to Implement Article 36 of Additional Protocol 1 of 1977. International Committee of the Red Cross, January. <http://www.icrc.org/eng/assets/files/other/icrc_864 _icrc/geneva.pdf> (accessed July 14, 2011).

Minsky, Marvin. 1968. *Semantic Information Processing*. Cambridge, MA: MIT Press.

Mohan, Ravi. 2007. Robotics and the Future of Warfare. <http://ravimohan.blogspot.com/2007/12/robotics-and-future-of-warfare.html> (accessed July 14, 2011).

O'Rourke, Ronald, 2007. Unmanned vehicles for U.S. Navy forces: Background and issues for Congress. CRS Report for Congress, updated April 12, 2007.

Shachtman, Noah, 2007. How technology almost lost the war: In Iraq, the critical networks are social, not electronic. *Wired* 15 (12) (November 27). <http://www.wired.com/polities/security/magazine/15-12/ff_futurewar> (accessed July 14, 2011).

Singer, P. W. 2009. *Wired for War. The Robotics Revolution and Conflict in the 21st Century.* New York: Penguin Group.

U.S. Department of Defense. 2009. *FY2009–2034 Unmanned Systems Integrated Roadmap.* Washington, DC: Department of Defense.

Verchio, Donna Marie. 2001. Just say no! The SIrUS Project: Well-intentioned, but unnecessary and superfluous. *Air Force Law Review* 21 (Spring): 224–225.

Wallach, Wendell, and Colin Allen. 2009. *Moral Machines: Teaching Robots Right From Wrong.* New York: Oxford University Press.

11 A Body to Kick, but Still No Soul to Damn: Legal Perspectives on Robotics

Peter M. Asaro

The continued advancement of robotic technologies has already begun to present novel questions of social and moral responsibility. While the overall aim of this collection is to consider the ethical and social issues raised by robotics, this chapter will focus on the legal issues raised by robotics. It starts from the assumption that we might better understand the social and moral issues surrounding robotics through an exploration of how the law might approach these issues. While it is acknowledged that there are instances where what is legal is not necessarily morally esteemed, and what is morally required may not be legal, in general, there is a significant overlap between what is legal and what is moral. Indeed, many of the crucial concepts are shared, and as such this chapter will explore how the law views responsibility, culpability, causality, intentionality, autonomy, and agency. As a philosopher, rather than a lawyer or legal scholar, my concern will be with these theoretical concepts, and how their justificatory frameworks can be used to interpret and apply law to the new kinds of cases, which teleoperated, semi-autonomous, and fully autonomous robotics have already, or, may soon, present. Insofar as some of the issues will also involve matters of industry and community standards, public opinions, and beliefs, as well as social values and public morals, the chapter will consider questions of value. While my concern will be primarily with the law as it is typically understood and applied in the United States, my aim is that these reflections will also prove useful to scholars and lawyers of other legal traditions.

Indeed, the legal issues raised by robotic technologies touch on a number of significant fundamental issues across far-ranging areas of law. In each of these areas, there can be found existing legal precedents and frameworks which either directly apply to robotics cases, or which might be extended and interpreted in various ways so as to be made applicable. My aim is to consider each in turn, as well as to identify the principles that might underlie a coherent legal understanding of the development and use of robotic systems. Furthermore, I will consider the means by which we might judge the potential of robots to have a legal standing of their own. It will thus be helpful to organize the discussion in terms of both the salient types of robots—teleoperated, semi-autonomous, and autonomous—as well as the principal areas of law, criminal and civil.[1]

The most obvious issues that arise for the application of the law to robotics stem from the challenge that these complex computational systems pose to our traditional notions of intentionality, as well as how and whom to punish for wrongful acts (Bechtel 1985; Moon and Nass 1998). Most of the scholarship on law and robotics to date has focused on treating robots as manufactured products (Asaro 2006, 2007; Schaerer, Kelley, and Nicolescu 2009), subject to civil liability, or on whether robots can themselves become criminally liable (Dennett 1997; Asaro 2007), or the challenges robotic teleoperation poses to legal jurisdiction (Asaro 2011). I will begin by considering the more straightforward cases of semi-autonomous robots, which can be treated much like other commercial products. For these cases, the law has a highly developed set of precedents and principles from the area of law known as *product liabilities*, which can be applied.

I will then consider the implications of increasingly autonomous robots, which begin to approach more sophisticated and human-like performances. At some point in the future, there may be good reasons to consider holding such robots to standards of criminal or civil liability for their actions, as well as compelling reasons to hold their owners and users to higher, or lower, standards of responsibility for the wrongdoings of their robots. These considerations will draw upon a variety of legal areas with similar structures of distributing intention, action, autonomy, and agency. There exist certain similarities between such robots and their owners and controllers, and the ways in which individuals have traditionally been held to account for the wrongdoings of other subordinate intelligent, sentient, conscious, autonomous, and semi-autonomous agents. Examples include laws pertaining to the assignment of responsibility between animals and their owners, employees and their bosses, soldiers and their commanders, and slaves and their masters, as well as *agency law*, in which agents are entrusted with even greater levels of responsibility than is the case with typical subordinates. There are also issues involving whether robots themselves are entitled to legal standing, redress, or even rights, including the ability to sign contracts, be subject to criminal liabilities, or the means by which they might be justly subjected to punishment for crimes. This will bring us to consider the punishments against other kinds of nonhuman legal agents, namely corporations, and what can be learned about robot punishments from corporate punishments.

11.1   Robots and Product Liability

Many of the most common potential harms posed by robotic systems will be covered by civil laws governing product liability. That is, we can treat robots as we do other technological artifacts—such as toys, weapons, cars, or airliners—and expect them to raise similar legal and moral issues in their production and use. In fact, the companies that currently manufacture robots are already subject to product liability, and retain lawyers who are paid to advise them on their legal responsibilities in producing, advertising, and selling these robots to the general public. Most of the public's current concerns about the possible harms that robots might cause would ultimately fall under this legal interpretation, such as a robotic lawnmower that runs over someone's foot, or a self-driving car that causes a traffic accident.

It will be helpful at this point to review the basic elements of product liability law.[2] Consider, for example, a toy robot that shoots small foam projectiles. If that toy were to cause several children to choke to death, the manufacturer might be held liable under civil law, and be compelled to pay damages to the families that lost children because of the toy. If it can be proven in court that the company was *negligent*, with regard to the defects, risks, and potential hazards arising from the use of their product, then the company could also be criminally, as well as civilly, liable for the damages caused to victims by their product. Legal liability due to negligence in product liability cases depends on either *failures to warn*, or *failures to take proper care* in assessing the potential risks a product poses. A failure to warn occurs when the manufacturer fails to notify consumers of a foreseeable risk, such as using an otherwise safe device in a manner that presents a potential for harm. For example, many power tools display warnings to operators to use eye protection or safety guards, which can greatly reduce the risks of using the device. The legal standard motivates manufacturers to put such warning labels on their products, and, in the preceding example, the manufacturer might avoid liability by putting a label on the package, stating that the robot toy contains parts that are a choking hazard to young children.

A failure to take proper care is more difficult to characterize. The idea is that the manufacturer failed to foresee a risk, which, if they had taken proper care, they would have likely foreseen. This counterfactual notion is typically measured against a somewhat vague community standard of reason, or an industry standard of practice, about just what proper care is expected among similar companies for similar products. In some sense, the more obvious the risk is, according to such a standard, the more likely that the negligence involved rises to the level of criminality.

The potential failure to take proper care, and the reciprocal responsibility to take proper care, is perhaps the central issue in practical robot ethics from a design perspective. What constitutes proper care, and what risks might be foreseeable, or in principle unforeseeable, is a deep and vexing problem. This is due to the inherent complexity of anticipating potential future interactions, and the relative autonomy of a robotic product once it is produced. It is likely to be very difficult or impossible to foresee many of the risks posed by sophisticated robots that will be capable of interacting with people and the world in highly complex ways—and may even develop and learn new ways of acting that extend beyond their initial design. Robot ethics shares this

problem with bioengineering ethics—both the difficulty in predicting the future interactions of a product when the full scope of possible interactions can at best only be
estimated, and in producing a product that is an intrinsically dynamic and evolving
system, whose behavior may not be easily controlled after it has been produced and
released into the world (Mitcham 1990).

A classic defense against charges of failures to warn and failures to take proper care
is the industry standard defense. The basic argument of the industry standard defense
is that the manufacturer acted in accordance with the stated or unstated standards of
the industry they are participating in. Thus, they were merely doing what other similar
manufacturers were doing, and were taking proper care as measured against their
peers. This appeal to a relative measure again points to the vagueness of the concept
of proper care, and the inherent difficulty of determining what specific and practical
legal and moral duties follow from the obligation to take proper care. This vague
concept also fails to tell us what sorts of practices *should* be followed in the design of
robots. An obvious role for robot ethics should be to seek to establish standards for
the robot industry, which will ensure that the relevant forms of proper care are taken,
and I believe this should be one of the primary goals for future robot ethics research.

If the company in question willfully sought to remain ignorant of the risks its
robotic products might pose, such as by refusing to test a product or ignoring warnings from designers, then its negligence could also be deemed criminal. This would
be a case of *mens rea*, in which the culpable state of mind is one of ignorance, either
willfully or unreasonably. That is, if the risks posed are so obvious that they would be
recognized by anyone taking the time to consider them, or knowledge of the risks had
to be actively avoided, then that ignorance is criminal. Beyond that, if it can be shown
that the manufacturer was actually aware of the risk, then this amounts to *recklessness*.
Reckless endangerment requires a mental state of foreseeing risks or possible dangers,
whether to specific individuals or an uncertain public, though not explicitly intending
that any potential victims actually be harmed.[3] In some cases, recklessness can also
be proved by showing that a "reasonable person" should have foreseen the risks
involved, even if it cannot be proven that the defendant actually had foreseen the
risks. An even more severely culpable state of mind would be if the company sold the
dangerous toys *knowingly,* in awareness of the fact that they would cause damages,
even though they did not intend the damages. And the most severe form of culpable
liability is that of having the mental intention to cause the harm, or otherwise *pur-
posely* causing harms. While these are all cases of criminal liability, as we will see later
in our discussion of corporate punishment, such cases are almost always settled by
awarding punitive monetary damages to victims and their legal advocates, rather than
penalties owed to the state, such as imprisoning the guilty parties.[4]

Another interesting aspect of liability is that it can be differentially apportioned.
That is to say, for example, one party might be 10 percent responsible, while another

is 90 percent responsible for some harmful event. This kind of analysis of the causal chains resulting in harms is not uncommon in cases involving traffic accidents, airliner crashes, and product liability. In many jurisdictions, there are laws that separate differential causal responsibility from the consequent legal liability. Among these are laws imposing *joint and several liability*, which holds all parties equally responsible for compensation, even if they are not equally responsible for the harm, or *strict liability*, which can hold a party fully responsible for compensation. These liability structures are meant to ensure that justice is done, in that the wronged individual is made whole (monetarily) by holding those most able to compensate them fully liable for paying all of the damages, even when they are not fully responsible for causing the harm. Nonetheless, these cases still recognize that various factors and parties contribute differentially to causing some event.

Differential apportionment could prove to be a useful tool when considering issues in robot ethics. For instance, a badly designed object recognition program might be responsible for some damage caused by a robot, but a bad camera could also contribute, as could a weak battery, or a malfunctioning actuator, and so on. This implies that engineers need to think carefully about how the subsystem they are working on could interact with other subsystems—whether as designed or in partial breakdown situations—in potentially harmful ways. That, in turn, would suggest that systems engineering approaches that can manage these complex interactions would become increasingly important for consumer robotics. It also means that manufacturers will need to ensure the quality of the components they use, including software, test the ways in which components interact with each other, as well as prescribe appropriate maintenance regimes to ensure the proper functioning of those components. This is typical of complex and potentially dangerous systems, such as in airliners and industrial robots, and may prove necessary for many consumer robots, as well.

There is, however, a limit to what robot manufacturers, engineers, and designers can do to limit the potential uses of, and harms caused by, their products. This is because other parties, namely the consumers and users of robots, will choose to do various sorts of things with them and will have to assume the responsibility for those choices. For instance, when one uses a product in a manner wholly unintended by its designers and manufacturers, such as using a tent as a parachute, we no longer hold the manufacturer liable for the harms that result. Schaerer, Kelley, and Nicolescu (2009) argue that users should be held liable only in those cases in which it can be shown that they acted with harmful intentions. I disagree with this argument because of the intrinsic flexibility of design inherent in the programmability of robots. Typically, we do not hold manufacturers responsible when the hardware has been tampered with or extensively modified, or when the hardware is running software developed by users or a third party, even when there is no malice involved. We also do not always hold the company that develops a piece of software responsible when it turns out to be vulnerable to a malicious third party, such as a hacker or virus. Again, the operative legal considerations are causal responsibility and culpable intent. However, in manufacturing a product that is programmable, and thus wildly customizable, a great deal of responsibility lies in the hands of those who do the programming, as well as those who use the robot by giving it various commands.

The challenge presented by programmable general-purpose robots is that it is unreasonable to expect their manufacturers to anticipate all the things their robots might be programmed to do or asked to do, and thus unreasonable to hold them liable for those things. At least, the less foreseeable the uses, the less responsible the manufacturer might be. But there is no clear and definitive line here. At one extreme are cases where the manufacturer ought to be held liable, and at the other extreme cases where the programmer or user ought to be held liable. At one extreme, we would find the narrowly specified applications of the robots for which its manufacturers intended the product to be used. At the other extreme, we might find a highly original custom application or program, which perhaps only that particular programmer or user might have dreamt up.

Like built-in programming, the context in which the robot has been placed and the instructions given to it by its owners may also be the determining, or contributing, causes of some harm, where the robot is the proximate cause. Orders and operator commands are like programming, in some sense, and as natural language processing grows more sophisticated, the two may become increasingly indistinguishable. And even a well-programmed robot can be ordered to do things that might unintentionally cause harms in certain situations. In short, there will always be risks inherent in the use of robots, and at some point the users will be judged to have knowingly assumed these risks in the very act of choosing to use a robot. Properly assessing responsibility in liability cases will be difficult and contested, and will depend on decisions in future cases that establish various legal precedents for interpreting such cases.

It also seems likely that robotic technologies will advance much like computer technologies, in that hackers and amateur enthusiasts will push the envelope of capabilities of new devices as much as commercial manufacturers do, especially in terms of the software and programming of robots. Even iRobot's mild-mannered Roomba vacuum-cleaning robot has a fully programmable version called Create (iRobot.com 2010), and hackers have created their own software development kits (SDKs) to customize the Roomba robot as they see fit, though at their own liability (Kurt 2006). As long as these robotic products have enough safe and legitimate uses, it would be difficult to prohibit or regulate them, just as it would be difficult to hold the manufacturers responsible for any creative, even if dangerous, uses of their products. Cars and guns are also very dangerous consumer products, but it is the users who tend to be held liable for most of the harms they cause, not the manufacturers, because the use of those potentially dangerous products place additional burdens of responsibility on the user. For manufacturers to be held responsible in those cases, it is usually necessary to show that there is some defect in the product, or that manufacturers misled consumers.

The crucial issue raised by Schaerer, Kelley, and Nicolescu (2009) is whether to hold the manufacturer strictly liable for all the damages (because they are better able to pay compensation and to ensure responsible design), or whether their limited ability to foresee a possible application of their technology should limit their liability in some way. One implication of applying strict liability, as Schaerer, Kelley, and Nicolescu argue, is that doing so may result in consumer robots being designed by manufacturers to limit their liability by making them difficult to be reprogrammed by users, or safeguarding them from obeying commands with hazardous implications. This could include making the open-ended programming of their robots more difficult, or incorporating safety measures intended to prevent harm to humans and property, such as ethical governors (Arkin 2009).[5] Conversely, by shielding individual users from liability, this could also encourage the reckless use of robotic systems by end-users. Cars are causally involved in many unintended harms, yet it is the drivers who are typically held responsible rather than manufacturers. This issue points to a fundamental tension between identifying the causal responsibility of original manufacturers, end-users, and third-parties, and the need for legal policies that can shape the responsible design and use of consumer robots, even if they run counter to our intuitions about causal responsibility.

An additional challenge that we may soon face is determining the extent to which a given robot has the ability to act "of its own accord," either unexpectedly or according to decisions it reaches independently of any person. As robot control programs become more capable of various forms of artificial reasoning and decision-making, these reasoning systems will become more and more like the orders and commands of human operators in terms of their being causally responsible for the robots' actions, and, as such, will tend to obscure the distinction just made between manufacturers and users. While some sophisticated users may actually design their own artificial intelligence systems for their robots, most will rely on the reasoning systems that come with these robots. Thus, liability for faults in that reasoning system might still revert to the manufacturer, except in cases where it can be shown that the user trained or reprogrammed the system to behave in ways in which it was not originally designed to behave. In its general form, the question of where commands and orders arise from is integral to the legal notions of autonomy and agency. There is a growing literature addressing the question of whether robots can be capable of moral autonomy, or even legal responsibility (Wallach and Allen 2009). But missing from these discussions is the recognition that the law does not always hold morally autonomous humans fully responsible for their own actions. The notable cases include those of diminished mental capacity, involuntary actions, or when agents are following orders of a superior.

The next section will consider the possibility that even if a robot could become, in some sense, fully autonomous, then we might not be inclined to hold it legally liable for all of the harms it might cause.

11.2 Vicarious Liability, Agents, and Diminished Responsibility

There are multiple areas of the law that deal with cases in which one independent, autonomous, rational being is acting on behalf of, or in subordination to, another. Often discussed in the robotics literature are laws governing the ownership of domesticated animals; however, there are also analogous cases involving the laws governing the liability of employees and soldiers following orders, as well as historical laws governing the liability of masters for their slaves, and the harms they cause when agents are carrying out the orders of their superiors. The laws governing animals are the simpler cases, as animals are not granted legal standing, though they may be entitled to protections from abuse in many jurisdictions. More complicated are cases where a person can act either on behalf of a superior or on their own behalf, and judging a specific act as being one or the other can have differing legal implications. The area of law dealing with these three-party relationships is called *agency law*,[6] and we will consider this after first considering the legal liabilities surrounding domesticated and wild animals.

It has been recognized that robots might be treated very much like domesticated animals, in that they clearly have some capacity for autonomous action, yet we are not inclined to ascribe to them moral responsibility, or mental culpability, or the rights that we grant to a human person (Caverley 2006; Schaerer, Kelly, and Nicolescu 2009). Domesticated animals are treated as property, and as such any harms to them are treated as property damages to the owner. Because they are domesticated, they are generally seen as not being particularly dangerous if properly kept. Despite this, it is recognized that animals sometimes act on their own volition and cause harms, and so their owners can be held liable for the damages caused by their animals, even though the owners have no culpable intentions. If, however, the owners' behavior was criminally negligent, reckless, or purposeful, then the owners can be held criminally liable for the actions of their animals. For instance, it can be criminal when someone fails to keep his or her animal properly restrained, trains an animal to be vicious, orders an animal to attack, or otherwise intends for the animal to bring about a harm.

We should note that in such cases the intention of the animal is rarely relevant—it does not matter much for legal purposes whether the animal intended the harms it caused or not. Rather, it is the owner's intention that is most relevant. Moreover, in those cases where the animal's intention runs counter to the owner's intention, this can have two different consequences. In cases of domestic animals, where the animal suddenly behaves erratically, unexpectedly, or disobeys its owner, then this tends to diminish the *mens rea* of the owner, though does not release them from liability, and often motivates the destruction of the animal. However, in cases of exotic or wild animals, such as big cats, nonhuman primates, and poisonous snakes, there is a certain presupposition of their having independent reasoning (i.e., being wild) and being more physically dangerous than domesticated animals. And with the recognition of the intrinsic danger they pose to other people, there is an additional burden of responsibility on the owner. Owning such animals has various restrictions in different states (Bornfreeusa.org 2010; Kelley et al. 2010), and the very act of owning them is recognized as putting other members of the community at risk, should the animal escape or someone accidentally happen upon them. Failing to properly keep such an animal can automatically constitute criminally reckless endangerment, based on the known dangerousness of the animal.

Such a standard might also be applied in robotics. A standard off-the-shelf robot might be considered as being like a domesticated animal, in that its manufacturer has been entrusted to design a robot that is safe to use in most common situations. However, a highly modified, custom programmed, or experimental robot might be seen as being more like a wild animal, which might act in dangerous or unexpected ways. Thus, someone who heavily modifies his or her robot, or builds a highly experimental robot, is also undertaking greater responsibility for potentially endangering the public with that robot. A good example would be someone who armed a robot with a dangerous weapon. Such an act could itself be seen as a form of reckless endangerment, subject to criminal prosecution, even if the robot did not actually harm anyone or destroy any property with the weapon. Similar principles apply in drunk-driving laws. By driving a car while drunk, an individual is putting others at risk, even if they do not actually have an accident. It is because of this increased risk that the activity is deemed criminal (as well as being codified in law as criminal). Building a robot that intentionally, knowingly, or recklessly endangers the public could be similarly viewed as a criminal activity, and laws to this effect should be established. More limited cases of negligent endangerment might be determined to be civilly or criminally liable.

With certain technologies that are known to be dangerous if misused—such as cars, planes, guns, and explosives—there are laws that regulate their ownership and use. This ensures both that the possession and use can be restricted to individuals who are trained and tested on the proper use of a technology, as well as to establish an explicit and traceable connection between a piece of technology and a responsible individual. Thus, the use of an airplane or automobile requires completing a regime of training and testing to obtain an operator's license. The ownership of a gun or explosives requires a license, which also aids in tracking individuals who might obtain such materials for illicit purposes, and in tracking the materials themselves. The ownership of dangerous exotic animals, and in many jurisdictions even certain particularly aggressive domesticated dog breeds, such as pit bulls, often requires a special license (Wikipedia.org 2010). It would not be unreasonable to expect that certain classes of robots, especially those that are deemed dangerous, either physically or because of their unpredictable behavior or experimental nature of their reasoning systems, might require special licenses to own and operate. Licenses might also be required to prevent children from being able to command dangerous robots, just as they are not allowed to drive cars, until they have reached a certain age and received training in the responsible uses of the technology.

The treatment of robots as animals is appealing because it does not require us to give any special rights or considerations to the robots—our only concern is with the owners. Another interesting area of legal history is the laws governing slavery. The history of these laws goes back to ancient Rome, and they have varied greatly in different times, places, and cultures, up to and including the slave laws of the United States. The U.S. slave laws ultimately treated slaves as property, but included numerous specialized clauses intended to manage the unique difficulties and dangers of enslaving human beings, as well as encoded specific racial aspects of slavery into the laws themselves. For the most part, slaves were treated as expensive animals, so that if a slave damaged the property of someone other than his or her owner, their owner was liable for the damage. Similarly, as property, the slave was protected from harm from individuals other than his or her owner, but such harms were viewed as property damage rather than crimes, such as assault or even murder. Indeed, the laws largely enshrined the ability of owners to harm their own slaves and not be subject to the criminal laws that might otherwise apply. Yet, it was also recognized that slaves exercised a will of their own, and so owners were not held liable for damages caused by their slaves if they had escaped. And, unlike animals, in the act of escaping, slaves were held liable for their own choice of whether to escape or not, though those who aided them were also held liable for assisting them in their escape. A full consideration of the implications of slave law for our understanding of robotics is beyond the scope of this chapter, but will be the subject of further research.

Agency law deals with cases in which one individual acts on behalf of another individual. In these cases, the *agent* acts on behalf of a *principal*. There are various circumstances where these relationships are established, but they generally involve some form of employment, often involving a contract.[7] Whether or not there is a written contract, the liability of the principal for the actions of its agents is derived from the doctrine of *respondeat superior*—that superiors are responsible for the action of their subordinates. Thus, if an employee causes a harm in the conduct of their job, and thus explicitly or implicitly at the discretion of their employer, the employer is liable. This is called *vicarious liability*—when one person or legal entity is liable for the actions of another. For instance, when a delivery truck damages a parked car, the delivery company, rather than the individual driver, can be held liable. There are exceptions to this, however, which recognize the independent autonomy of employees. One of the employer's defenses against such liability claims is to argue that the employee was acting on their own behalf, and not that of the employer—which generally means showing they were not doing their job in the typical manner. Courts make a distinction between *detours*, in which an employee must digress from the usual manner of carrying out their job in order to achieve the purposes of their employer, and *frolics*, in which the employee is acting solely for their own purposes. Thus, when a driver finds an intersection blocked and must take a different route to make a delivery, the employer would still be liable for the damage to the parked car. However, if the driver had decided to take a different route in order to visit a friend before making the delivery, then the court may decide that this constitutes a frolic and the driver is responsible for the damage to the parked car because the driver was not carrying out their duties as an employee, or fulfilling the will or purpose of the employer, at the time of the accident.

These ideas might be usefully applied to many kinds of service robots. It would seem that, for most uses of a robot to assist a person in daily life, such as driving them around, shopping for them, cleaning and maintaining their home, running errands, and so on, the robot would be little different than a human servant or employee. As such, vicarious liability would apply, and the owner would therefore be liable for any harm caused by that robot in the conduct of their owner's business. This would also include cases of detour, in which the robot was unable to carry out its duties in the normal or directed manner, and sought alternative routes, plans, or strategies for achieving its given goals.

As robots grow more sophisticated and autonomous, we might eventually be tempted to argue that they actually are capable of developing their own purposes. For such a robot, an owner might seek a defense from liability for the actions of a robot which was on a frolic of its own—a robot which, though employed in the service of its owner, caused some harm while pursuing its own purpose. Depending on the ways in which such robots might be programmed, and our ability to review its reasoning and planning processes, we might actually be able to determine from its internal records which purposes it was actually in pursuit of when it caused a particular harm. Of course, it might be that it was pursuing a dual purpose, or that the purposes were obscure, in which case the courts would have to make this determination in much the same manner as they do for human agents. However, this raises several issues regarding whether robots might themselves have legal standing, especially if they are capable of frolicking, or whether they might be subject to penalties and punishments, and it is to these issues that we now turn.

11.3 Rights, Personhood, and Diminished Responsibility

Modern legal systems were established on the presupposition that all legal entities are persons. While a robot might someday be considered a person, we are not likely to face this situation any time soon. However, over time the law has managed to deal with several kinds of nonpersons, or quasi-persons, and we can look to these for some insights on how we might treat robots that are nonpersons, or quasi-persons. Personhood is a hotly debated concept, and many perspectives in that debate are based in strongly held beliefs from religious faith and philosophical dispositions. Most notably, the status of unborn human fetuses, and the status of severely brain damaged and comatose individuals have led to much debate in the United States over their appropriate legal consideration and rights. Yet, despite strongly differing perspectives on such issues, the legal systems in pluralistic societies have found ways to deal practically with these and several other borderline cases of personhood.

Minor children are a prime example of quasi-persons. Minors do not enjoy the full rights of personhood that adults do. In particular, they cannot sign contracts or become involved in various sorts of legal arrangements, because they do not have the right to do so as minors. They can become involved in such arrangements only through the actions of their parents or legal guardians. In this sense, they do not have full legal standing. Of course, the killing of a child is murder in the same way that the killing of an adult is, and so a child is still a legal person in this sense—and, in fact, is entitled to more protections than an adult. Children can thus be considered a type of quasi-person, or legal quasi-agent. The case of permanently mentally impaired people can be quite similar to children. Even full-fledged legal persons can claim temporary impairment of judgment, and thereby diminished responsibility for their actions, given certain circumstances, for example, temporary insanity, or being involuntarily drugged. The point is that some aspects of legal agency can apply to entities that fall short of full-fledged personhood and full responsibility, and it seems reasonable to think that some robots will eventually be granted this kind of quasi-agency in the eyes of the law before they achieve full legal personhood.

11.4 Crime, Punishment, and Personhood in Corporations and Robots

Criminal law is concerned with punishing wrongdoers, whereas civil law is primarily concerned with compelling wrongdoers to compensate those harmed. There is an important principle underlying this distinction: crimes deserve to be punished, regardless of any compensation to those directly harmed by the crime. Put another way, the harmed party in a crime is the whole of society. Thus, the case is prosecuted by the state, or, by "the people," and the debt owed by the wrongdoer is owed to the society. While the punishments may take different forms, the point of punishment is traditionally conceived of as being corrective in one or more senses: that the wrong-doer pays their debt to society (retribution); that the wrongdoer is to be reformed so as not to repeat the offense (reform); or that other people in society will be deterred from committing a similar wrong (deterrence).

There are two fundamental problems with applying criminal law to robots: (1) criminal actions require a moral agent to perform them, and (2) it is not clear that it is possible to punish a robot. While moral agency is not essential to civil law, moral agency is essential to criminal law, and is deeply connected to our concepts of punishment (retribution, reform, and deterrence). Moral agency might be defined in various ways, but, in criminal law, it ultimately must serve the role of being an autonomous subject who has a culpable mind, and who can be punished. Without moral agency, there can be harm (and hence civil liability), but not guilt. Thus, there is no debt incurred to society unless there is a moral agent to incur it—it is merely an accident or act of nature, but not a crime. Similarly, only a moral agent can be reformed, which implies the development or correction of a moral character—otherwise it is merely the fixing of a problem. And finally, deterrence only makes sense when moral agents are capable of recognizing the similarity of their potential choices and actions to those of other moral agents who have been punished for the wrong choices and actions—without this reflexivity of choice by a moral agent, and recognition of similarity between and among moral agents, punishment cannot possibly result in deterrence.

We saw in the previous section that it is more likely that we will treat robots as quasi-persons long before they achieve full personhood. Solum (1991–1992) has given careful consideration to the question of whether an artificial intelligence (AI) might be able to achieve legal personhood, using a thought experiment in which an AI acts as the manager of a trust. He concludes that while personhood is not impossible in principle for an AI to achieve, it is also not clear how we would know that any particular AI has achieved it. The same argument could be applied to robots. Solum imagines a legal Turing Test in which it comes down to a court's determination whether an AI could stand trial as a legal agent in its own right, and not merely as a proxy or agent of some other legal entity. He argues that a court would ultimately base its decision on whether the robot in question has moral agency, and whether it is possible to punish it—in other words, could the court fine or imprison an AI that mismanages a trust? In cases of quasi-personhood and diminished responsibility, however, children and the mentally impaired are usually shielded from punishment as a result of their limited legal status, specifically because they lack proper moral agency.

There is another relevant example in law of legal responsibility resting in a nonhuman entity, namely corporations. The corporation is a nonhuman entity that has been effectively granted many of the legal rights and responsibilities of a person. Corporations can (through the actions of their agents) own property, sign contracts, and be held liable for negligence. In certain cases, corporations can even be punished for criminal activities, such as fraud, criminal negligence, and causing environmental damage. A crucial aspect of treating corporations as persons depends on the ability to punish them, though this is not nearly so straightforward as it is for human persons. As a seventeenth-century Lord Chancellor of England put it, corporations have "no soul to damn and no body to kick" (Coffee 1981), so how can they be expected to have a moral conscience? Of course, corporations exist to make money, for themselves or stockholders, and as such can be given monetary punishments, and in certain cases, such as antitrust violations, split apart, or dissolved altogether. Though they cannot be imprisoned in criminal cases, responsible agents within the corporation can be prosecuted for their individual actions. As a result of this, and other aspects of corporations being complex sociotechnical systems in which there are many stakeholders with different relations to the monetary wealth of a corporation, it can be difficult to assign a punishment that achieves retribution, reform, and deterrence, while meeting other requirements of justice, such as fairness and proportionality.[8]

Clearly, robots are different in many important respects from corporations. However, there are also many important similarities, and it is no coincidence that Coffee's (1981) seminal paper on corporate punishment draws heavily on Simon's (1947) work on organizational behavior and decision making, and in particular how corporate punishment could influence organizational decision making through deterrence. Nonetheless, a great deal of work needs to be done in order to judge just how fruitful this analogy is. While monetary penalties work as punishments for corporations, this is because they target the essential purpose for the existence of corporations—to make money. The essential purposes of robots may not be so straightforward, and, if they exist at all, they will vary from robot to robot and may not take a form that can be easily or fairly penalized by a court.

The most obvious difference from corporations is that robots do have bodies to kick, though it is not clear that kicking them would achieve the traditional goals of punishment. The various forms of corporal punishment presuppose additional psychological desires and fears central to being human that may not readily apply to robots—concerning pain, freedom of movement, mortality, and so on. Thus, torture, humiliation, imprisonment, and death are not likely to be effective in achieving retribution, reform, or deterrence in robots. There could be a policy to destroy any robots that do harm, but, as is the case with animals that harm people, it would essentially be a preventative measure to avoid future harms by an individual, rather than a true punishment. Whether it might be possible to build in a technological means to enable genuine punishment in robots is an open question. In short, there is little sense in trying to apply our traditional notions of punishment to robots directly. This appears to me to be a greater hurdle to ascribing moral agency to robots than other hurdles, such as whether it is possible to effectively program moral decision making.

11.5 Conclusion

I hope that this brief overview of how certain legal concepts might be applied to current and future robots has convinced the reader that jurisprudence is a good place to begin framing some of the issues in robot ethics. I do not claim that this is the only viable approach, or that it will be capable of resolving every issue in robot ethics. Rather, I maintain that we can delineate different classes of ethical problems, some of which will have straightforward solutions from a legal perspective, while other classes of problems will remain unresolved. In terms of thinking about robots as manufactured products, many of the most practical and pressing issues facing robotics engineers can be seen as being essentially like those facing other engineers. In these cases, it is necessary to take proper care in imagining, assessing, and mitigating the potential risks of a technology. Just what this means for robotics will, of course, differ from other technologies, and should be the focus of further discussion and research. It is my belief that robot ethics will have its greatest influence by seeking to define and establish expectations and standards of practice for the robotics industry.

There remain a host of metaethical questions facing robot ethics that lie beyond the scope of the legal perspective. While moral agency is significant to the legal perspective, jurisprudence alone cannot determine or define just what moral agency is. Similarly, the ethical questions facing the construction of truly autonomous technologies demand special consideration in their own right. While there was no room to discuss it in this chapter, the legal perspective can also contribute to framing issues in the use of robots in warfare, though it offers little in the way of determining what social values we should aspire to enshrine in the laws governing the use of lethal robots. In particular, international law, humanitarian law, uniform codes of military conduct, the Geneva Conventions, the Nuremberg Principles, and international laws banning antipersonnel mines and limiting biological, chemical, and nuclear weapons, are all starting points for theorizing the ethics of using robot technologies in warfare, but may fall short in suggesting new standards for the ethical conduct of the kind of warfare that robots might make possible.

Notes

1. In the system of Anglo-American law, a distinction is drawn between criminal and civil law, and within civil law there is a further distinction between the laws of torts and contracts. Tort law deals with property rights and infringements outside of, or in addition to, contractual obligations and crimes, and is primarily concerned with damage to one's person or property and other harms. Thus, one has the right to sue responsible parties for damages that one has suffered, even

if one is not engaged in an explicit legal contract with the other party, and in addition to or regardless of whether the other party also committed a criminal act when causing the damages in question. Tort law seeks justice by compelling wrongdoers to compensate, or "make whole," those who were harmed for their loss (Prosser et al. 1984). Criminal law deals with what we tend to think of as moral wrongdoing or offenses against society, such as theft, assault, murder, etc., and seeks justice by punishing the wrongdoer.

There are several crucial differences between the concepts of criminal damages and civil damages and their accordant penalties. Most generally, for something to be a crime, there must be a law that explicitly stipulates the act in question as being criminal, whereas civil damages can result from a broad range of acts, or even inaction, and need not be explicitly specified in written law. Criminal acts are usually distinguished by their having criminal intent—a culpable state of mind in the individual committing the crime, known in Latin as *mens rea*. While certain forms of negligence can rise to the level of criminality, and can be characterized as nonmental states of ignorance, judgments of criminality typically consider mental states explicitly. Civil law, in comparison, is often indifferent to the mental states of the agents involved. And finally, there are differences in the exactment of punishments for transgressions. Under civil law, the damages are repaired by a transfer of money or property from the liable transgressor to the victim, while in criminal law the debt of the guilty transgressor is owed to the general public at large or the state, for the transgressor has violated the common good. A criminal penalty owed to society need not be evaluated in monetary terms, but might instead be measured in the revocation of liberty within society, expulsion from society, and in cultures of corporeal punishment, the revocation of bodily integrity or life, or the infliction of pain, humiliation, and suffering. In some instances, both frameworks apply, and criminal penalties may be owed over and above the restorative monetary damages owed to the victim of a crime.

2. For more on product liability law, see chapter 17 of Prosser et al. 1984.

3. It is in this way very much like the *doctrine of double effect* in Just War Theory (Walzer 1977), in that it separates knowledge of the possible harms of one's actions from the intention to actually bring those harms about. According to the doctrine of double effect, the killing of innocent civilians is permissible if the intended effect is on a militarily valid target, whereas the killing of civilians is not permissible if the intended effect is actually to harm the civilians.

4. For more on criminal negligence and liability, see chapter 5 of Prosser et al. 1984.

5. This notion is also popular in science fiction, starting with Isaac Asimov's "Three Laws of Robotics" (later four), and the "restraining bolts" in *Star Wars* droids, all of which aim to prevent robots from doing harm, despite maintaining their willingness to obey orders.

6. For more on the legal theory of agency, see Gregory 2001.

7. The principal can also be a corporation, in which case it is unable to act without its agents, which raises certain issues for corporate punishment, as we will see.

8. As Coffee (1981) argues, typical monetary fines against a company hurt shareholders and low-level employees more directly than they hurt the managers and decision makers in a

company, which diminishes their ability to deter or reform those who made the bad decisions and thus the fairness of imposing such fines.

Acknowledgments

I would like to thank Lawrence Solum for his thoughts on an early draft of this chapter, Malcolm MacIver for his thoughtful questions and comments, and special thanks go to John Evers for his numerous helpful discussions and legal insights on these topics.

References

Arkin, Ronald. 2009. *Governing Lethal Behavior in Autonomous Robots*. New York: CRC Press.

Asaro, Peter. 2006. What should we want from a robot ethic? *International Review of Information Ethics* 6 (12): 9–16.

Asaro, Peter. 2007. Robots and responsibility from a legal perspective. <http://peterasaro.org/writing/ASARO%20Legal%20Perspective.pdf> (accessed July 14, 2011).

Asaro, Peter. 2011. Remote-control crimes: Roboethics and legal jurisdictions of tele-agency. *IEEE Robotics and Automation Magazine* 18 (1): 68–71.

Bechtel, William. 1985. Attributing responsibility to computer systems. *Metaphilosophy* 16 (4): 296–306.

Bornfreeusa.org. 2010. Exotic animals summary. <http://www.bornfreeusa.org/b4a2_exotic_animals_summary.php> (accessed September 11, 2010).

Caverley, Daniel. 2006. Android science and animal rights: Does an analogy exist? *Connection Science* 18 (4): 403–417.

Coffee, John. 1981. "No soul to damn: No body to kick": An unscandalized inquiry into the problem of corporate punishment. *Michigan Law Review* 79 (3): 386–459.

Dennett, Daniel. 1997. When HAL kills, who's to blame?: Computer ethics. In *HAL's Legacy: 2001's Computer as Dream and Reality*, ed. D. G. Stork, 351–365. Cambridge, MA: MIT Press.

Gregory, William. 2001. *The Law of Agency and Partnership*, 3rd ed. New York: West Group.

iRobot.com. 2010. <http://store.irobot.com/shop/index.jsp?categoryId=3311368> (accessed September 11, 2010).

Kelley, Richard, Enrique Schaerer, Micaela Gomez, and Monica Nicolescu. 2010. *Advanced Robotics* 24 (13): 1861–1871.

Kurt, Tod E. 2006. *Hacking Roomba: ExtremeTech*. New York: Wiley. See also <http://hackingroomba.com/code/> (accessed November 26, 2010).

Mitcham, Carl. 1990. Ethics in bioengineering. *Journal of Business Ethics* 9 (3): 227–231.

Moon, Y., and Clifford Nass. 1998. Are computers scapegoats? Attributions of responsibility in human-computer interaction. *International Journal of Human-Computer Studies* 49 (1): 79–94.

Prosser, William Lloyd, and W. Page Keeton, Dan B. Owens, Robert E. Keeton, and David G. Owen, eds. 1984. *Prosser and Keeton on Torts.* 5th ed. New York: West Group.

Schaerer, Enrique, Richard Kelley, and Monica Nicolescu. 2009. Robots as animals: A framework for liability and responsibility in human-robot interactions. *Proceedings of RO-MAN 2009: The 18th IEEE International Symposium on Robot and Human Interactive Communication,* Toyama, Japan, Sept. 27–Oct. 2, 72–77. <http://ieeexplore.ieee.org/stamp/stamp.jsp?tp=&arnumber=5326244> (accessed July 14, 2011).

Simon, Herbert. 1947. *Administrative Behavior: A Study of Decision-Making Processes in Administrative Organizations.* New York: Free Press.

Solum, Lawrence. 1991–1992. Legal personhood for artificial intelligences. *North Carolina Law Review* (April): 1231–1287.

Wallach, Wendell, and Colin Allen. 2009. *Moral Machines: Teaching Robots Right from Wrong.* Oxford, UK: Oxford University Press.

Walzer, Michael. 1977. *Just and Unjust Wars: A Moral Argument with Historical Illustrations.* New York: Basic Books.

Wikipedia.org. 2010. Breed-specific legislation. <http://en.wikipedia.org/wiki/Breed-specific _legislation> (accessed November 26, 2010).

12 Robots and Privacy

M. Ryan Calo

Robots are commonplace today in factories and on battlefields. The consumer market for robots is rapidly catching up. A worldwide survey of robots by the United Nations in 2006 revealed 3.8 million in operation, 2.9 million of which were for personal or service use. By 2007, there were 4.1 million robots working just in people's homes (Singer 2009, 7–8; Sharkey 2008, 3). Microsoft founder Bill Gates has gone so far as to argue in an opinion piece that we are at the point now with personal robots that we were in the 1970s with personal computers, of which there are many billions today (Gates 2007). As these sophisticated machines become more prevalent—as robots leave the factory floor and battlefield and enter the public and private sphere in meaningful numbers—society will shift in unanticipated ways. This chapter explores how the mainstreaming of robots might specifically affect privacy.[1]

It is not hard to imagine why robots raise privacy concerns. Practically by definition, robots are equipped with the ability to sense, process, and record the world around them (Denning et al. 2009; Singer 2009, 67).[2] Robots can go places humans cannot go, see things humans cannot see. Robots are, first and foremost, a human instrument. And, after industrial manufacturing, the principle use to which we've put that instrument has been surveillance.

Yet increasing the power to observe is just one of ways in which robots may implicate privacy within the next decade. This chapter breaks the effects of robots on privacy into three categories—direct surveillance, increased access, and social meaning—with the goal of introducing the reader to a wide variety of issues. Where possible, the chapter points toward ways in which we might mitigate or redress the potential impact of robots on privacy, but acknowledges that, in some cases, redress will be difficult under the current state of privacy law.

As stated, the clearest way in which robots implicate privacy is that they greatly facilitate *direct surveillance*. Robots of all shapes and sizes, equipped with an array of sophisticated sensors and processors, greatly magnify the human capacity to observe. The military and law enforcement have already begun to scale up reliance on robotic technology to better monitor foreign and domestic populations. But robots also present corporations and individuals with new tools of observation in arenas as diverse as security, voyeurism, and marketing. This widespread availability is itself problematic, in that it could operate to dampen constitutional privacy guarantees by shifting citizen expectations.

A second way in which robots implicate privacy is that they introduce new points of *access* to historically protected spaces. The home robot in particular presents a novel opportunity for government, private litigants, and hackers to access information about the interior of a living space. Robots on the market today interact uncertainly with federal electronic privacy laws and, as at least one recent study has shown, several popular robot products are vulnerable to technological attacks—all the more dangerous in that they give hackers access to objects and rooms instead of folders and files.

Society can likely negotiate these initial effects of surveillance and unwanted access with better laws and engineering practices. But there is a third, more nuanced category of robotic privacy harm—one far less amenable to reform. This third way by which robots implicate privacy flows from their unique *social meaning*. Robots are increasingly human-like and socially interactive in design, making them more engaging and salient to their end-users and the larger community. Many studies demonstrate that people are hardwired to react to heavily anthropomorphic technologies, such as robots, as though a person were actually present, including with respect to the sensation of being observed and evaluated.

That robots have this social dimension translates into at least three distinct privacy dangers. First, the introduction of social robots into living and other spaces historically reserved for solitude may reduce the dwindling opportunities for interiority and self-reflection that privacy operates to protect (Calo 2010, 842–849). Second, social robots may be in a unique position to extract information from people (cf. Kerr 2004). They can leverage most of the same advantages of humans (fear, praise, etc) in information gathering. But they also have perfect memories, are tireless, and cannot be embarrassed, giving robots advantages over human persuaders (Fogg 2003, 213).

Finally, the social nature of robots may lead to new types of highly sensitive personal information—implicating what might be called "setting privacy." It says little about an individual how often he runs his dishwasher or whether he sets it to auto dry.[3] It says a lot about him what "companionship program" he runs on his personal robot. Robots exist somewhere in the twilight between person and object and can be exquisitely manipulated and tailored. A description of how a person programs and interacts with a robot might read like a session with a psychologist—except recorded, and without the attendant logistic or legal protections.

These categories of surveillance, access, and social meaning do not stand apart—they are contingent and interrelated. For example: reports have surfaced of insurgents hacking into military drone surveillance equipment using commonly available software. One could also imagine the purposive introduction by government of social machines into private spaces in order to deter unwanted behavior by creating the impression of observation. Nor is the implication of robots for privacy entirely negative—vulnerable populations, such as victims of domestic violence, may one day use robots to prevent access to their person or home and police against abuse. Robots could also carry out sensitive tasks on behalf of humans, allowing for greater anonymity. These and other correlations between privacy and robotics will no doubt play out in detail over the next few decades.

12.1   Robots that Spy

Robots of all kinds are increasing the military's already vast capacity for direct surveillance (Singer 2009). Enormous, unmanned drones can stay aloft, undetected, for days and relay surface activity across a broad territory. Smaller drones can sweep large areas, as well as stake out particular locations by hovering nearby and alerting a base upon detecting activity. Backpack-size drones permit soldiers to see over hills and scout short distances. The military is exploring the use of even smaller robots capable of flying up to a house and perching on a windowsill.

Some of the concepts under development are stranger than fiction. Although not developed specifically for surveillance, Shigeo Hirose's Ninja is a robot that climbs high-rises using suction pads. Other robots can separate or change shape in order to climb stairs or fit through tight spaces. The Pentagon is reportedly exploring how to merge hardware with live insects that would permit them to be controlled remotely and relay audio (Shachtman 2009).

In addition to the ability to scale walls, wriggle through pipes, fly up to windows, crawl under doors, hover for days, and hide at great altitudes, robots may come with programming that enhances their capacity for stealth. Researchers at Seoul National University in South Korea, for instance, are developing an algorithm that would assist a robot in hiding from, and sneaking up on, a potential intruder. Wireless or satellite networking permits large-scale cooperation among robots. Sensor technology, too, is advancing. Military robots can be equipped with cameras, laser or sonar range finders, magnetic resonance imaging (MRI), thermal imaging, GPS, and other technologies.

The use of robotic surveillance is not limited to the military. As Noel Sharkey has observed, law enforcement agencies in multiple parts of the world are also deploying more and more robots to carry out surveillance and other tasks (Sharkey 2008). Reports have recently surfaced of unmanned aerial vehicles being used for surveillance in the United Kingdom. The drones are "programmed to take off and land on their own, stay airborne for up to 15 hours and reach heights of 20,000 feet, making them invisible from the ground" (Lewis 2010). Drone pilot programs have been reported in Houston, Texas, and other border regions within the United States.

Nor is robotic surveillance limited to the government. Private entities are free to lease or buy unmanned drones or other robotic technology to survey property, secure premises, or monitor employees. Reporters have begun to speculate about the possibility of robot paparazzi—air or land robots "assigned" to follow a specific celebrity. Artist Ken Renaldo built a series of such "paparazzi bots" to explore human–computer interaction in the context of pop culture.

The replacement of human staff with robots also presents novel opportunities for data collection by mediating commercial transactions. Consider robot shopping assistants now in use in Japan. These machines identify and approach customers and try to guide them toward a product. Unlike ordinary store clerks, however, robots are capable of recording and processing every aspect of the transaction. Face-recognition technology permits easy reidentification. Such meticulous, point-blank customer data could be of extraordinary use in both loss prevention and marketing research.[4]

Much has been written about the dangers of ubiquitous surveillance. Visible drones patrolling a city invoke George Orwell's *Nineteen Eighty-Four*. But given the variety in design and capabilities of spy robots and other technologies, Daniel Solove's vision may be closer to the truth. Solove rejects the Big Brother metaphor and describes living in the modern world by invoking the work of Franz Kafka, where an individual never quite knows whether information is being gathered or used against her (Solove 2004, 36–41). The unprecedented surveillance robots permit implicates each of the common concerns associated with pervasive monitoring, including the chilling of speech and interference with self-determination (Schwartz 2000). As the Supreme Court has noted, excessive surveillance may even violate the First Amendment's prohibition on the interference with speech and assembly (*United States v. United States District Court*; Solove 2007).

The potential use of robots to vastly increase our capacity for surveillance presents a variety of specific ethical and legal challenges. The ethical dilemma in many ways echoes Joseph Weizenbaum's discussion of voice recognition technology in his seminal critique of artificial intelligence, *Computers, Power, and Human Reason*. Weizenbaum wondered aloud why the U.S. Navy was funding no fewer than four artificial intelligence labs in the 1970s to work on voice recognition technology. He asked, only to be told that the Navy wanted to be able to drive ships by voice command. Weizenbaum suspected that the government would instead use voice recognition technology to make monitoring communications "very much easier than it is now" (Weizenbaum 1976, 272). Today, artificial intelligence permits the automated recognition and data mining that underpin modern surveillance.

Roboticists might similarly ask questions about the uses to which their technology will be put—in particular, whether the only conceivable use of the robot is massive or covert surveillance. As is already occurring in the digital space, roboticists might simultaneously begin to develop privacy-*enhancing* robots that could help individuals preserve their privacy in tomorrow's complex world. These might include robots that shield the home or person from unwanted attention, robotic surrogates, or other innovations for now found only in science fiction.

The unchecked use of drones and other robotic technology could also operate to dampen the privacy protections enjoyed by citizens under the law. Well into the twentieth century, the protection of the Fourth Amendment of the U.S. Constitution against unreasonable government intrusions into private spaces was tied to the common law of trespass. Thus, if a technique of surveillance did not involve the physical invasion of property, no search could be said to occur. The U.S. Supreme Court eventually "decoupled violation of a person's Fourth Amendment rights from trespass violations of his property" (*Kyllo v. United States*). Courts now look to whether the government has violated a citizen's expectation of privacy that society was prepared to recognize as reasonable (*Kyllo v. United States*).

Whether a given expectation of privacy is reasonable has come to turn in part on whether the technology or technique the government employed was "in general public use"—the idea being that if citizens might readily anticipate discovery, any expectation of privacy would be unreasonable. The bar for "general" and "public" has proven lower than these words might suggest on their face. Although few people have access to a plane or helicopter, the Supreme Court has held the use of either to spot marijuana growing on a property not to constitute a search under the Fourth Amendment (*California v. Ciraolo*; *Florida v. Riley*). Under the prevailing logic, it should be sufficient that "any member of the public" could legally operate a drone or other surveillance robot to obviate the need for law enforcement to secure a warrant to do so.[5]

Due to their mobility, size, and sheer, inhuman patience, robots permit a variety of otherwise untenable techniques. Drones make it possible routinely to circle properties looking for that missing roof tile or other opening thought to be of importance in *Riley*. A small robot could linger on the sidewalk across from a doorway or garage and wait until it opened to photograph the interior. A drone or automated vehicle could peer into every window in a neighborhood from such a vantage point that an ordinary officer on foot could see into the house without even triggering the prohibition on "enhancement" of senses prohibited in pre-*Kyllo* cases such as *United States v. Taborda*, which involved the use of a telescope. Such practices greatly diminish privacy; if we came to anticipate them, it is not obvious under the current state of the law that these activities would violate the Constitution.

One school of thought—introduced to cyberlaw by Lawrence Lessig and championed by Richard Posner, Orin Kerr, and other thoughts leaders—goes so far as to hold that no search occurs under the Fourth Amendment unless and until a human being actually accesses the relevant information. This view finds support in cases like *United States v. Place* and *Illinois v. Caballes*, where no warrant was required for a dog to sniff a bag on the theory that the human police officer did not access the content of the bag and learned only about the presence or absence of contraband, in which the defendant could have no privacy interest. One can at least imagine a rule permitting robots to search for certain illegal activities by almost any means—for instance, x-ray, night vision, or thermal imaging—and alert law enforcement only should contraband be detected. Left unchecked, these circumstances combine to diminish even further the privacy protections realistically available to citizens and consumers.

12.2   Robots: A Window into the Home

Robots can be designed and deployed as a powerful instrument of surveillance. Equally problematic, however, is the degree to which a robot might inadvertently grant access to historically private spaces and activities. In particular, the use of a robot capable of connecting to the Internet within the home creates the possibility for unprecedented access to the interior of the house by law enforcement, civil litigants, and hackers. As a matter of both law of technology, such access could turn out to be surprisingly easy.

With prices coming down and new players entering the industry, the market for home robots—sometimes called personal or service robots—is rapidly expanding. Home robots can come equipped with an array of sensors, including potentially standard and infrared cameras, sonar or laser rangefinders, odor detectors, accelerometers, and global positioning systems (GPS). Several varieties of home robots connect wirelessly to computers or the Internet, some to relay images and sounds across the Internet in real time, others to update programming. The popular WowWee Rovio, for instance, is a commercially available robot used for security and entertainment. It can be controlled remotely via the Internet and broadcasts both sound and video to a website control panel.

12.2.1   Access by Law

What does the introduction of mobile, networked sensors into the home mean for citizen privacy? At a minimum, the government will be able to secure a warrant for recorded information with sufficient legal process, physically seizing the robot or gaining live access to the stream of sensory data. Just as law enforcement is presently able to compel in-car navigation providers to turn on a microphone in one's car (Zittrain 2008, 110) or telephone companies to compromise mobile phones, so could the government tap into the data stream from a home robot—or even maneuver the robot to the room or object it wishes to observe.

The mere fact that a machine is making an extensive, unguided record of events in the home represents a privacy risk. Still, were warrants required to access robot sensory data in all instances, robot purchasers would arguably suffer only an incremental loss of privacy. Police can already enter, search, and plant recording devices in the home with sufficient legal process. Depending on how courts come to apply electronic privacy laws, however, much data gathered by home robots could be accessed by the government in response to a mere subpoena or even voluntarily upon request.

Commercially available robots can patrol a house and relay images and sounds wirelessly to a computer and across the Internet. The robot's owner needs only travel to a website and log in to access the footage. Depending on the configuration, images and sounds could easily be captured and stored remotely for later retrieval or to establish a "buffer" (i.e., for uninterrupted viewing on a slow Internet connection). Or consider a second scenario: a family purchases a home robot that, upon introduction to a new environment, automatically explores every inch of house to which it has access. Lacking the onboard capability to process all of the data, the robot periodically uploads it to the manufacturer for analysis and retrieval.[6]

In these existing and plausible scenarios, the government is in a position to access information about the home activities—historically subject to the highest level of protection against intrusion by the government (*Silverman v. United States*)—with relatively little process. As a matter of constitutional law, individuals that voluntarily commit information to third parties lose some measure of protection for that information (*United States v. Miller*). Particularly where access is routine, such information is no longer entitled to Fourth Amendment protection under what is known as the "third-party doctrine" (Freiwald 2007, 37–49).

Federal law imposes access limitations on certain forms of electronic information. The Electronic Communications Privacy Act lays out the circumstances under which entities can disclose "electronic communications" to which they have access by virtue of providing a service (18 USC § 2510). How this statute might apply to a robot provider, manufacturer, website, or other service, however, is unclear. Depending on how a court characterizes the entity storing or transmitting the data—for instance, as a "remote computing service"—law enforcement could gain access to some robot sensory data without recourse to a judge.

Indeed, a court could conceivably characterize the relevant entity as falling out of the statute's protection altogether, in which case the service provider would be free to turn over details of customers' homes voluntarily upon request. Private litigants could also theoretically secure a court order for robot sensory data stored remotely to show, for instance, that a spouse had been unfaithful. Again, due to the jealousy with which constitutional, federal, and state privacy law has historically guarded the home, this level of access to the inner workings of a household with so little process would represent a serious departure.

12.2.2 Access by Vulnerability

Government and private parties might access robot data transmitted across the Internet or stored remotely through relatively light legal process, but the state of current technology also offers practical means for individuals to gain access to, even control of, robots in the home. If, as Bill Gates predicts, robots soon reach the prevalence and utility that personal computers possess today, less than solid security could have profound implications for household privacy.

Recent work by Tamara Denning, Tadayoshi Kohno, and colleagues at University of Washington has shown that commercially available home robots are insecure and could be hijacked by hackers. The University of Washington team researchers looked at three robots—the WowWee Rovio, the Erector Spykee, and the WowWee RobotSapien V2—each equipped with cameras and capable of wireless networking. The team uncovered numerous vulnerabilities. Attackers could identify Rovio or Spykee data streams by their unique signatures, for instance, and eavesdrop on nearby conversation or even operate the robot.[7] Attacks could be launched within wireless range (e.g., right outside the home) or by sniffing packets of information traveling by Internet protocol. A sophisticated hacker might even be able to locate home robot feeds on the Internet using a search engine (Denning et al. 2009).[8]

The potential to compromise devices in the home is, in a sense, an old problem; the insecurity of webcams has long been an issue of concern. The difference with home robots is that they can move and manipulate, in addition to record and relay. A compromised robot could, as the University of Washington team points out, pick up spare keys and place them in a position to be photographed for later duplication. (Or it could simply drop them outside the door through a mail slot.) A robot hacked by neighborhood kids could vandalize a home or frighten a child or elderly person. These sorts of physical intrusions into the home compromise security and exacerbate the feeling of vulnerability to a greater degree than was previously feasible.

12.3   Robots as Social Actors

The preceding sections identified two key ways in which robots implicate privacy. First, they augment the surveillance capacity of the government or private actors. Second, they create opportunities for legal and technical access to historically private spaces and information. Responding to these challenges will be difficult, but the path is relatively clear from the perspective of law and policy. As a legal matter, for instance, the Supreme Court could uncouple Fourth Amendment protections from the availability of technology, hold that indiscriminate robotic patrols are unreasonable, or otherwise account for new forms of robotic surveillance.

The Federal Trade Commission (FTC), the primary federal agency responsible for consumer protection, could step in to regulate what information a robotic shopping assistant could collect about consumers. The FTC could also bring an enforcement proceeding against a robot company for inadequate security under Section 5 of the Federal Trade Commission Act (as it has for websites and other companies). Congress could amend the Electronic Communications Privacy Act to require a warrant for video or audio footage relayed from the interior of a home. As of this writing, coalitions of nonprofits and companies have petitioned the government to reform this Act, along a number of relevant lines.

Beyond these regulatory measures, roboticists could follow the lead of Weizenbaum and others and ask questions about the ethical ramifications of building machines capable of ubiquitous surveillance. Roboethicists urge formal adoption by roboticists of the ethical code known as PAPA (privacy, accuracy, intellectual property, and access) developed for computers (Veruggio and Operto 2008, 1510–1511). Various state and federal law enforcement agencies could establish voluntary guidelines and limits on the use of police robots. And robotics companies could learn from Denning and her colleagues and build in better protections for home robots to ensure they are less vulnerable to hackers.

This section raises another dimension of robots' potential impact on privacy, one that is not as easy to remedy as a legal or technical matter. It explores how our reactions to robots as social technologies implicate privacy in novel ways. The tendency to anthropomorphize robots is common, even where the robot hardly resembles a living being. Technology forecaster Paul Saffo observes many people name their robotic vacuum cleaners and take them on vacation. Reports have emerged of soldiers treating bomb-diffusing drones like comrades and even risking their lives to rescue a "wounded" robot.

Meanwhile, robots increasingly are designed to interact more socially. Resemblance to a person makes robots more engaging and increases acceptance and cooperation. This turns out to be important in many early robot applications. Social robots will be deployed to care for the elderly and disabled, for example, and to diagnosis autism and other issues in children. They need to be accepted by people in order to do so. At the darker end of the spectrum, some roboticists are building robots with an eye toward sexual gratification; others predict that "love and sex with robots" is just around the corner (Levy 2007). Robots' social meaning could have a profound effect on privacy and the values it protects, one that is more complex and harder to resolve than anything mentioned thus far in this chapter.

12.3.1 Robots and Solitude

An extensive literature in communications and psychology demonstrates that humans are hardwired to react to social machines as though a person were really present.[9] Generally speaking, the more human-like the technology, the greater the reaction will be. People cooperate with sufficiently human-like machines, are polite to them, decline to sustain eye-contact, decline to mistreat or roughhouse with them, and respond positively to their flattery (Reeves and Nass 1996). There is even a neurological correlation to the reaction; the same "mirror" neurons fire in the presence of real and virtual social agents.

Importantly, the brain's hardwired propensity to treat social machines as human extends to the sensation of being observed and evaluated. Introducing a simulated person (or simply a face, voice, or eyes) into an environment leads to various changes in behavior. These range from giving more in a charity game, to paying for coffee more often on the honor system, to making more errors when completing difficult tasks. People disclose less and self-promote more to a computer interface that appears human. Indeed, the false suggestion of person's presence causes measurable physiological changes, namely, a state of "psychological arousal" that does not occur when one is alone (Calo 2010, 835–842).

The propensity to react to robots and other social technology as though they were actually human has repercussions for privacy and the values it protects (Calo 2010, 842–849). One of privacy's central roles in society is to help create and safeguard moments when people can be alone. As Alan Westin famously wrote in his 1970 treatise on privacy, people require "moments 'off stage' when the individual can be himself." Privacy provides "a respite from the emotional stimulation of daily life" that the presence of others inevitably engenders (Westin 1967, 35). The absence of opportunities for solitude would, many believe, cause not only discomfort and conformity, but also outright psychological harm.

Social technology, meanwhile, is beginning to appear in more—and more private—places. Researchers at both MIT and Stanford University are working on robotic companions in vehicles, where Americans spend a significant amount of their time. Robots wander hospitals and offices. They are, as described, showing up in the home with increasing frequency. The government of South Korea has an official goal of one robot per household by 2015. (The title of Bill Gates's op-ed referenced at the outset of this chapter?—"A Robot In Every Home.") The introduction of machines that our brains understand as people into historically private spaces may reduce already dwindling opportunities for solitude. We may withdraw from the actual whirlwind of daily life only to reenter its functional equivalent in the car, office, or home.[10]

12.3.2 Robot Interrogators

For reasons already listed, robots could be as effective as humans in eliciting confidences or information.[11] Due to our propensity to receive them as people, social robots—or, more accurately, their designers and operators—can employ flattery, shame, fear, or other techniques commonly used in persuasion (Fogg 2003). But unlike humans, robots are not themselves susceptible to these techniques. Moreover, robots have certain built-in advantages over human persuaders. They can exhibit perfect recall, for instance, and, assuming an ongoing energy source, have no need for interruptions or breaks. People tend to place greater trust in computers, at least, as sources of information (Fogg 2003, 213). And robotic expression can be perfectly fine-tuned to convey a particular sentiment at a particular time, which is why they are useful in treating certain populations, such as autistic children.

The government and industry could accordingly use social robots to extract information with great efficiency. Setting aside the specter of robotic CIA interrogators, imagine the possibilities of social robots for consumer marketing. Ian Kerr has explored the use of online "bots" or low-level artificial intelligence programs to gather information about consumers on the Internet (Kerr 2004). As one example, Kerr points to the text-based virtual representative ELLEgirlBuddy, developed by ActiveBuddy, Inc. to promote *Elle Girl* magazine and its advertisers. This software interacted with thousands of teens via instant messenger before it was eventually retired. ELLEgirlBuddy mimicked teen lingo and sought to foster a relationship with its interlocutors, all the while collecting information for marketing use (Kerr 2004). Social robots—deployed in stores, offices, and elsewhere—could be used as highly efficient gatherers of consumer information and, eventually, tuned to deliver the perfect marketing pitch.

12.3.3 Setting Privacy

Many contemporary privacy advocates worry that a "smart" energy grid connected to household devices, though probably better for the environment, will permit guesses about the interior life of a household. Indeed, one day soon it may be possible to determine an array of habits—when a person gets home, whether and how long they play video games, whether they have company—merely by looking at an energy meter. This important, looming problem echoes the issues discussed earlier in reference to access to the historically private home.

The privacy issues of smart grids are in a way cabined, however, by the sheer banality of our interaction with most household devices. Notwithstanding Supreme Court Justice Anton Scalia's reference to how a thermal imagining device might reveal the "lady in her sauna" (*Kyllo v. United States*), the temperature to which we set the thermostat or how long we are in the shower does not say all that much about us. Even the books we borrow from the library or the videos we rent (each protected, incidentally, under privacy law) permit at most inferences about our personality and mental state.

Our interactions with social robots could be altogether different. Consumers ultimately will be able to program robots not only to operate at a particular time or accomplish a specific task, but also to adopt or act out a nearly infinite variety of personalities and scenarios with independent social meaning to the owner and the community. If the history of other technologies is any guide, many of these applications will be controversial. Already people appear to rely on robots with programmable personalities for companionship and gratification. Additional uses will simply be idiosyncratic, odd, or otherwise private.

In interacting with programmable social robots, we stand to surface our most intimate psychological attributes. As David Levy predicts, "robots will transform human notions of love and sexuality," in part by permitting humans to better explore themselves (Levy 2007, 22). And even as we manifest these interior reflections of our subconscious, a technology will be *recording* them. Whether through robot sensory equipment, or embedded as an expression of code, the way we use human-like robots will be fixed in a file. Suddenly our appliance settings will not only matter, they also will reveal information about us that a psychotherapist might envy. This arguably novel category of highly personal information could, as happens with any other type of information, be stolen, sold, or subpoenaed.[12]

12.3.4 The Challenge of Social Meaning

Again, we can imagine ways to mitigate these harms. But the law is, in a basic sense, ill equipped to deal with the robots' social dimension. This is so because notice and consent tend to defeat privacy claims and because harm is difficult to measure in privacy cases. Consider the example of a robot in the home that interrupts solitude. The harm is subconscious, variable, and difficult to measure, which is likely to give any court or regulator pause in permitting recovery. Insofar as consent defeats many privacy claims, the robot's presence in the home is likely to be invited, even purchased. Similarly, it is difficult enough to measure which commercial activities rise to the level of deception or unfairness, without having to parse human reactions to computer salespeople. Rather than relying on legal or technological fixes, the privacy challenges of social robots will require an in-depth examination of human–robot interaction within multiple disciplines over many years.

12.4 Conclusion

According to a popular quote by science fiction writer William Gibson, "the future is already here. It just hasn't been evenly distributed yet." Gibson's insight certainly appears to describe robotics. One day soon, robots will be a part of the mainstream, profoundly affecting our society. This chapter has attempted to introduce a variety of ways in which robots may implicate the set of societal values loosely grouped under the term "privacy." The first two categories of impact—surveillance and access—admit of relatively well-understood ethical, technological, and legal responses. The third category, however, tied to social meaning, presents an extremely difficult set of challenges. The harms at issue are hard to identify, measure, and resist. They are in many instances invited. And neither law nor technology has obvious tools to combat them. Our basic recourse as creators and consumers of social robots is to proceed very carefully.

Notes

1. For the purposes of this chapter, a robot is a stand-alone machine with the ability to sense, process, and interact physically with the world. The term "home robot" or "personal robot" is used to indicate machines consumers might buy and to distinguish them from military, law enforcement, or assembly robots. This leaves out a small universe of robotic technologies—"smart" homes, embedded medical devices, prosthetics—that also have privacy implications not fully developed here. Artificial intelligence, in particular, whether or not it is "embodied" in a robot, has deep repercussions for privacy, for instance, in that it underpins data mining.

2. This is not to minimize the privacy risks associated with smart energy grids or the "Internet of things," namely, embedded computing technology into everyday spaces and products. Information stemming from such technology can be leveraged, particularly in the aggregate, in ways that negatively impact privacy.

3. One of the chief benefits of Internet commerce is the ability to target messages and perform detailed analytics on advertising and website use. As several recent reports have cataloged, outdoor advertisers are finding ways to track customers in real space. Billboards record images of passersby, for instance, and change on the basis of the radio stations to which passing cars are tuned. Robotics will only accelerate this trend by further mediating consumer transactions offline.

4. Surveillance may not automatically be lawful merely because the tools that were used are available to the public. In *United States v. Taborda*, for instance, the U.S. Court of Appeals for the Second Circuit suppressed evidence secured on the basis of using a telescope to peer into a home on the theory that "the inference of intended privacy at home is [not] rebutted by a failure to obstruct telescopic viewing by closing the curtains." But following the Supreme Court opinion in *Kyllo*—the Fourth Amendment case involving thermal imaging of a home—general availability appears to support a presumption that the tool can be used without a warrant.

5. This is how at least two robots—SRI International's Centibots and Intel's Home Exploring Robotic Butler—already function.

6. An earlier study found similar vulnerabilities in one version of iRobot's popular Roomba, which moves slowly, cannot grasp objects, and is not equipped with a camera.

7. As discussed previously, terrorist insurgents have also hacked into military drones.

8. The standard explanation is that we evolved at a time when cooperation with other humans conferred evolutionary advantages and, because of the absence of media, what appeared to be human actually was. There are reasons to be skeptical of explanations stemming from evolutionary psychology—namely, it can be used to prove multiple conflicting phenomena. Whatever the explanation, however, the evidence that we do react in this way is quite extensive.

9. Communications scholar Sam Lehman-Wilzig criticizes this idea on the basis that, if we treat robots like other people, we can simply shut the door on them as we do with one another in order to gain solitude. People may not consciously realize that robots have the same impact on

us as another person does, however, and robots and other social machines and interfaces can and do go many places—cars, computers, etc.—that humans cannot.

10. It could also be argued that we will get used to robots in our midst, thereby defeating the mechanism that interrupts solitude. What evidence there is on the matter points in the other direction, however. For instance, a study of the effect on participants of a picture of eyes when paying for coffee on the honor system saw no diminishment in behavior over many weeks. Nor is it clear that people will come to trust robots in the same way they might intimates, relatives, or servants—assuming we even already do.

11. Of course, artificial intelligence is not at the point where a machine can routinely trick a person into believe it is human—the so-called Turing Test. The mere belief that the robot is human is not necessary in order to leverage the psychological principles of interrogation and other forms of persuasion.

12. This is somewhat true already with respect to virtual worlds and open-ended games. Human–robot interactions stand to amplify the danger in several ways. There is likely to be a greater investment and stigma attached to physical rather than virtual behavior, for instance (or so one hopes, given the content of many video games). Ultimately our use of robots may reveal information we do not even want to know about ourselves, much less risk others discovering.

References

Calo, M. Ryan. 2010. People can be so fake: A new dimension to privacy and technology scholarship. *Penn State Law Review* 114: 809.

Denning, Tamara, Cynthia Matuszek, Karl Koscher, Joshua Smith, and Tadayoshi Kohno. 2009. A spotlight on security and privacy risks with future household robots: Attacks and lessons. *Proceedings of the 11th International Conference on Ubiquitous Computing*, September 30–October 3.

Fogg, B. J. 2003. *Persuasive Technologies: Using Computers to Change What We Think and Do*. San Francisco: Morgan Kaufmann Publishers.

Freiwald, Susan. 2007. First principles of communications privacy. *Stanford Technology Law Review* 3: 1.

Gates, Bill. 2007. A robot in every home. *Scientific American* 296 (1) (January): 58–65.

Kerr, Ian. 2004. Bots, babes, and Californication of commerce. *University of Ottawa Law and Technology Journal* 1: 285.

Levy, David. 2007. *Love + Sex with Robots*. New York: Harper Perennial.

Lewis, Paul. 2010. CCTV in the sky: Police plan to use military-style spy drones. *The Guardian* (January).

Reeves, Byron, and Cliff Nass. 1996. *The Media Equation*. Cambridge, UK: Cambridge University Press.

Schwartz, Paul. 2000. Internet privacy and the state. *Connecticut Law Review* 32: 815.

Shachtman, Noah. 2009. Pentagon's cyborg beetle spies take off. *Wired.com* (January). <http://www.wired.com/dangerroom/2009/01/pentagons-cybor/> (accessed March 22, 2011).

Sharkey, Noel. 2008. "2084: Big robot is watching you." A commissioned report. <http://staffwww.dcs.shef.ac.uk/people/N.Sharkey/> (accessed September 12, 2010).

Singer, Peter Warren. 2009. *Wired for War*. New York: The Penguin Press.

Solove, Daniel. 2004. *The Digital Person: Technology and Privacy in the Digital Age*. New York: New York University Press.

Solove, Daniel. 2007. The First Amendment as criminal procedure. *New York University Law Review* 82: 112.

Veruggio, Gianmarco, and Fiorella Operto. 2008. Roboethics: Social and ethical implications of robotics. In *Springer Handbook of Robotics*, ed. Bruno Siciliano and Oussama Khatib, 1499–1524. Berlin, Germany: Springer-Verlag.

Weizenbaum, Joseph. 1976. *Computers Power and Human Reason: From Judgment to Calculation*. San Francisco: W. H. Freeman and Company.

Westin, Allen. 1967. *Privacy and Freedom*. New York: Atheneum.

Zittrain, Jonathan. 2008. *The Future of the Internet: And How to Stop It*. New Haven, CT: Yale University Press.

V Psychology and Sex

The anthropomorphization of robots is an important trend, not merely for the privacy implications noted in chapter 12, but also for increasing public acceptance, even affinity, toward robots. But this betides a new danger of "too much of a good thing": Can one become emotionally *over*-invested in robots? Are there potential harms due to emotional and psychic dependence that raise serious moral concerns, either to the human users of robots or to the public at large?

Matthias Scheutz discusses the dangers of emotional bonds with robots in chapter 13; he argues that social robots will differ from industrial or military robots in appearance, environment, programming, mobility, autonomy, and perceived agency. As humans have a tendency to personify and become emotionally dependent on social robots, opportunities will abound for malicious exploitation of such unidirectional emotional bonds by the creators or purveyors of robots. Scheutz recommends regulations to forestall such worries, including the possibility of creating robots with emotions of their own.

David Levy in chapter 14 and Blay Whitby in chapter 15 investigate aspects of one of the most notorious, widely publicized, and most intense types of psychological and emotional experiences humans will have with robots: sex.

Levy's chapter focuses on the idea that robot prostitutes, or "sexbots," will soon become widely accepted alternatives to human sex workers, and he takes up five aspects of the ethics of robot prostitution. He considers the ethical issues concerning the general use of robot prostitutes, effects on an individual's self-respect in using a robot in this way, how such use affects other human intimate relationships (e.g., is it infidelity?), and the impact of robotic prostitutes on human sex workers and (eventually) on the sexbots themselves.

Whitby examines robot lovers within the general context of the ethics of caring technologies. In Japan and South Korea, robots are widely assumed to have a future significant role in elder care and babysitting. But wishful thinking and hype can obscure both what is actually possible, and what should—or should not—be allowed. Whitby notes that Masahiro Mori's hypothesis of the "Uncanny Valley" poses a difficult technical barrier to creating realistic-looking robot lovers, but robots may soon be able to better human companions' ability to retain intimate knowledge of one's own quirks, and respond to (and even anticipate) one's feelings. He notes people unable to find lovers, or prevented from doing so (e.g., criminals), are obvious markets for robotic companions, but notes a disquieting further possibility: people may seek robots in order to do things to them that would be abhorrent when done to another human. As such, he considers the possibility of love for (or by) a robot, and reflects on Levy's arguments. He ends with a call for public discussion and the possible development of professional ethics codes to guide the responsible development of robotic companions. Then, in part VI, we focus on the broader notion of robots as caregivers.

13 The Inherent Dangers of Unidirectional Emotional Bonds between Humans and Social Robots

Matthias Scheutz

The early twenty-first century is witnessing a rapid advance in social robots. From vacuum cleaning robots (like the Roomba), to entertainment robots (like the Pleo), to robot pets (like KittyCat), to robot dolls (like Baby Alive), to therapy robots (like Paro), and many others, social robots are rapidly finding applications in households and elder-care settings. In 2006, the number of service robots worldwide alone outnumbered industrial robots by a factor of four, and this gap is expected to widen to a factor of six by 2010, fueled by ambitious goals like those of South Korea, to put one robot into each household by the year 2013, or by the Japanese expectation that the robot industry will be worth ten times the present value in 2025 (Gates 2007).

From these expectations alone, it should be clear that social robots will soon become an integral part of human societies, very much like computers and the Internet in the last decade. In fact, using computer technology as an analogy, it seems likely that social robotics will follow a similar trajectory: once social robots have been fully embraced by societies, life without them will become inconceivable.

As a consequence of this societal penetration, social robots will also enter our personal lives, and that fact alone requires us to reflect on what exactly happens in our interactions with these machines. For social robots are specifically designed for personal interactions that will involve human emotions and feelings: "a sociable robot is able to communicate and interact with us, understand, and even relate to us, in a personal way. It is a robot that is socially intelligent in a human-like way" (Breazeal 2002). And while social robots can have benefits for humans (e.g., health benefits as demonstrated with Paro [Shibata 2005]), it is also possible that they could inflict harm—emotional harm, that is. And exactly herein lies the hitherto underestimated danger: the potential for humans' emotional dependence on social robots.

As we will see shortly, such emotional dependence on social robots is different from other human dependencies on technology (e.g., different both in kind and quality from depending on one's cell phone, wrist watch, or PDA). To be able to understand the difference and the potential ramifications of building complex social robots that are freely deployed in human societies, we have to understand how social robots are

Table 13.1
Industrial robots versus computers versus social robots

Aspect/device	Industrial robots	Computers	Social robots
application	industrial production	any	personal/service
environment	restricted	any	any
appearance	machine-like	machine-like	(often) life-like
programming	task-specific	open-ended	(sometimes) open-ended
actuation	yes	no	yes
mobility	limited	none	(often) unlimited
autonomy	no	no	yes (limited)
agency	no	no	?

different from other related technologies and how they, as a result, can affect humans at a very basic level.

13.1 Social Robots Are Different

Start by comparing social robots to related technologies, namely computers and industrial robots (see table 13.1). These two kinds of machines are particularly relevant, because social robots contain computers (for their behavior control) and share with industrial robots the property of being robots (in the sense of being machines with motion or manipulation capabilities or both). And computers and industrial robots have been around for decades, while social robots are a recent invention.

Very much like industrial robots, social robots have the capability to initiate motion (of actuators or themselves) and thus exhibit behavior (compared to stationary objects like computers). Different from industrial robots, which are typically confined to factories, social robots are directly targeted at consumers for service purposes (like the Roomba vacuum cleaner) or for entertainment (like the AIBO robo-dog).

Very much like computers, social robots have managed to enter individuals' homes and thus their private lives, and increasingly are becoming part of people's daily routines (Forlizzi and DiSalvo 2006). Different from computers, robots can interact with their owners at various levels of sophistication, and they can even initiate and terminate those interactions on their own.

And, unlike industrial robots and computers, social robots are often mobile, and their mobility is driven by different forms of preprogrammed or learned behaviors. Even if behaviors are predetermined and allow for very limited variability (e.g., as in various robotic toys or the Roomba), current social robots nevertheless change their position in the world. And despite the fact that these behavioral repertoires are very simple, social robots nevertheless can make (limited) decisions about what action to take or what behaviors to exhibit. They base these decisions on their perceptions of the environment and their internal states, rather than following predetermined action sequences based on preprogrammed commands, as is usually the case with robots in industrial automation (Parasuraman, Sheridan, and Wickens 2000).

The simple rule-governed mobility of social robots, especially when robots are able to adapt and change their behaviors (e.g., by learning from experience), has far-reaching consequences. For—as will become clear—it enables robots to affect humans in very much the same way that animals (e.g., pets) or even other people affect humans. In particular, the rule-governed mobility of social robots allows for, and ultimately prompts, humans to ascribe intentions to social robots in order to be able to make sense of their behaviors (e.g., the robot did not clean in the corner because it thought it could not get there). The claim is that the autonomy of social robots is among the critical properties that cause people to view robots differently from other artifacts such as computers or cars.

13.2 Autonomy + Mobility = Perceived Agency?

There are several intuitions behind applying the notion of autonomy—which has its roots in the concept of human agency—to artifacts like robots. These intuitions are derived from ideas about what it means for a human being to be autonomous: "To be autonomous is to be a law to oneself; autonomous agents are self-governing agents. Most of us want to be autonomous because we want to be accountable for what we do, and because it seems that if we are not the ones calling the shots, then we cannot be accountable" (Buss 2002). Clearly, current robots (and those in the near future) will neither be self-governing agents that want to be autonomous, nor will they be in a position where they could be accountable or held accountable for their actions. This is because they will not have the reflective self-awareness that is prerequisite for accountable, self-governing behavior. Yet, there is a sense in which some robots are, at least to some extent, "self-governing," and can thus be said, again, in a weak sense, to be autonomous—a robot, for example, that is capable of picking up an object at point A and dropping it off at point B without human supervision or intervention is, at least to some extent, "self-governing."

A much stronger and richer sense of autonomy, one that comes closest to the notion of human autonomy, is centered on an "agent's active use of its capabilities to pursue its goals, without intervention by any other agent in the decision-making processes used to determine how those goals should be pursued" (Barber and Martin 1999). This notion stresses the idea of decision making by an artificial system or agent to pursue its goals and, thus, requires the agent to at least have mechanisms for decision making and goal representations, and ideally also additional representations of other intentional states (such as desires, motives, etc.), as well as nonintentional states (such as task representations, models of other agents, etc.).

Yet, there is also an independent sense in which the autonomy of an artificial system is a matter of degrees: "for example, consider an unmanned rover. The command, 'find evidence of stratification in a rock' requires a higher level autonomy than, 'go straight 10 meters'" (Dorais et al. 1998). The degrees or levels of autonomy can depend on several factors: for example, how complex the commands are that it can execute, how many of its subsystems can be controlled without human intervention, under what circumstances the system will override manual control, and the overall duration of autonomous operation (Dorais et al. 1998; see also Huang 2004).

There is yet another dimension of robot autonomy, orthogonal to the preceding conceptual distinctions that focus on functional, behavioral, and architectural aspects, but of clear relevance to human–robot interactions. This dimension concerns a human's perception of the (level of) autonomy of an artificial system and the impact the perceived autonomy has on that human's behavior.

The relationship among these different characterizations of robot autonomy has been summarized as a robot's "ability of sensing, perceiving, analyzing, communicating, planning, decision-making, and acting, to achieve its goals as assigned by its human operator(s) through designed human-robot interaction. Autonomy is characterized as involving levels demarcated by factors including mission complexity, environmental difficulty, and level of HRI to accomplish the missions" (Huang 2004).

There is converging evidence that the degree of autonomy that a robot exhibits is an important factor in determining the extent to which it will be viewed as human-like, where the investigated robots are typically able to move freely, respond to commands, recognize objects, understand human speech, and make decisions (Kiesler and Hinds 2004; Scheutz et al. 2007). Perceived autonomy is so critical because it implies capabilities for self-governed movement, understanding, and decision-making (Kiesler and Hinds 2004), capabilities that together comprise important components of how we define the qualities of "humanness" or "human-like" (Friedman, Kahn, and Hagman 2007).

The distinguishing features of mobility and autonomy, therefore, set autonomous social robots apart from other types of robots, computers, and artifacts, and are ultimately a critical factor for shaping the human perceptions of autonomous robots as "social agents."

13.3   Evidence from HRI Studies

Over the last few years, we have conducted several human–robot interaction experiments to investigate the degree to which humans perceive robots as autonomous agents and to isolate the effects that perceived autonomy can have both on human attitudes toward robots and human behavior. To be able to gain a better understanding of people's true beliefs about robots, we developed a rigorous evaluation framework that encompasses both subjective and objective methods and measures (Rose, Scheutz, and Schermerhorn 2010). Here, we briefly summarize the results from three studies.

13.3.1 Study 1: Dynamic Autonomy

We investigated the extent to which robot autonomy based on independent decision making and behavior by the robot can affect the objective task performance of a mixed human–robot team while being subjectively acceptable to the human team leader (Schermerhorn and Scheutz 2009; Scheutz and Crowell 2007). In this task, a human subject worked together with a robot to accomplish a team goal within a given time limit. While both human and robot had tasks to perform, neither robot nor human could accomplish the team goal alone. In one of the task conditions (the "autonomy condition"), the robot was allowed to act autonomously when time was running out in an effort to complete the team goal. As part of this effort, it was able to refuse human commands that would have interfered with its plans. In the other condition (the "no autonomy condition"), the robot would never show any initiative on its own and would only carry out human commands. Human subjects were tested in both conditions (without knowing anything about the conditions) and then asked to rate various properties of the robot. Overall, subjects rated the "autonomous robot" as more helpful and capable, and believed that it made its own decisions and acted like a team member. There was also evidence that they found the autonomous robot to be more cooperative, easier to interact with, and less annoying than the nonautonomous robot. Surprisingly, there was no difference in the subjects' assessment of the degree to which the robot disobeyed commands (even though it clearly disobeyed commands in almost all subject runs in the autonomy condition while it never disobeyed any command in the no-autonomy condition). We concluded that subjects preferred the autonomous robot as a team partner.

13.3.2 Study 2: Affect Facilitation

We also investigated the utility of affect recognition and expression by the robot in a similar team task (Scheutz et al. 2007; Scheutz et al. 2006). Here, instead of making autonomous decisions, the robot always carried out human orders. However, in one condition (the "affect condition") it was allowed to express urgency in its voice or respond to sensed human stress with stress of its own (again expressed in its voice), compared to the "no-affect condition," where the robot's voice was never modulated. Each subject was exposed to only one condition and comparison was made among subject groups. The results showed that allowing the robot to express affect and respond to human affect with affect expressions of its own—in circumstances where humans would likely do the same and where affective modulations of the voice thus make intuitive sense to humans—can significantly improve team performance, based on objective performance measures. Moreover, subjects in the "affect condition" changed their views regarding robot autonomy and robot emotions from their pre-experimental position based on their experience with the robot in the experiment. While they were neutral before the experiment as to whether robots should be allowed to act autonomously and whether robots should have emotions of their own, they were slightly in favor of both capabilities after the experiments. This is different from subjects in the no-affect group who did not change their positions as a result of the experiment. We concluded that appropriate affect expression by the robot in a joint human–robot task can lead to better acceptability of robot autonomy and other human-like features like emotions in robots.

13.3.3   Study 3: Social Inhibition and Facilitation

While the previous two studies attempted to determine human perceptions and agreement with robot autonomy indirectly through human participation in a human–robot team task (where the types of interactions with the robot were critical for achieving the goal, and thus for the subjects' views of the robot's capabilities), the third study attempted to determine the human-likeness of the robot directly. Specifically, the study investigated people's perceptions of social presence in robots during a sequence of different interactions, where the robot functioned as a survey taker as well as an observer of human task performance (Crowell et al. 2009; Schermerhorn, Scheutz, and Crowell 2008). The experimental design used well-known results in psychology about social inhibition and facilitation that occurs in humans when they are observed performing tasks by other humans (Zajonc 1965). Our experimental results showed that robots can have effects on humans and human performance that are otherwise only observed with humans. Interestingly, there was a gender difference in subjects' perception of the robot, with only males showing "social inhibition effects" caused by the presence of the robot while they were performing a math task. Postexperimental surveys confirmed that male subjects viewed the robot as more human-like than did the female subjects.

Together, these laboratory studies provide experimental evidence about human perceptions of autonomous robots. In particular, they show that humans seem to prefer autonomous robots over nonautonomous robots when they have to work with them, that humans prefer human-like features (e.g., affect) in robots and that those features are correlated with beliefs about autonomy, and that a robot's presence can affect humans in a way that is usually only caused by the presence of another human. The question then arises whether the findings also apply to "robots in the wild," outside of the well-controlled laboratory environment. As the next section will demonstrate, there is already ample evidence for people's susceptibility to the lure of social robots outside the lab, especially when they have repeated, longer-term interactions with robots.

13.4 The Personification of Robots

An increasing body of evidence demonstrates how humans anthropomorphize robots, project their own mentality onto them, and form what seem like deep emotional yet unidirectional relationships with them. Documented examples, which we will summarize in the stories that follow, range from interviews with soldiers that worked with robots on defusing improvised explosive devices (IEDs), to ethnographic studies with robot-pet owners (of the AIBO robot dog) and with owners of the robotic Roomba vacuum cleaner.

13.4.1 From Garreau's "Bots on the Ground"

The first story is about a robot developed by roboticist Mark Tilden for the purpose of defusing land mines. The robot achieves the task by stepping on them, which causes the mine to detonate and destroy the robot's leg. Hence, the robot was designed with several legs to be able to detonate several mines before becoming useless. Here is the story:

At the Yuma Test Grounds in Arizona, the autonomous robot, 5 feet long and modeled on a stick-insect, strutted out for a live-fire test and worked beautifully, he [Tilden] says. Every time it found a mine, blew it up and lost a limb, it picked itself up and readjusted to move forward on its remaining legs, continuing to clear a path through the minefield. Finally it was down to one leg. Still, it pulled itself forward. Tilden was ecstatic. The machine was working splendidly. The human in command of the exercise, however—an Army colonel—blew a fuse. The colonel ordered the test stopped. Why? asked Tilden. What's wrong? The colonel just could not stand the pathos of watching the burned, scarred, and crippled machine drag itself forward on its last leg. This test, he charged, was inhumane. (Garreau 2007)

Whether or not "inhumane" was an appropriate attribution, the fact remains that the only explanation for not wanting to watch a mindless, lifeless machine, purposefully developed for blowing up mines, destroy itself, is that the human projected some agency onto the robot, ascribing to it some inner life, and possibly even feelings.

Another example, recounted by a Marine sergeant running a robot repair shop in Iraq, is the technician who returned his IED-defusing robot, which he had named "Scooby-Doo," for repair. While it is well known that humans have a tendency to name inanimate things they like and use frequently (e.g., their car), naming comes at a price: it automatically generates a kind of intimacy with and connectedness to the named object. And, in the case of robots, it only reinforces what the self-propelled behavior of a robot already does: prompting the inscription of intentionality into an artifact and thus implicating granting it agency! Here is a story recounted by a robot technician named Bogosh:

"There wasn't a whole lot left of Scooby," Bogosh says. The biggest piece was its 3-by-3-by-4-inch head, containing its video camera. On the side had been painted "its battle list, its track record. This had been a really great robot." The veteran explosives technician looming over Bogosh was visibly upset. He insisted he did not want a new robot. He wanted Scooby-Doo back. "Sometimes they get a little emotional over it," Bogosh says. "Like having a pet dog. It attacks the IEDs, comes back, and attacks again. It becomes part of the team, gets a name. They get upset when anything happens to one of the team. They identify with the little robot quickly. They count on it a lot in a mission." (Garreau 2007)

In fact, soldiers take pictures of their robots, introduce robots to their friends and family abroad, and even promote them, all indications of treating robots as if they were intentional creatures. "When we first got there, our robot, his name was Frankenstein, says Sgt. Orlando Nieves, an EOD [Explosive Ordnance Disposal techni-cian] from Brooklyn. 'He'd been in a couple of explosions and he was made of pieces and parts from other robots.' Not only did the troops promote him to private first class, they awarded him an EOD badge—a coveted honor. 'It was a big deal. He was part of our team, one of us. He did feel like family' (Garreau 2007).

13.5 Robot Dogs Are Pets, Too

Even if these examples seem hardly believable, one might be lenient and justify the soldiers' attribution of human qualities to robots by pointing to the extraordinary circumstances that these soldiers encounter in combat, and the huge emotional toll it takes on the human psyche. But surprisingly, being in a deserted remote location, dealing with life-threatening situations, is not necessary to elicit the kinds of reactions to robots we saw with soldiers in Iraq. Ordinary citizens living in the United States seem to fall prey to suggestive behaviors of social robots. For example, Peter Kahn and colleagues (Kahn, Friedman, and Hagman 2002) examined the postings of users in AIBO news groups, where robo-dog owners share their experiences with AIBO freely, and identified four categories of postings:

Essences refer to the presence or absence of technological, biological, or animistic underpinnings of AIBO (e.g., "He's resting his eyes"). *Agency* refers to the presence or absence of mental states for AIBO, such as intentions, feelings, and psychological characteristics (e.g., "He has woken in the night very sad and distressed"). *Social standing* refers to ways in which AIBO does or does not engage in social interactions, such as communication, emotional connection, and companion-ship (e.g., "I care about him as a pal, not as a cool piece of technology"). *Moral standing* refers to ways in which AIBO may or may not engender moral regard, be morally responsible, be blameworthy, have rights, or deserve respect (e.g., "I actually felt sad and guilty for causing him pain!"). (Kahn, Friedman, and Hagman 2002)

While they found relatively few references to AIBO's moral standing (12 percent), people made very frequent references to essences (79 percent), agency (60 percent), and social standing (59 percent). It seems clear that AIBO owners have a strong tendency to form (false) beliefs about (possible) mental states of their robots.

13.6 Even the Roomba Does the Trick

Another example group are owners of Roomba vacuum cleaners that have been interviewed in a variety of studies over the last several years, given that the Roomba is one of the most widely sold autonomous robots. While at first glance it would seem that the Roomba has no social dimension (neither in its design nor in its behavior) that could trigger people's social emotions, it turns out that humans, over time, develop a strong sense of gratitude toward the Roomba for cleaning their home. The mere fact that an autonomous machine keeps working for them day in and day out seems to evoke a sense of, if not urge for, reciprocation. Roomba owners seem to want to do something nice for their Roombas, even though the robot does not even know that it has owners (it treats humans as obstacles in the same way it treats chairs, tables, and other objects that it avoids while driving and cleaning). The sheer range of human responses is mind blowing (e.g., see Sung et al. 2007). Some will clean for the Roomba, so that it can get a rest, while others will introduce their Roomba to their parents, or bring it along when they travel because they managed to develop a (unidirectional) relationship: "I can't imagine not having him any longer. He's my BABY! . . . When I write emails about him, which I've done that, as well, I just like him, I call him Roomba baby. . . . He's a sweetie" (Sung et al. 2007).

13.7 Not Even Experienced Roboticists Are Always Spared

Somewhat surprisingly, it is even possible for an experienced roboticist to be affected by the suggestive force of apparent autonomous behavior. In our own lab, for example, we found our humanoid robot CRAMER disturbing when it was left on (by accident) and started shifting attention from speaker to speaker (as if it understood what was being said). And, according to Garreau, graduate students at MIT working in the lab with the Kismet robot put up a curtain between themselves and the robot at times because the robot's gaze was breaking their concentration. In fact, even the creator of Kismet, Cynthia Breazeal, seems to have developed a very personal relationship with her own creation:

Breazeal experienced what might be called a maternal connection to Kismet; she certainly describes a sense of connection with it as more than "mere" machine. When she graduated from MIT and left the AI laboratory where she had done her doctoral research, the tradition of academic property rights demanded that Kismet be left behind in the laboratory that had paid for its development. What she left behind was the robot "head" and its attendant software. Breazeal described a sharp sense of loss. (Turkle 2006)

13.8 The Dangers Ahead

These accounts are only a small set of the ever-mounting evidence that humans are becoming increasingly attached to robots. From seemingly innocuous facts such as the naming of their robots, to more worrisome episodes such as promoting robots to military ranks, calling robots "pals," and exhibiting "shameful" reactions (such as the woman who shut her bedroom door because she was getting undressed and felt that her AIBO was watching her), the personification of social robots is widespread and is becoming a testimony for the human willingness to form unidirectional emotional bonds with these machines.

It is important in this context to note how little is required on the robotic side to cause people to form relationships with robots. Consider the case of the AIBO. Clearly, it is modeled after a real dog in that its physical shape resembles that of a dog and its behaviors bear some resemblance to dog behaviors (wagging tail, barking, etc.). Hence, one might argue that it is really a robotic substitute for what otherwise would be legitimate companion. But then, consider the PackBot, which is not even a fully autonomous robot; rather, it is under tight remote control from its operator. Moreover, it has tracks and does not resemble any particular biological creature. Yet, it does play a critical role in the soldiers' daily routines and fight for survival. Hence, one might argue that these special circumstances make humans forget the very machine-like appearance and lack of autonomy of PackBot. And PackBot has another unique feature that might contribute to the soldier's identification with the robot: soldiers are able to see the world from the robot's perspective (through visual real-time streams from the robot's cameras). This could easily blur the distinction between the robot itself and the human operating it, at least for the human operator (there is evidence from cognitive science that humans view sensory or actuator augmentations as part of their bodies when they have gained sufficient experience using them).

For further contrast, consider now the Roomba, which neither has animal-like appearance, nor allows the human to see the world from its perspective. It is a mere disc that drives around in certain patterns, to avoid bumping into things. Yet, it manages to instill the idea of agency in people, and can cause them to even experience gratitude for its service, so much so that they will clean in its stead. One would hardly be able to make that point for dishwashers!

It is also interesting to note how little these robots have to contribute on their end to any relationship, in other words, how inept and unable they are to partake as a genuine partner: neither the Roomba nor the PackBot, for example, have any notion of "other"; there are no built-in algorithms for detecting and recognizing people.

Rather, anything that causes their contact sensors to be triggered is treated in the same way, namely as an "obstacle" that needs to be avoided.

13.9   The False Pretense: Robots Are Agents

None of the social robots available for purchase today (or in the foreseeable future, for that matter) care about humans, simply because they cannot care. That is, these robots do not have the architectural and computational mechanisms that would allow them to care, largely because we do not even know what it takes, computationally, for a system to care about anything (cf. Haugeland 2002). Yet, this fact is clearly getting lost in the increasing hype about social robots. It almost seems as if industry is trying hard to make the case for the opposite, thus enforcing the personification of social robots.

Take, for example, one of the new Hasbro robot dolls, called Baby Alive, which can say simple phrases like "I'm hungry," "oh oh, I made a stinky," and "mommy, I love you." The commercial advertising for the robot emphasizes, "how real it is" by explicitly using the phrase "a baby so real." Other companies have been advertising their toys as "recreating the emotions" of a cat, a dog, an infant, and so on (see also Scheutz 2002).

Even companies like iRobot that are clearly aware of the computational and cognitive limitations of their products, find it useful, for whatever reason, to create a Facebook page for their PackBot product, where PackBot stories and news are recounted in first-person narratives, as if there were a single entity called "PackBot" that had experienced all these situations and events.

And, finally, academics themselves are often less careful than they ought to be when presenting their research. For example, researchers who work on emotions often say loosely that their robots have emotions, implement emotions, use emotions, and so on. This kind of suggestive language (e.g., during research presentations or even in published research papers) makes it easy for nonexpert readers to conflate the control processes in these artifacts with similarly labeled, yet substantively very different control processes in natural organisms, particularly humans (e.g., Scheutz 2002). The repeated labeling of control states in robotic architectures and of behaviors exhibited by robots with terms familiar from human and animal psychology helps to create, maintain, and sustain the false belief that "somebody is at home" in current robots. And while people, when asked explicitly, might deny that they think of the robot as a person, an animal, or an otherwise alive agent, this response generated at the conscious level might be forgotten at the subconscious level at which robots can affect humans so deeply. Social robots are clearly able to push our "Darwinian buttons," those mechanisms that evolution produced in our social brains to cope with the dynamics and complexities of social groups, mechanisms that automatically trigger inferences about other agents' mental states, beliefs, desires, and intentions.

13.10 The Potential for Abuse

The fact alone that humans are already anthropomorphizing existing social robots in ways that clearly overstate the robots' capabilities is a sufficient indication that the personification of social robots is moving forward quickly, and that more sophisticated future robots will likely be even more anthropomorphized. Features of future robots, like human-like appearance, natural language interactions, and so on, might prompt people to be even more trusting in them or develop attitudes toward robots that could and likely would be exploited. For example, if it turns out that humans are reliably more truthful with robots than they are with other humans, it will only be a matter of time before robots will interrogate humans. And if it turns out that robots are generally more believable than humans, then it will only be a matter of time before robots are used as sales representatives.

Moreover, it will become even easier and more natural for humans to establish unidirectional emotional bonds with more sophisticated robots, often without noticing, akin to becoming addicted, where one's realization of one's addiction always comes after the fact. And with more sophisticated robots that are specifically programmed to exhibit behavior that could easily be misinterpreted as showing social emotions such as sympathy and empathy, it will become increasingly difficult for people to even realize that their social emotional bonds are unidirectional, aside from a basic emotional resistance that we are already seeing today (e.g., when people insist that they get back the very same robot that they sent in for repair and not another copy).

What is so dangerous about unidirectional emotional bonds is that they create psychological dependencies that could have serious consequences for human societies, because they can be exploited at a large scale. For example, social robots that appear "lovable" might be able to get people to perform actions that the very same people would not have performed otherwise, simply by threatening to end their relation with the human (e.g., an admittedly futuristic sounding request of a robo-dog to dispose of a real dog: "please get rid of this animal, he is scaring me, I don't want him around any longer"). More important, social robots that cause people to establish emotional bonds with them, and trust them deeply as a result, could be misused to manipulate people in ways that were not possible before. For example, a company might exploit the robot's unique relationship with its owner to make the robot convince the owner to purchase products the company wishes to promote. Note that unlike human relationships where, under normal circumstances, social emotional mechanisms such as empathy and guilt would prevent the escalation of such scenarios; there does not have to be anything on the robots' side to stop them from abusing their influence over their owners.

13.11 We Need to Act, Now!

Despite our best intentions to build useful robots for society, thereby making the case for robo-soldiers, robo-pets, robo-nurses, robo-therapists, robo-companions, and so forth, current and even more so future robot technology poses a serious threat to humanity. And while there is clearly a huge potential for robots to do a lot of good for humans (from elder care to applications in therapy), any potential good cannot be discussed without reflecting any potentially detrimental consequences of allowing machines to enter our personal social and emotional lives.

Some have warned us for quite some time about the dangers of producing increasingly human-like robots: "it is also practically important to avoid making robots that are reasonable targets for either human sympathy or dislike. If robots are visibly sad, bored or angry, humans, starting with children, will react to them as persons. Then they would very likely come to occupy some status in human society. Human society is complicated enough already" (McCarthy 1999). Yet, it is clear that, as a research community, the fields of artificial intelligence, robotics, and the nascent field of human–robot interaction have not reflected enough on the social and ethical implications of their artifacts. Such a reflection, if considered soon enough, might be able to inform future robotics research in useful ways, for example, on how research should proceed with respect to questions such as the slowly crystallizing perspective of future robotic soldiers (Moshkina and Arkin 2007) or robotic sex partners (Levy 2007).

Different from the first discussions about robot consciousness and robot rights in the 1960s, in which philosophers thought it opportune to begin reflecting on these subjects, since the existence of such robots was still far off (Putnam 1964), we are now running out of time. We need to start right away to investigate the potential dangers of social robots, find ways to mitigate them, and possibly develop principles that future lawmakers can use to impose clear restrictions on the types of social robots that can be deployed.

For example, one could simply prohibit and stop all research and development on social robots. While this option would certainly solve some of the problems, by avoiding them altogether, it seems completely unreasonable to believe that research and development of social robots could be prohibited and stopped, while other research in robotics and artificial intelligence continues.

Another option might be to require, by law, that all commercially available robots have some form of ethical reasoning built in. For example, some researchers have argued that ethical principles will need to be integrated into the decision-making algorithms in the robotic architecture in such a way that the robot will not be able to alter, ignore, or turn off these mechanisms (e.g., Arkin 2009). While this option might work for limited domains, where the number of possible actions is clearly constrained and the ethical implications of all actions can be determined ahead of time, it is unclear how general ethical principles could be devised that would work for an unknown number of situations, largely because philosophy in all of its history has not been able to agree on the right set of universal ethical principles, aside from being computationally feasible in real time given the computational constraints of the robotic platform. Even if there were a way to encode ethics in a set of universal laws, very much like Asimov conceived of the Three Laws of Robotics (in his short story "Runaround" from 1942), there are strong logical reasons why such as system cannot work—it would be straightforward to present a robot with logical paradoxes that would render any rational reasoning system ineffective, for example by ordering it to "not obey any orders, including this one," an order that, by simply stating it, automatically makes the robot disobedient no matter how sophisticated its control system may be.

Another option might be, again, required by law, to make it part of a social robot's design, appearance, and behavior, that the robot continuously signal, unmistakably and clearly, to the human that it is a machine, that it does not have emotions, that it cannot reciprocate (very similar to the "smoking kills" labels on European cigarette packs). Of course, these reminders that robots are machines are no guarantee that people will not fall for them, but it might reduce the likelihood and extent to which people will form emotional bonds with robots. And it will present the challenge of walking a fine line between making interactions with robots easier and more natural, while clearly instilling in humans the belief that robots are human-made machines with no internal life (at least the present ones). It is currently unclear how effective such mechanisms could be, although empirically testing their effectiveness would be straightforward (e.g., add a particular mechanism to a particular generation of Roombas, repeat the previous ethnographic studies, and compare the extent to which people engage in the same behaviors as before).

In the end, what we need is a way to ensure that robots will not be able to manipulate us in ways that would not be possible for other (normal) human beings. And a radical step might be necessary to achieve this: to endow future robots with human-like emotions and feelings. Specifically, we need to do for robots what evolution did for us, namely to equip us with an emotional system that strikes a balance between individual well-being and socially acceptable behavior. By having the same "unalterable affective evaluation" as those realized in humans, future social robots will be able to function in human societies in human-like ways (for all the reasons we are now investigating in HRI and AI/robotics), with the side effect of having "genuine feelings" that make them just as vulnerable and manipulable as humans.

Some have voiced their reservations about endowing robots with emotions arguing that it would take extra effort to implement human-like emotions in robots (e.g., McCarthy 1999), while others have maintained that certain types of emotions will necessarily be possible (and even instantiated) in complex robotic architectures with particular architectural properties (Sloman and Croucher 1981). Without taking a stance on whether emotions have to be explicitly built in or result as emergent phenomena in certain types of architectures, it is important to appreciate that this suggestion does not apply to any type of robot, but only to certain types of social robots. We certainly do not need a space exploration robot to be emotional, and nobody would set foot on a plane with an automatic flight controller that can get depressed, if not suicidal. However, if we had a choice between a *Terminator 3*-type scenario, where intelligent robots take control, despite human efforts to prevent it, and a grouchy household robot that is tired of cleaning up the kitchen floor, the choice is obvious.

References

Arkin, R. 2009. *Governing Lethal Behavior in Autonomous Robots*. Boca Raton, FL: Chapman and Hall.

Barber, K., and C. Martin. 1999. Agent autonomy: Specification, measurement, and dynamic adjustment. In *Proceedings of the Autonomy Control Software Workshop at Autonomous Agents* (Agents'99), 8–15. Seattle: Association for Computing Machinery.

Breazeal, C. L. 2002. *Designing Sociable Robots*. Cambridge, MA: MIT Press.

Buss, S. 2002. Personal autonomy. *Stanford Encyclopedia of Philosophy* (Winter ed.), ed. E. N. Zalta. Metaphysics Research Lab, CSLI, Stanford University. <http://plato.stanford.edu/entries/personal-autonomy/> (accessed July 14, 2011).

Crowell, C., M. Scheutz, P. Schermerhorn, and M. Villano. 2009. Gendered voice and robot entities: Perceptions and reactions of male and female subjects. In *Proceedings of the 2009 IEEE/RSJ International Conference on Intelligent Robots and Systems*, 3735–3741. St. Louis, MO: IEEE.

Dorais, G., R. P. Bonasso, D. Kortenkamp, B. Pell, and D. Schreckenghost. 1998. Adjustable autonomy for human-centered autonomous systems on Mars. Mars Society Conference, University of Colorado at Boulder, Colorado, August.

Forlizzi, J., and C. DiSalvo. 2006. Service robots in the domestic environment: a study of the roomba vacuum in the home. In *Proceedings of the 1st ACM SIGCHI/SIGART Conference on Human-Robot Interaction* (HRI '06), 258–265. New York: ACM.

Friedman, B., Jr., Peter H. Kahn, and J. Hagman. 2003. Hardware companions? What online Aibo discussion forums reveal about the human-robotic relationship. In *Proceedings of the SIGCHI Conference on Human Factors in Computing Systems*, 273–280. Ft. Lauderdale, FL: SIGCHI.

Garreau, J. 2007. Bots on the ground in the field of battle (or even above it): Robots are a soldier's best friend. *Washington Post*. May 6.

Gates, Bill. 2007. A robot in every home. *Scientific American* (January): 58–65.

Haugeland, J. 2002. *Computationalism: New Direction*, ed. M. Scheutz, 159–174. Cambridge, MA: MIT Press.

Huang, H. M., ed. 2004. *Autonomy Levels for Unmanned Systems (ALFUS) Framework, Volume I: Terminology*. Gaithersburg, MD: National Institute of Standards and Technology.

Kahn, Peter H., Jr., B. Friedman, and J. Hagman. 2002. "I care about him as a pal": Conceptions of robotic pets in online Aibo discussion forums. In *Proceedings of CHI Extended Abstracts '2002*, 632–633. Minneapolis, MN: ACM.

Kiesler, S., and P. Hinds. 2004. Introduction to the special issue on human-robot interaction. *Human-Computer Interaction* 19 (1): 1–8.

Levy, D. 2007. *Love and Sex with Robots: The Evolution of Human-Robot Relationships*. New York: Harper.

McCarthy, J. 1999. Making robots conscious of their mental state. In *Machine Intelligence 15*, ed. Koichi Furukawa, Donald Michie, and Stephen Muggleton, 3–17. Oxford: Clarendon Press.

Moshkina, L., and R. Arkin. 2007. *Lethality and Autonomous Systems: Survey Design and Results*. GVU Technical Report; GIT-GVU-07-16. Atlanta: Georgia Institute of Technology.

Parasuraman, R., T. Sheridan, and C. Wickens. 2000. A model for types and levels of human interaction with automation. *IEEE Transactions on Systems, Man, and Cybernetics, Part A: Systems and Humans* 30 (3): 286–297.

Putnam, H. 1964. Robots: Machines or artificially created life? *Journal of Philosophy* 61 (November): 668–691.

Rose, R., M. Scheutz, and P. Schermerhorn. 2010. Towards a conceptual and methodological framework for determining robot believability. *Interaction Studies* 11 (2): 314–335.

Schermerhorn, P., and M. Scheutz. 2009. Dynamic robot autonomy: Investigating the effects of robot decision-making in a human-robot team task. In *Proceedings of the 2009 International Conference on Multimodal Interfaces* (ICMI-MLMI '09), 63–70. New York: ACM.

Schermerhorn, P., M. Scheutz, and C. Crowell. 2008. Robot social presence and gender: Do females view robots differently than males? In *Proceedings of the 3rd ACM/IEEE International Conference on Human Robot Interaction* (HRI '08), 263–270. New York: ACM.

Scheutz, M. 2002. Agents with or without emotions? In *Proceedings of the Fifteenth International Florida Artificial Intelligence Research Society Conference*, ed. Susan M. Haller and Gene Simmons, 89–94. Pensacola Beach, FL: AAAI Press.

Scheutz, M., and C. Crowell. 2007. The burden of embodied autonomy: Some reflections on the social and ethical implications of autonomous robots. ICRA 2007 Workshop on Roboethics, Rome, Italy, April.

Scheutz, M., P. Schermerhorn, J. Kramer, and D. Anderson. 2007. First steps toward natural human-like HRI. *Autonomous Robots* 22 (4): 411–423.

Scheutz, M., P. Schermerhorn, and J. Kramer. 2006. The utility of affect expression in natural language interactions in joint human-robot tasks. In *Proceedings of the 1st ACM SIGCHI/SIGART Conference on Human-robot interaction* (HRI '06), 226–233. New York: ACM.

Shibata, T. 2005. Human interactive robot for psychological enrichment and therapy. In *Proceedings of AISB 2005: Social Intelligence and Interaction in Animals, Robots and Agents*, 98–107. Hatfield, UK: University of Hertfordshire.

Sloman, A., and M. Croucher. 1981. Why robots will have emotions. In *Proceedings of the 7th International Joint Conference on Artificial Intelligence* (IJCAI '81), ed. Patrick J. Hayes, 84–86. Vancouver, BC: William Kaufmann.

Sung, J. Y., L. Guo, R. E. Grinter, and H. I. Christensen. 2007. "My Roomba is Rambo": Intimate home appliances. In *Proceedings of UbiComp 2007*, ed. John Krumm, Gregory D. Abowd, Aruna Seneviratne, and Thomas Strang, 145–162. Innsbruck, Austria: Springer.

Turkle, S. 2006. *A Nascent Robotics Culture: New Complicities for Companionship*. AAAI Technical Report Series, July.

Zajonc, R. B. 1965. Social facilitation. *Science* 149: 269–274.

14 The Ethics of Robot Prostitutes

David Levy

I pay for sex because that is the only way I can get sex. I am not ashamed of paying for sex. I pay for food. I pay for clothing. I pay for shelter. Why should I not also pay for sex? Paying for sex does not diminish the pleasure I derive from it.
—Hugh Loebner[1] (1998)

Recent discussions on roboethics have introduced the subject of sex with robots (Levy 2006a, 2006b, 2006c). In particular, one authoritative statement on this topic received worldwide media publicity during 2006—the prediction by Henrik Christensen, chairman of EURON, the European Robotics Research Network, that "people will be having sex with robots within five years."

The arrival of sexbots[2] seems imminent when one considers recent trends in the development of humanoids, sex dolls, and sex machines of various types. Sophisticated humanoids such as the Repliée Q1 (Minato et al. 2005) have already been developed that are humanlike in appearance. Advances in materials science have enabled sex doll manufacturers to improve significantly on the inflatable products of the preceding decades, creating dolls with prices in the region of $5,000 to $7,000 (Levy 2007). Low-cost sexual devices, designed mostly for use by women, now sell tens of millions annually in the United States (Good Vibrations 2005). Far more intricate and more expensive machines are manufactured that actually simulate sexual intercourse, and are sold on websites such as <www.orgasmalley.com>, whose prices range from $140 to $1,800.

It takes little imagination to appreciate that it is already technically possible to construct a robot that combines the look, feel, and functions of humanoids, sex dolls, and sex machines. When sexbots first appear on the market, they will most likely be beyond the pockets of all but the wealthy. The current cost of constructing a sophisticated humanoid dwarfs the cost of purchasing an upmarket sex doll, as can be seen from the $130,000 starting price of the robot heads designed by industry leader David Hanson, and manufactured by his company Hanson Robotics Inc. With the first sexbots costing a six-figure (dollar) sum, or possibly more, their hire will be the only way for most of us who want to experiment with the joys of robot sex to do so (Levy 2007).

14.1   Sex Dolls for Hire

In terms of sales volumes, Japan leads the way with the current generation of high-priced sex dolls (Levy 2007). Their popularity on the retail market has also spawned a doll variant of the more traditional form of "escort" service. In a 2004 newspaper article entitled "Rent-a-Doll Blows Hooker Market Wide Open," (Connell 2004) the *Mainichi Daily News* explains how one leading purveyor, *Doll no Mori* (Forest of Dolls), started their 24/7 doll-escort service in southern Tokyo and the neighboring Kanagawa prefecture: "We opened for business in July this year," said Hajime Kimura, owner of Doll no Mori. "Originally, we were going to run a regular call girl service, but one day while we were surfing the Net we found this business offering love doll deliveries. We decided the labor costs would be cheaper and changed our line of business."

Outlays are low, he explains, with the doll's initial cost the major investment, and wages are never a problem for employers. "We've got four dolls working for us at the moment. We get at least one job a day, even on weekdays, so we made back our initial investment in the first month," Kimura says. "Unlike employing people, everything we make becomes a profit and we never have to worry about the girls not turning up for work." *Doll no Mori* charges start at 13,000 yen (around $110) for a seventy-minute session with the dolls, which is about the same price as a regular call girl service. The company boasts of many repeat customers. "Nearly all our customers choose our two-hour option."

Within little more than a year of the doll-for-hire idea taking root in Japan, sex entrepreneurs in South Korea also started to cash in. Upmarket sex dolls were introduced to the Korean public at the Sexpo exhibition in Seoul in August 2005, and were immediately seen as a possible antidote to Korea's Special Law on Prostitution that had been placed on the statute books the previous year. Before long, hotels in Korea were hiring out "doll experience rooms" for around 25,000 won per hour ($25), a fee that included a bed, a computer to enable the customer to visit pornographic websites, and the use of a doll. This initiative quickly became so successful at plugging the gap created by the antiprostitution law that, before long, establishments were opening up that were dedicated solely to the use of sex dolls, including at least four in the city of Suwon. These hotels assumed, quite reasonably, that there was no question of them running foul of the law, since their dolls were not human. But the Korean police were not so sure. The news website Chosun.com (Chosun.com 2006, now at soompi.com) reported, in October 2006, that the police in Gyeonggi Province were "looking into whether these businesses violate the law . . . Since the sex acts are occurring with a doll and not a human being, it is unclear whether the Special Law on Prostitution applies."

The early successes of these sex-doll-for-hire businesses are a clear indicator of things to come. If static sex dolls can be hired out successfully, then sexbots with moving components seem certain to be even more successful. If vibrators can be such a huge commercial success, then malebots with vibrating penises would also seem likely to have great commercial potential.

14.2 Paying a (Human) Sex Worker

Prostitution is known as "the world's oldest profession," and is one that continually attracts controversy because of the ethical issues involved in selling sex. On the one hand, there are arguments such as: prostitution harms women, exploits women, demeans women, spreads sexual diseases, fuels drug problems, leads to an increase in organized crime, breaks up relationships, and more (Ericsson 1980). In contrast, there are those, including many of the clients themselves, who acknowledge and praise the social benefits of prostitution and the valuable services performed by the profession for its clients. These supporters employ arguments such as: prostitutes have careers based on giving pleasure, they can teach the sexually inexperienced how to become better lovers, they make people less lonely, they relieve millions of people of unwanted stress and tension, and they provide sex without commitment for those who want it (Pateman 1988, 2003). The ethical issues surrounding all these and other arguments related to prostitution have been debated for centuries.

In order to gain some insight as to why people will be willing, even eager, to hire the services of malebots and fembots, it is useful first to investigate the reasons for paying for the services of human sex workers. A comprehensive analysis of the principal reasons is given by Levy (2007), discussing not only men hiring female sex workers, but also the far less prevalent but increasing phenomenon of women hiring men.

Several reasons have been identified as to why men pay women for sex—what the men want or expect from these sexual encounters. The reasons most commonly stated by male clients include:

Variety Here, variety, means the opportunity to have sex with a range of different women (McKeganey and Barnard 1996; Plumridge et al. 1997). A robot will be able to provide variety in terms of its conversation, its voice, its knowledge and virtual interests, its virtual personality, and just about every other aspect of its being, including its appearance and size. While variety in these characteristics of sex workers is one major reason for men paying for sex, variety in the sexual experience itself is, for many clients, another important factor, often *the* most important. Many clients are interested in sexual practices to which they do not otherwise have access, such as oral sex, often because their partners are unable or unwilling to accommodate their desires (Monto 2001). An electromechanically sophisticated robot that can indulge in oral sex will be able to satisfy this particular human motivation.

Lack of Complications and Constraints The literature has identified a small group of motivations that might collectively be described as a lack of complications and constraints. For many clients, the principal benefits of the commercial sex exchange include the clear purpose and bounded nature of the arrangement, as well as its anonymity, its brevity, and the lack of emotional involvement (Bernstein 2005; McKeganey 1994). Sexbots, almost by definition, will be able to satisfy these particular human motivations.

Lack of Success with the Opposite Sex For a variety of reasons, many men experience difficulty in developing relationships with women. In some cases, this is because the man is ugly, physically deformed, psychologically inadequate, a stranger in another town or a foreign land, or simply lacking in the necessary social skills or sexual assurance or both. Such men, with normal male desires, have a need for sexual intimacy that they cannot satisfy because of their lack of sexual effectiveness—they simply cannot attract a mate, or are afraid to try, or suffer from a combination of both. By paying for sex, they reduce the risk of rejection to an absolute minimum, thereby almost guaranteeing themselves sex on a plate. For these men, prostitution is the only sex available, a reason for paying for sex that was indicated by almost 40 percent of the clients in one study (Xantidis and McCabe 2000). None of these problem categories will present any difficulty to sexbots, which will be immune to any ugliness or physical deformity in their clients, and to their clients' psychological inadequacies.

In contrast to the relatively well-researched topic of men paying for sex, there is almost no systematic published research on the reasons why women pay, or what exactly they are seeking. But what little published evidence there is on this topic suggests that the reasons are close to those that motivate the male clients of sex workers, principally, a lack of complications and constraints and a lack of success with men (Levy 2007).

In summary, sexbots for hire will be able to satisfy the motivational as well as the sexual needs for individuals (of both sexes) who would otherwise be the clients of sex workers—to provide variety, to offer sex without complications or constraints, and to meet the needs of those who have no success in finding human sex partners. In addition, there is one significant health benefit for the clients in hiring a sexbot instead of a sex worker, namely the relative ease with which hirers can assure themselves of freedom of infection from sexually transmitted diseases. The sexual hygiene of a robot could and should be undertaken by the clients, as a case reported in *Genitourinary Medicine* testifies (Kleist and Moi 1993).

14.3 Some Ethical Aspects of Robot Prostitution

In the subsections that follow, we consider five aspects of the ethics of robot prostitution.

14.3.1 The Ethics of Making Robot Prostitutes Available for General Use

The prime purpose of a sexbot is to assist the user in achieving orgasm, without the necessity of having another human being present. This is the same purpose as vibrators for women, which are now so popular that they are openly sold on the shelves on some of the biggest and most reputable drug store and pharmacy chains in the United States and Europe. It would seem anomalous, in view of this widespread tacit acceptability of vibrators, to brand their use immoral, just as it is difficult to argue that the design, development, manufacture, and sale of sexbots is unethical.

14.3.2 The Ethics, vis à vis Oneself and Society in General, of Using Robot Prostitutes

With most of the clients of sex workers, self-respect is an important issue. There are those, like Hugh Loebner, who are so proud of the use they make of the services of sex workers that they happily publicize their commercial sex activities, but they represent a small minority. The majority feel that there is still a moral stigma attached to their encounters, and they will go to some length in their attempts to avoid being found out by those close to them, or, even worse, being named and shamed in public, as some police forces do. For this majority, the issue of self-respect will be much better catered to by hiring robot prostitutes instead of sex workers, because robots are not generally perceived as living beings but as artifacts, and the same moral stigma does not therefore apply. Yet there will, at least for some time, be a moral stigma of a different sort. We understand sex with a person, but most people do not appreciate the concept of sex with a robot, and what we do not understand we tend to stigmatize.

In contemplating how the use of robot prostitutes might affect society, it is also important to consider the legal issues. Most of us in a free-thinking society are unlikely to feel that the use of sexbots by adults in private is a practice that should be prevented by legislation. Yet in Alabama, Texas, and some other U.S. jurisdictions, the sale of vibrators has been deemed illegal (Levy 2007), so it is hard to predict how the law will view the sale and hire of sexbots in the more conservative-minded states. Among those who have argued that people should have the right to avail themselves of the services of sex workers, David Richards (1979) makes a strong case: "We are able to understand the humane and fulfilling force of sexuality *per se* in human life, the scope of human autonomous self-control in regulating its expression, and the implications of these facts for the widening application of the concept of human rights to the sexual area

. . . sexual autonomy appears to be a central aspect of moral personality, through which we define our ideas of a free person who has taken responsibility for her or his life." Clearly, Richards's arguments carry even more force when related to robot prostitutes rather than to human sex workers.

14.3.3 The Ethics, vis à vis One's Partner or Spouse, of Using Robot Prostitutes

How the use of a robot prostitute is perceived by a spouse or partner is open to many possibilities. Will a spouse or partner who considers infidelity with another human to be reasonable behavior be likely to be upset by the hire of a sexbot? Certainly there will be many who feel that the sexual demands placed on them within their relationship are excessive, and who will therefore appreciate a night off now and then, in the knowledge that what is taking place is nothing "worse" than a form of masturbation. There will also be some who positively relish the idea of robots, programmed to be sexually adept, teaching their partner to improve their lovemaking skills. And there will be couples, both of whom derive pleasure and sexual satisfaction from a threesome in which the third participant is willing to indulge in whatever sexual activity is asked of it (subject of course to its programming and engineering). In contrast, there will be some partners and spouses who find the very idea of sex with a robot to be anathema. The ethics of using a robot prostitute within a relationship will depend very much on the sexual ethics of the relationship itself when robots do not enter the picture.

14.3.4 The Ethics, vis à vis Human Sex Workers, of Using Robot Prostitutes

It is a common perception that prostitution is a "bad thing" for the sex workers. This is because it is seen, inter alia, as degrading them, encouraging them into a lifestyle in which an addiction to hard drugs often forms an integral part, and strongly increasing the likelihood of their catching AIDS or some other possibly fatal sexually transmitted disease. If this is so, and not all sex workers agree with this perception of their profession as a bad thing, then the introduction of robot prostitutes can only be a "good thing," because it will most likely cause a dramatic drop in the numbers who ply their trade in whichever countries robot prostitutes are made available. This eventuality was predicted as long ago as 1983, when *The Guardian* reported (Weatherby 1983) that New York prostitutes "share some of the fears of other workers—that technology developments may put them completely out of business. All the peepshows now sell substitutes—dolls to have sex with, vibrators, plastic vaginas and penises—and as one woman groused in New York 'It won't be long before customers can buy a robot from the drug-store and they won't need us at all.'" This problem, the compulsory redundancy of sex workers, is an important ethical issue, since in many cases those who turn to prostitution as their occupation do so because they have literally no other way to earn the money they need.

14.3.5   The Ethics, vis à vis the Sexbots Themselves, of Using Robot Prostitutes

Up to now the discussion in this chapter has been based on the assumption that sexbots will be mere artifacts, without any consciousness and therefore with no rights comparable to those of human beings. Recently, however, the study of robotics has taken on a new dimension, with the emergence of ideas relating to artificial consciousness (AC).[3] This area of research is concerned with "the study and creation of artifacts which have mental characteristics typically associated with consciousness such as (self-)awareness, emotion, affect, phenomenal states, imagination, etc." (AISB 2005).

Without wishing to prejudice what will undoubtedly be a lively and long-running debate on robot consciousness, this author considers it appropriate to raise the issue of how AC, when designed into robots, should affect our thinking regarding robot prostitutes. Should they then be considered to have legal rights and ethical status, and therefore worthy of society's concern for their well-being and their behavior, just as our view of sex workers is very much influenced by our concern for *their* well-being and behavior? David Calverley asserts (2005) that natural law mitigates in favor of an artificial consciousness having intrinsic rights, and therefore, simply by virtue of having an artificial consciousness, a robot should be ascribed *legal* rights. If this is held to be so, then concomitant with those legal rights will come legal responsibilities, and robot prostitutes might therefore become subject to some of the same or similar legal restrictions that currently apply to sex workers.

The legal status and rights of robots are but one aspect of their ethical status. Torrance (2006) discusses our responsibility in, and the ethical consequences of, creating robots that are considered to possess conscious states, and he introduces the notion of artificial ethics (AE)—the creation of "systems that perform in ways which confer or imply the possession of ethical status when humans perform in those ways. For example, having a right to life, or a right not to be treated merely as an instrument of someone else's needs or desires, are properties which are part of the ethical status of a human being, but a person doesn't acquire such rights just because of what they *do*. This may extend to ethics when applied to artificial agents."

These questions from Calverley, Torrance, and others in this recent but already fascinating field are certainly issues that will form part of the coming debate on the ethics of robot sex and robot prostitution. This author does not pretend to have any answers as yet, but for the time being rests content to have raised the profile of these issues for the awareness of the roboethics community.

14.4   Conclusion

With the advent of robot sex, robot prostitution inevitably becomes a topic for discussion. The author believes that the availability of sexual robot partners will be of significant social and psychological benefit for society, but accepts that there are important ethical issues to be considered relating to robot prostitutes. This chapter has highlighted some of these issues. The debate is just beginning.

Notes

1. Hugh Loebner is the founder and sponsor of the annual Loebner Prize in Artificial Intelligence, a Turing Test contest to find the best conversational computer program.

2. In common with accepted practice this chapter employs the term "sexbot" to mean *any* robot with sexual functionality, and "malebot" or "fembot" to indicate a sexbot with artificial genitalia corresponding to a particular sex.

3. Sometimes referred to as "machine consciousness" (MC).

References

AISB. 2005. Symposium overview. In *Proceedings of the Symposium on Next Generation Approaches to Machine Consciousness*, ed. R. Chrisley, R. W. Clowes, and S. Torrance, iv. Hatfield, UK: University of Hertfordshire; London: AISB.

Bernstein, E. 2005. Desire, demand, and the commerce of sex. In *Regulating Sex: The Politics of Intimacy and Identity*, ed. E. Bernstein and L. Schaffner, 101–128. New York: Routledge.

Calverley, D. 2005. Toward a method for determining the legal status of a conscious machine. In *Proceedings of the Symposium on Next Generation Approaches to Machine Consciousness*, ed. R. Chrisley, R. W. Clowes, and S. Torrance, 75–84. Hatfield, UK: University of Hertfordshire; London: AISB.

Chosun.com. 2006. Do the anti-prostitution laws protect sex dolls? <http://www.soompi.com/forums/topic/90967-sex-doll-brothels-springing-up-in-korea/>.

Connell, R. 2004. Rent-a-doll blows hooker market wide open. *Mainichi Daily News*, December 16.

Ericsson, L. 1980. Charges against prostitution: An attempt at philosophical assessment. *Ethics* 90 (3): 335–366.

Good Vibrations. 2005. Personal communication. <http://www.goodvibes.com/> (accessed July 14, 2011).

Kleist, E., and H. Moi. 1993. Transmission of gonorrhoea through an inflatable doll. *Genitourinary Medicine* 69 (4): 322.

Levy, D. 2006a. Marriage and sex with robots. EURON Workshop on Roboethics, Genoa, March.

Levy, D. 2006b. Emotional relationships with robotic companions. EURON Workshop on Roboethics, Genoa, March.

Levy, D. 2006c. A history of machines with sexual functions: Past, present and robot. EURON Workshop on Roboethics, Genoa, March.

Levy, D. 2007. *Love and Sex with Robots*. New York: Harper Collins.

Loebner, H. 1998. Being a john. In *Prostitution: On Whores, Hustlers, and Johns*, ed. J. Elias, V. Bullough, V. Elias, and G. Brewer, 221–225. Amherst, NY: Prometheus Books.

McKeganey, N. 1994. Why do men buy sex and what are their assessments of the HIV-related risks when they do? *AIDS Care* 6 (3): 289–301.

McKeganey, N., and M. Barnard. 1996. *Sex Work on the Streets: Prostitutes and Their Clients*. Buckingham, UK: Open University Press.

Minato, T., M. Shimada, S. Itakura, K. Lee, and H. Ishiguro. 2005. Does gaze reveal the human likeness of an android? Paper presented at 2005 4th IEEE International Conference on Development and Learning, Osaka, Japan.

Monto, M. 2001. Prostitution and fellatio. *Journal of Sex Research* 58 (2): 140–145.

Pateman, C. 1988. *The Sexual Contract*. Stanford, CA: Stanford University Press.

Pateman, C. 2003. Defending prostitution: Charges against Ericsson. *Ethics* 93 (3): 561–565.

Plumridge, E., J. Chetwynd, A. Reed, and S. Gifford. 1997. Discourses of emotionality in commercial sex: The missing client voice. *Feminism and Psychology* 7 (2): 165–181.

Richards, D. 1979. Commercial sex and the rights of the person: A moral argument for the decriminalization of prostitution. *University of Pennsylvania Law Review* 127: 1195–1287.

Torrance, S. 2006. The ethical status of artificial agents—With and without consciousness. *Ethics of Human Interaction with Robotic, Bionic, and AI Systems: Concepts and Policies*, workshop proceedings, ed. G. Tamburrini and E. Datteri, <http://ethicbots.na.infn.it/meetings/firstworkshop/abstracts/torrance.htm> (accessed July 14, 2011).

Weatherby, W. J. 1983. Hard times on the street walk. *The Guardian*, February 23, p. 21.

Xantidis, L., and M. McCabe. 2000. Personality characteristics of male clients of female commercial sex workers in Australia. *Archives of Sexual Behavior* 29 (2): 165–176.

15 Do You Want a Robot Lover? The Ethics of Caring Technologies

Blay Whitby

Do you want a robot lover? You might perhaps think that you do and that it is nobody else's business but yours, but the widespread use of robots in intimate and caring roles will bring about important social changes. We need to examine these changes now and consider them from an ethical standpoint. Robotic carers and artificial companions are a technology that is likely to be available in the near to mid-term future. In Japan and South Korea, robots are seen as potential carers for the elderly and as baby-sitters. Many researchers are looking to make their products display emotion and respond to emotional displays by users. At least one writer has predicted marriage to robots will be accepted in progressive countries by 2050. This chapter examines some of the implications of these possibilities—both technical and social. Do they represent socially and ethically acceptable developments? What is likely to be technically feasible, and just what should we allow?

15.1 The Debate

It is a truth universally acknowledged that a young man (or woman) in possession of a good fortune must be in want of a robotic companion.[1] So do *you* want a robot lover? Maybe not: perhaps you would prefer instead a robot to act as a domestic servant or as a personal care assistant in your declining years. Perhaps a more interesting question for you might be: Would you leave your children in the care of a robot nanny? All these considerations highlight immediate problems for robot ethics, and it is urgent to address them because this sort of caring technology is about to enter widespread use. The technology discussed in this chapter is absolutely not science-fiction technology. The discussion is about technology that is already in use or under development.

It's important to read the chapter title as a *question* because many futurologists, industrialists, and investors have already decided that you do indeed want something along these lines. There are advantages and disadvantages to the use of robots in personal settings. Many people will want the sort of technology under consideration here for both good and bad reasons, but there will be costs—and not just the financial costs of promoting research and development in this area at the expense of other areas. There are social dangers that ought to be avoided and about which such unrestrained commercial interests may need to be made aware. It is unlikely that the social problems of robot ethics will be solved by allowing markets to decide freely.

There is also coherent and powerful opposition to robot lovers, perhaps also to technology employed in other companionship roles. The Roman Catholic Church, the world's largest religious organization, has clear and profound doctrinal opposition to sexual acts other than within marriage for the purpose of procreation.[2] For this reason, the very idea of a robot lover, and maybe even a robot companion, will be completely unacceptable to them. Many other religious groups are likely to take similar positions.

A position of general and complete opposition to the technology, however, pays no attention to the real human benefits that this technology might bring. Robot carers, and, in particular, "smart homes," could enable older people to remain independent longer, and this may well be something they would freely choose. Robot companions, too, may have many social benefits. The ethical issues are nuanced. It seems highly likely that a more balanced ethical response will need to be both technically and philosophically informed. Explicit ethical principles are needed for the design and introduction of this sort of technology. These ethical principles need urgent exploration and discussion.

15.2   What Is a Robot?

It is natural for people to see novel technologies in terms of those they replace. That is why automobiles were once referred to as "horseless carriages" and radio as "wireless." That is also why science-fiction accounts of robots have tended to make robots human-like in appearance and size. In fact, very few robots have turned out to be human-like in appearance or size. Real robots are now commonplace, but because they don't look like the ones in the movies, they have not always been recognized as robots.

Contemporary robots range from bits of software that autonomously perform activities (both good and bad) on the Internet, to the post-A320 range of Airbus airliners, which are so highly automated that the pilots effectively give them executive commands, rather than actually flying them; the aircraft itself automatically takes care of the flying. One of the most successful modern robots has been the BGM-109 Tomahawk family of cruise missiles.[3] However, the major employment opportunity for robots is still in assembly-line industrial production, particularly automobile manufacturing, where they could be said to represent about 10 percent of the workforce.

These everyday robots are for the most part extremely nonhuman-like in appearance. However, there are three main dimensions along which robots can be human-like. It is important to distinguish them. The first, and perhaps, least important dimension is that of physical appearance. Rather more important is the fact that robots can also be human-like in behavior—deliberately imitating some human behaviors without looking particularly human. Just as importantly, robots can be human-like along the dimension of the tasks they perform. This latter dimension is clearly the most important when we consider the ethics of robots in caring and companionship roles.

One very important fictional image for robots has been that of domestic servant. Obviously, this is another example of people seeing the new technology in terms of previous technologies. Indeed, the first use of the word "robot," in Karel Čapek's 1921 play *Rossum's Universal Robots* ([1921] 2004) coincided perhaps significantly, with a period in Europe when human domestic servants were becoming hard to find[4] The possibility of building some sort of mechanical butler has preoccupied artists and, to a not-insignificant extent, technologists ever since the 1920s. This is despite the fact that in a modern house a great deal of a butler's traditional work has been automated. Dishwashing machines, central heating timers, easy-care fabrics, telephone answering machines, and similar technologies do much of the work once done by domestic servants.

However, there still remains a manifest desire to find further technological replacements for humans in certain roles. Whether this desire is prompted by real human need or instead by uncaring commercial imperatives is another issue raised by the question in the title. That there exists a profitable market for expensive new technology is not of itself sufficient moral defense for allowing widespread sales. There are ethical questions—in particular, who gains and who loses—that need to be examined at an early stage. We will consider, in detail, the question "why would anyone want a robot lover?" in a subsequent section. In conclusion of this section, it is worth remarking that there seems some ethical ambiguity about the answers (or lack of answers) to questions such as "why would anybody want a robot butler, or teacher, or physician?" There may be morally good responses to these questions, but they are rarely, if ever, stated. The field of robot ethics has some immediate and urgent groundwork to do.

15.3 What Is a Robot Lover?

Just as the creations of artists may have misled us about the nature of robots in general, so we may have been misled about the appearance and nature of a robot lover. It is not particularly difficult to employ an actress (or actor) to play an on-screen robot. Audiences tend not to object on the grounds that a machine so human in appearance and behavior, or indeed so physically attractive, is technically impossible for the foreseeable future.

In reality, robots cannot yet achieve anything like this standard of physical resemblance to humans. One significant problem among many is that of the "Uncanny Valley" (Mori 1970). This is a phenomenon first documented by Masahiro Mori in 1970, but much talked of in more recent years as technology has advanced to the point where it has much more immediate relevance. The Uncanny Valley involves severe revulsion on the part of humans when observing things such as robots that look and move in a way that is almost, but not quite, human-like. It is hypothesized that this phenomenon is an evolved human response—maybe to prompt the avoidance of very ill or incapacitated humans. Another plausible theory is that observing the unnatural movements of a very human-like robot triggers our fears of our own mortality.

Whatever the biological antecedents, the Uncanny Valley is a major problem for those designing realistic cinema animations, as well as for robot builders. It is also a major technological hurdle preventing the building of any robot that could produce the same kind of feelings that might be engendered by an attractive actress pretending to be a robot, at least for the foreseeable future. One can certainly buy sex dolls, but animating them in a way that does not disgust is likely to prove rather more difficult than was once anticipated. The problem of the Uncanny Valley is an important reason why it is highly unlikely that *physical* robots will be adopted as artificial sexual companions by those with mainstream sexual preferences, at least for the immediate future.

It is necessary to remember, however, that a gynoid- or android-style robot companion is only one possible technological development among many—albeit the main possibility that has been portrayed in artistic examinations of the future. The other two dimensions, mentioned in the preceding section, along which a robot can be human-like (behavior and role) are important here. If the robot *or other automated technology* performs at least some of the tasks of a human lover, then its introduction may well be analogous to the way in which household automation has taken over much of the role of domestic servants over the last fifty years.

Work, which might contribute to the development of robot lovers, is now proceeding in a number of technological areas. Boden (2006, 1094) lists thirteen technologies now working or under development, which could help move toward the sort of artificial companions under discussion. These include such things as the monitoring and manipulation of users' emotions by artificial systems, the detection of lying by users, and the realistic simulation of emotion by artificial systems.

A real robot lover might not be much like the pretty, human-sized, very human-resembling robot of the movies, therefore. However, it is likely to be a slightly different, but just as effective, sort of artificial companion. It could have intimate knowledge of its user. It could respond to and perhaps even anticipate its user's feelings. This line of development would be harder to recognize as anything like a robot lover because it will be integrated into other technologies. For example, it might emerge as a "user interface" to other caring technologies such as a smart home. It is certainly not inconceivable that such systems could provide more worthwhile companionship than humans provide in some cases.

All of the examples on Boden's list are technologies that could be developed (and indeed are now being developed) with the best of intentions. Those seeking to make computers more emotionally aware, for example, declare objectives such as making their systems more usable, more helpful, or better tutors. However, the same developments could equally be employed to make systems more seductive, more sexual, or more emotionally indispensable. These dimensions are likely to be of great interest to those intending to make profits from the technology.

As ethicists, we should not be concerned primarily with the pretty gynoid of the movies. We should probably worry much more about the use of advanced AI technology in a wide variety of caring roles. This is a way in which robot love is much more likely to emerge by stealth than in any obvious fashion. In discussing the ethics of technological developments we need always to be aware of the force of Kranzberg's First Law (1986, 545). In brief, this insists that technology is never of itself good, or bad, or neutral. It is always all three, depending on how we use it. We have choices about how we use technology in intimate and caring settings. These merit calm, informed, and thoughtful discussion.

15.4 Why Would Anyone Want a Robot Lover?

The joke in the opening sentence of section 15.1 reflects the humor in unquestioning acceptance of the need for robotic companions. It is worth briefly examining this frequently assumed need. There is no shortage of humans and while there may be many jobs suitable for robots, the provision of human companionship would seem to be a most unlikely area for automation.

The obvious answer to the question "Why would anyone want a robot lover?" is because a person is unable to find a human lover. There could be many reasons for this. However, from an ethical standpoint it is clear that we should, ceteris paribus, prefer to try to remedy or ameliorate the human problems, rather than substitute an artificial device. It may be that there are some individuals who are so extremely unattractive, or socially unskilled, or troubled in some other way, that human society is impossible for them. However, this is rarely claimed as a justification for the technological developments under discussion here.

For example, the category of people who most obviously are considered unfit for human society is convicted violent criminals. They would seem an ideal target market for robot lovers. Interestingly, they are a market rarely, if ever, mentioned by the enthusiasts for technology. However, for those who possibly can find human companionship, it would seem morally better to arrange this companionship than to substitute a technological solution. There are a number of reasons for this.

First, it clearly can be argued that peaceful, even loving, interaction among humans is a moral good in itself. Second, we should probably distrust the motives of those who wish to introduce technology in a way that tends to substitute for interaction between humans. Third, for a social mammal such as a human, companionship and social interaction are of crucial psychological importance. Ultimately, it may perhaps be that we can scientifically analyze all of these psychological needs. It may also be possible one day to build technology that completely fulfills these needs. However, as things stand, we cannot be sure that our caring technologies are capable of meeting all the relevant psychological needs. In advance of any certainty about this, there is clearly a risk of severe psychological damage. To a greater or lesser extent, these three moral reservations apply to all technology that is employed to substitute for humans in intimate caring roles. They are apparent when applied to the case of a robot lover, but apply equally to other caring technologies.

A very different view is taken by David Levy. Levy is an enthusiast for the use of robots in caring and loving roles (2007). Although he does not state it directly, his answer to the question "Why would anyone want a robot lover?" is essentially that it is the result of an inevitable process of technological development. He identifies three routes by which people might come to love robots (Levy 2007, 127–159). The first is claimed to be similar, if not identical, to the normal development of love between humans. The second route is best described as technophilia—people preferring a robot lover mainly because it is a robot. The third route is the way in which some people are so socially isolated that the love of a robot is a preferred option to normal human companionship.

Levy's third route may be ethically dubious precisely because of the three moral reservations stated above. It also represents an attempt to fix a potentially serious set of human problems by the proposed use of a yet-to-be-developed technology. In ethical terms, this may be a smokescreen to distract us from what we really need to fix. The supposed need for this technology is something that should not be accepted on trust—especially as there are profits to be made from selling the technology whereas the financial benefits of fixing social problems are not so obvious. Once again, the urgent need for widespread public debate should be clear.

What can be said of Levy's other two routes? Human sexuality is best described as a highly creative exercise. There is no doubt that sexuality directed at robots, rather than humans, is already practiced. These two routes are not only open they are already well traveled. The primary question in this section is not whether this is good, healthy, or in need of a response. It is rather: is sex with robots a route to love with robots?

Levy is perfectly correct in that it is a route to familiarity and dependence, but most writers would require a good deal more before calling this love. For example, Mark Fisher's excellent definition of love as a process rather than a state or an emotion rightly emphasizes the importance of reciprocity (1990, 23–35). Reciprocity from a robot is clearly different from reciprocity from a human. Levy asserts simply, "reciprocal liking is another attribute that will be easy to replicate in robots" (Levy 2007, 147). This is a highly debatable assertion.

Whether or not we could accept that Levy's sort of intimacy and familiarity with robots is actually "love" will be analyzed further in the next section. For now we can allow that there will almost certainly be increasing use of robots in intimate settings— and especially for sexual purposes. If we heed the lessons from previous technologies, then it would seem that there would be a ready market for robotic sex toys of various types. This is despite the widespread opposition of religious groups. For example, the continuing profitability of the pornography industry, despite effective opposition, and even legislative prohibition in some cultures, suggests that there is a strong underlying demand for pornographic material. It is reasonable to expect that there will also be a strong underlying demand for robotic sex toys.

The ethics of the sexual use of robots is, as has previously been remarked, nuanced and complex. The simple arguments portraying such use as all good or all bad should be quickly dismissed. In particular, the fact that there is a strong underlying demand is not any sort of moral justification. The fact that there are people who may be unable to find any lover other than a robotic one has been shown to be inadequate as a justification for the widespread use of such technology.

On the one hand, to allow a completely free market in robot lovers (and by the same token, robot carers and robot companions of all sorts) would be unforgivably rash. On the other hand, the case in favor of a free market in this technology could be based on the traditions of personal freedom set out most clearly in John Stuart Mill's definitive essay "On Liberty." As Mill put it, "The only part of the conduct on anyone, for which he is amenable to society, is that which concerns others. In the part which merely concerns himself, his independence is, of right, absolute. Over himself, over his own body and mind, the individual is sovereign" ([1859] 1966, 14).

Those following this very influential view will claim that if people want to involve themselves with robots, in various ways, then they have Mill's "absolute" right to do so. It is worth remarking that this sovereignty over oneself has never really been absolute in practice. Mill allows that it does not apply to children and "backward states of society" ([1859] 1966, 14). Societies under threat of violence—for example, those engaged in wars or under terrorist threat—find it necessary to constrain individual private behavior.

Nevertheless, an argument might be made that individuals have the right to purchase robots as sex toys or as other forms of caring technology if by doing so they harm no one else. Indeed, for those following Levy's second route—that of techno-philia, it is a win-win situation. Not only are they likely to be happier with their robotic lover than they would be with a human lover, but also the rest of society is spared any consequences of having to deal with their paraphilic urges.

This argument is valid as stated, but some limitations must be pointed out. It may not always be the case that no one else is harmed by this sort of behavior. An individual who consorts with robots, rather than humans, may become more socially isolated. Even if they are happier with their robotic companion, the reduction in human contact may make them less socially able, and therefore, not so effective as a citizen. If the practice becomes widespread, then society as a whole may suffer, and morally may be entitled to take steps to prevent this sort of breakdown.

It is important to stress that these are limitations, not a counter-argument. The exact point of balance between the rights of the individual and the rights of society to protect established social order is a familiar area of debate in political philosophy. There is not space to consider these debates here, nor would it be accurate to say that there is any useful consensus. The immediate conclusion urged is that the availability of robot lovers and caring technologies raises these political debates and should be discussed in the political area, rather than simply as technology. There is a need for more scientific research into these social effects. There is also a need for balanced general public debate on the moral question of whether or not such social effects are to be held more important than individual liberty.

Unfortunately, the lessons from previous technologies suggest an even more worrying possibility. One important reason why people would choose a robot lover is in order to be able to do things to it that would be unacceptable if done to humans.

At present, the main explicit interaction that nonexperts have with artificial intelligence (AI) is in computer games. Although it is a technology that has many successful applications, at present computer gaming would be how the majority of people encounter and discuss AI. In this application area, generally speaking, AI is used to provide more interesting and elusive targets for people to shoot at. In short, the main reason people seem to buy AI technology is to play at killing it. Since computer gaming is so commercially successful and has led us to accept extreme levels of simulated violence, we should anticipate extreme levels of violence toward robots.

The ethical implications of this are complex and controversial. Some discussion has been initiated elsewhere, for example in Whitby 2008. It is not clear that the arguments from liberty, such as Mill's, will justify the abuse of robots. There are several questions to be considered about the private abuse of robots. First, are people who do this sort of thing in simulation more or less likely to do it to humans in reality? The evidence is not clear. There has been much discussion and a certain amount of useful research on whether the use of violent computer games desensitizes users to violence in reality. The balance of evidence is at least worrying (see, e.g., Anderson and

Bushman 2001). Second, is there some sort of cathartic release through this sort of private activity, which might make people better behaved in human–human relationships? Third, what is the ethical role of designers of the technology? It is obviously possible to design robots or caring technology that responds positively to, and actively encourages, abuse at one extreme. At the other extreme, it is just as possible to design the technology to summon the authorities at the slightest hint of abusive behavior or to log every expletive or angry word issued by the user as possible evidence in a prosecution.

There is a distinct lack of guidance on these design questions in existing professional and legal codes. This needs to be remedied because designers with different views on the ethics of abuse may build very different systems for the mass market, with totally unpredictable ethical consequences.

15.5 Love

There are two distinct questions to be considered about robot love: "Can you love a robot?" and "Can the robot love you?" The second question generates a great deal of philosophical interest. This interest is most unfortunate for anyone concerned with the ethics of robot love. One might suspect that some readers will be pursuing this chapter hoping primarily for the expression of a position on this long-standing philosophical debate. Any such readers may well be disappointed. It is not necessary to answer this second question to progress the arguments of this chapter. Indeed the philosophical focus on this question is a serious and unfortunate distraction from the immediate ethical issues. In short, it does not matter whether or not the robot is *really* capable of loving someone. What matters is how humans behave.

Of course, how people behave depends partly on their beliefs about the technology. If people come to believe that their robot or caring system is *really* in love with them, then they will probably be a good deal more likely to describe themselves as loving it in return. For this reason, a convincing simulation of love is just as ethically dangerous as anything approaching the real thing. Even, perhaps especially, if the simulation is not particularly convincing, over-enthusiastic marketing by those who wish to sell such technology may deliberately set out to foster such false beliefs. This is not an area where we can trust the free market.

Despite Levy's optimism, at present there is no technology under development that would enable any artifact in itself to experience genuine love. There are, by contrast, a number of technologies—for example, those cited by Boden—which would enable it to perform a fairly adequate simulation of loving a human (2006, 1094–1095). In the private and intimate contexts under consideration, the word "adequate" will have much weaker requirements than it would in a double-blind scientific trial or in a Turing Test situation.

To be detained by the philosophical question of to what extent an effective simulation is *really* love, is to be misdirected from the immediate ethical issues: Should we permit the use of effective simulations of love? If so, under what circumstances and to what extent? There are no easy answers to these questions, but they are portentous.

What, then of the first question: "Can you love a robot?" Although there is not the same level of philosophical controversy, this question, too, needs a good deal of unpacking. One writer who gives a clear affirmative answer is Levy (2007, 105–112). Levy has no doubt that you can love a robot. Indeed, he predicts that progressive states will recognize marriage to robots by 2050 (155). The sort of love that Levy imagines occurring stems precisely from the familiarity, indispensability, and intimate association with the technology that we have been considering in this chapter. However, whether or not we are ready to call this phenomenon "love" is highly debatable. Most people would hear this use of the word "love" as metaphorical.

If however, a significant proportion of people eventually come to talk of loving their robots in a way that at least closely resembles the way in which we use the word in the case of personal human relationships, then it is reasonable to assume that the word "love" is undergoing a change of definition. Love is a concept that has been defined in widely differing ways over recorded history. The discussions in Plato's *Symposium* ([385–380 BCE] 1999, 9–50), though still celebrated in modern English in expressions such as "a platonic relationship," differ significantly from modern views on love.[5] Approaches to the definition of love for much of the period between Plato and the modern era center on the notion of "agape"—the Christian principle demanding love for all.

The concept of love implicit in Austen's tongue-in-cheek work misquoted at the beginning of this chapter proved very influential for the nineteenth-century view of love (Austen [1813] 2006). However, her insistence on the central importance of material wealth often seemed unacceptable, or at least highly unromantic, to twentieth-century audiences. It is worth briefly mentioning the importance of material wealth because it may be an important factor in deciding who can have a robot lover or a robot nanny or a smart home, and who cannot. Even if Levy's account is too simplistic, it is quite possible that the sort of technology under discussion in this chapter will cause a great deal of rethinking of the definition of love.

If the definition of love is undergoing, or about to undergo, yet another major change, why should we care? This is not purely an esoteric academic issue. The definition of love is central to our view of human relationships. Changes in the definition of love, caused by the widespread use of caring technology, are certainly possible. They may even represent an improvement in human happiness. What they are not is something that can be ignored or avoided. We might feel the need for caution about the introduction of technology that brings about such changes.

What is essential is that these decisions should be more widely debated. If there is the possibility of such large social impacts as changes in the definition of, and even, the nature of basic human relationships, then there should be informed public debate. It is not acceptable to leave such important decisions solely in the hands of unaccountable, and almost always anonymous, technologists and designers.

15.6 Robot Carers

The notion of a robot spouse may seem too far-fetched to deserve serious discussion by contemporary technologists. The notion of love with robots may also be seen as not of great interest to the designers of present technology. However, this is most certainly not true of the notion of care. Robots, or more accurately, a wide variety of automated systems, are already entering into the field of personal and intimate care. In this case, less seems to hang on the philosophical question of whether or not a caring technology *really* takes care of someone. It is sufficient to say that it performs a wide variety of tasks that would previously have been performed by a human acting in the role of carer. There are technological developments taking place now that fall under this heading.

Among such technological developments are so-called smart homes. These take the form of a fully automated apartment. Among the technologies used are CCTV (closed-circuit television), motion detectors, heat sensors, intelligent refrigerators that monitor their contents, and an AI system that monitors the activities of the occupant. Such technology is designed to at least partly fulfill the role of a human carer or a team of human care assistants. It is a technology that will be in large-scale use within the next few years.

Another technology, which is close to market, is that of so-called robot nannies. These are mobile robots intended to entertain and monitor infants. The potential dangers of the misuse of robot nannies have been extensively discussed elsewhere (Sharkey and Sharkey 2010; Whitby 2010), so only general remarks will be made here. What is important about both smart homes and robot nannies is that they are technologies that exemplify the problems discussed earlier, and are not remote or science-fiction possibilities. In the case of these technologies, the need for ethical codes that give guidance is immediate, if not already overdue.

It might seem, at first glance, that technologies such as robots and other intelligent systems, which have more human-like interactions with users, should generally be welcomed. Indeed, most researchers in the relevant areas unquestioningly assume that achieving a greater number of interactions and making them more human-like are desirable research goals. Similarly, the development of domestic robots and other caring technologies to care for the elderly and the very young seems, at first glance, a thoroughly laudable goal. However, as we have seen, there are a number of important ethical issues involved in such developments that require careful consideration.

There is clear scientific evidence that humans adapt to technology to a far greater extent than technology can adapt to humans. The way that this can happen with even very crude AI technology was demonstrated by Weizenbaum's ELIZA (1984, 188–189). Although this famous early AI program only gave the appearance of a conversation by outputting phrases in response to key words in the user input, it was on occasion taken seriously as a conversational partner. This response was unexpected by Weizenbaum, and caused him great concern.

More specific studies have indicated that this process of adaptation will be especially noticeable in cases where AI technology and robots are used in everyday and intimate settings, such as the care of children and the elderly. For example, Fogg and Tseng (1999, 80–87) claim that empirical studies have shown that humans give more credibility to computer products after they have failed to solve a problem for themselves or in situations where the human has a strong need for information. This is particularly likely to emerge in applications where robots are employed in intimate and caring roles. Smart homes and robot nannies are prime examples of such applications.

When technology is placed in an intimate setting—for example, caring for a human in a smart home—it is also likely that the tendency of humans to see their interactions with machines in anthropomorphic terms will be increased, as demonstrated by the extensive studies of Reeves and Nass (Reeves and Nass 1996). Because of this, the interaction designs of such systems need to be handled in an ethically sensitive manner.

Interaction designers have mixed feelings about anthropomorphism. Some view it as facilitating good interaction but, crucially for present purposes, others take the view that it is ethically dubious. For example, Ben Schneiderman describes the human portrayal of a computer as "morally offensive to me" (qtd. in Don et al. 1992, 69). It is not easy to rule on this debate. To assume that it is always beneficial to exploit human emotional and social instincts in designing interfaces is simplistic, but so is assuming that it is never beneficial. From what has been said earlier, it should be clear that it is not an issue that can be left solely in the hands of designers, however sensitive their methods. It is an ethical issue that needs to be resolved now.

A further set of ethical issues stems from the tendency of designers to unthinkingly force their view of what constitutes an appropriate interaction onto users. In the field of information technology (IT) in general there have been many problems caused by this tendency. Some writers (e.g., Norman 1999) argue that there is a systematic problem. Even if we do not grant the full force of Norman's arguments, there would seem to be cause for ethical worries about human–robot interactions in such intimate contexts. Largely unaccountable technical experts may well force their views (both explicit and implicit) of what is appropriate and inappropriate on vulnerable users via this technology. In other fields, such as law and politics, we might reasonably expect decisions with such impacts to be taken in a fully informed and accountable manner including open public debate.

This is despite clear warnings having been offered (e.g, in Picard 1998 and Whitby 1988) that there are potential hazards to be avoided. The principles of user-centered design–more usually cited than actually followed in current software development— are generally based on the notion of creating tools for the user. In the case of the technologies under discussion here, by contrast, the goal is the creation of companions, or carers, for the user. This requires comparatively far more attention to the ethical dimensions of the interaction. What is needed is both technically and ethically informed debate on these issues with the ultimate goal of being able to provide a code of conduct for designers. It is important to consider these ethical issues with an appropriate urgency.

15.7 Conclusion

This chapter has a question as a title. The fact that it has raised more questions than answers should not be too surprising, therefore. The exact codes of ethics appropriate for this area have yet to be fully formed. It would be easy to create some sort of moral panic about robot lovers and automated caring technologies. The problems outlined in this chapter might make some people feel that total prohibition is a valid approach. This would be a serious mistake. Building caring systems of all sorts has great potential benefits. Prohibition would, on balance, be morally wrong. What is morally right is building and employing such systems in an ethical manner.

Similarly, work aimed at improving human–robot interaction in intimate contexts should not be outlawed or heavily restricted. However, despite the tremendous usefulness of this sort of technology, failure to address the various ethical issues entailed would bring serious dangers. Among these are the unintended consequences of limiting human freedom and dignity. This will be particularly the case with respect to vulnerable users—for example, very young infants cared for by robot nannies and old people with declining cognitive capacities cared for by a smart home.

To build and use such technology, in an ethical manner, requires a deliberate attempt to avoid forcing on to vulnerable users the designers' views and prejudices as to what is appropriate behavior. When building caring systems for especially vulnerable humans, sensitivity to their dignity and, in most cases, their autonomy is essential. The code of good practice of BCS, The Chartered Institute for IT, and the code of ethics of the Association for Computing Machinery do not provide specific guidance on the issues discussed in this chapter. This is not a criticism of these codes since they were designed for an era in which the typical user of computer technology was a businessman. Caring technologies move the goal posts of such codes.

It would be possible to rework these professional codes to cover many of the problems raised in this chapter. Among other things, the revised codes would need to safeguard human dignity—something the IT industry has not had to worry about much until now.

Should we let you have a robot lover? This is probably a question that will divide public opinion. Some people will defend Mill's liberal thesis that it is an entirely private matter. Others may see the very possibility as unforgivably perverse or as blasphemous. The debate should be started now.

We need to avoid a headlong rush into adopting technology driven only by uncaring commercial imperatives. It is worth remarking that there is a good deal less profit in persuading people to care personally for their elderly relatives than there is in selling smart homes. In blunt terms: if everybody chose a human lover, the market for robot lovers would be very small. The market for robot lovers and other caring technologies is maximized in the situation where nobody chooses human companionship.

We need professional codes, guidelines, and possibly, eventually, legislation to direct this technology in an ethical direction. We need designers and technologists who have appropriate ethical values and conduct their work in an ethical manner. But, above all, we need informed public discussion. To wait until these technologies are in widespread general use would be a serious mistake.

Notes

1. This is a slight misquotation of the opening sentence of *Pride and Prejudice* by Jane Austen, first published in 1813 (Austen [1813] 2006). Just as Austen sought to poke fun at the cultural assumptions of her time, so today it remains necessary to challenge the contemporary cultural assumptions behind the desirability of robots in caring and companionship roles.

2. Paul VI (1968, par. 13).

3. The Block IV Phase II Tomahawk Land Attack Missile produced by Raytheon has enhanced capabilities, including being able to locate and pursue a moving target.

4. The robots in Čapek's play are more like what we would now call androids or clones in the sense that they are biological, rather than mechanical.

5. The accounts of love given in Plato's *Symposium* cover a wide range. For the present discussion we should note that many accounts regard homosexual love as a higher form of love than heterosexual love and at least one, that of Pausanias, sees no possibility of reciprocity in love between a man and a woman—presumably because women are held to be incapable of rationality (Plato [385–380 BCE] 1999, 13–17).

References

Anderson, C. A., and B. J. Bushman. 2001. Effects of violent video games on aggressive behavior, aggressive cognition, aggressive affect, psychological arousal, and prosocial behavior: A meta-analytic review of the scientific literature. *Psychological Science* 12 (5): 353–359.

Austen, J. [1813] 2006. *Pride and Prejudice*. London: Headline Review.

Boden, M. A. 2006. *Mind as Machine: A History of Cognitive Science*. Oxford, UK: Oxford University Press.

Čapek, K. [1921] 2004. *Rossum's Universal Robots (R.U.R.)*. London: Penguin.

Don, A., S. Brennan, B. Laurel, and B. Shneiderman. 1992. Anthropomorphism: From ELIZA to *Terminator 2*. In *CHI 92, Proceedings of the SIGCHI Conference on Human Factors in Computing Systems*, ed. P. Bauersfield, J. Bennet, and G. Lynch, 67–70. New York: ACM.

Fisher, M. 1990. *Personal Love*. London: Duckworth.

Fogg, B. J., and H. Tseng. 1999. The elements of computer credibility. In *CHI 99, Proceedings of the SIGCHI Conference on Human factors in Computing Systems*, 80–87. New York: ACM.

Kranzberg, M. 1986. Technology and history: Kranzberg's laws. *Technology and Culture* 27 (3): 544–560.

Levy, D. 2007. *Love and Sex with Robots: The Evolution of Human-Robot Relations*. New York: HarperCollins.

Mill, J. S. [1859] 1966. On liberty. In *John Stuart Mill: A Selection of His Works*, ed. J. M. Robson. Toronto: Macmillan.

Mori, M. 1970. Bukimi no tani: The uncanny valley, trans. K. F. MacDorman and T. Minato. *Energy* 7 (4): 33–35.

Norman, D. 1999. *The Invisible Computer*. Cambridge, MA: MIT Press.

Paul VI. 1968. Encyclical Letter Humanae Vitae, par. 13.

Picard, R. 1998. *Affective Computing*. Cambridge, MA: MIT Press.

Plato. [385–380 BCE] 1999. *Symposium*, trans. C. Gill. London: Penguin Classics.

Reeves, B., and C. Nass. 1996. *The Media Equation*. Cambridge, UK: Cambridge University Press.

Sharkey, N., and Sharkey, A. 2010. The crying shame of robot nannies: An ethical appraisal. *Interaction Studies: Social Behaviour and Communication in Biological and Artificial Systems* 11 (2): 161–190.

Weizenbaum, J. 1984. *Computer Power and Human Reasoning*. Harmondsworth, UK: Pelican.

Whitby, B. 1988. *Artificial Intelligence: A Handbook of Professionalism.* Chichester, UK: Ellis Horwood.

Whitby, B. 2008. Sometimes it's hard to be a robot: A call for action on the ethics of abusing artificial agents. *Interacting with Computers, Special Issue: On the Abuse and Misuse of Social* 20 (3): 326–333.

Whitby, B. 2010. Oversold, unregulated, and unethical: Why we need to respond to robot nannies. *Interaction Studies: Social Behaviour and Communication in Biological and Artificial Systems* 11 (2): 290–294.

VI Medicine and Care

While the robots of part V provide intimate relationships, we will now look at robots that today provide more serious interaction: companionship and medical care, such as to the elderly, persons with disabilities, and children. Indeed, this is a major potential application for robotics and is receiving extensive attention and funding internationally, particularly in South Korea, Japan, and several European countries, although there is less funding for such projects in the United States.

Clearly, robots can provide round-the-clock care and increased safety. However, there are a number of risks and ethical issues associated with such applications for robots, and several of these are discussed in part VI. Two of the following chapters are concerned with robot caregivers that either exist now or can be expected to become available within the next decade. The third chapter looks at the ethical implications of human–robot relationships by imagining the existence of machines that can create "artificial people," either from organic or inorganic elements.

In chapter 16 Jason Borenstein and Yvette Pearson examine ethical issues associated with using robot caregivers. They suggest that continued interaction with robots is likely to change both human-to-human behavior as well as human–robot interactions. The authors also consider the ethical implications of situations in which the recipients of care might prefer their robot caregivers to human ones. These behavioral and psychological changes might be influenced by such factors as the robot's appearance, its degree of autonomy, and its ability to express emotions.

Noel Sharkey and Amanda Sharkey discuss issues of privacy, safety, and personal liberty associated with robot caregivers of the very young and the elderly. They ask in chapter 17 whether invasion of privacy should extend to robots as well as to human caregivers. Other questions they raise include limits on permitted robot behavior to ensure the safety of the people cared for and the reduced human contact experienced by children who are left with only robotic supervision.

Steve Petersen discusses in chapter 18 the ethical issues in a more speculative way, in contrast to the two preceding chapters. Distinguishing between "humans" and "people," he postulates that a "Person-o-Matic" machine, as he calls it, could be programmed to produce artificial people (or APs), but not humans. The APs could be manufactured from metal, plastic, electronics, and so on, or they could be synthesized from artificial DNA. He then discusses provocative ethical issues about the possible servitude of APs, as well as the possibility of programming them so that their major goal is to make humans happy. This chapter segues into part VII, which focuses on robot rights.

16 Robot Caregivers: Ethical Issues across the Human Lifespan

Jason Borenstein and Yvette Pearson

One of the distinct challenges associated with designing robots is which, if any, ethical theory should be incorporated into their programming. Yet instead of focusing on how to integrate a particular ethical theory into robots, another strategy for developing an ethically sound technology is to focus on whether a technological intervention is likely to advance or hinder human flourishing. In making design decisions, scientists and engineers should consider not only the technical dimensions and potential uses of the technology, but also the ethical implications of introducing a novel use of technology into a specific context.

In this case, the primary concern is about how the existence of robots may positively or negatively affect the lives of care recipients. Because incorporating robots into our lives may be motivated by the drive for efficiency in terms of time and resources, it is imperative to make a concerted effort to focus on the promotion and maintenance of central human capabilities as a primary goal of robot caregiver intervention. If the use of robot caregivers is also efficient and convenient for professional and "informal" human caregivers, those are acceptable side effects, but having them as the sole or main impetus for using robot caregivers is likely to produce undesirable ethical and social outcomes. Drawing from the capabilities approach, this chapter examines key ethical considerations that may help to determine whether the use of robotic caregivers is consistent with the promotion of human flourishing at different life stages.

Though care provided across the lifespan may have some common features, for example, the ability to monitor care recipients, what constitutes even basic care will vary from one life stage to the next. For example, the fact that infants and toddlers are just beginning their cognitive, physical, and social development means that caregivers should ensure that their actions do not interfere with or delay this development. This requires, among other things, that caregivers facilitate a child's ability to play, with play understood as pleasurable activity that allows for the "intermingling of emotional, intellectual, social, and physical development" (Lane and Mistrett 2008, 413). Although adults should have opportunities to engage in play as well, the purpose served by them engaging in play is not the same as it is with young children. Assuming an adult has already developed certain skills, including ways of communicating with caregivers and interfacing with his or her surrounding environment, fostering the actualization of these capacities is not as pertinent. Instead, robot caregivers should be designed to respond to care recipients' attempts to communicate their needs and to detect whether certain interventions, for example, a reminder to take medication, are necessary. In short, some types of functions are going to be more relevant than others depending on the care recipient's life stage and abilities.

In order to develop and deploy the technology ethically, it is necessary to consider the various facets of life that may be altered by the intervention of robot caregivers. This chapter will explore the likely effects of robot caregiver intervention on human–human interaction as well as the ethics of human–robot interaction (HRI). Included is an evaluation of the concern that the intervention of robot caregivers will lead to a reduction in human contact or increase isolation for members of society that tend to be marginalized as a result of their impairments. Another issue that will be examined is how care recipients might react to robot caregivers, including the possibility that a preference might arise for them over their human counterparts.

Whether robot caregivers will function as "extensions of us"[1] or outright replacements—for example, because care recipients grow to trust robot caregivers more than fellow humans—presents multiple concerns about expectations. In addition to the potential impact on relationships among humans, there are ethical issues related to the effects of robots on care recipients and caregivers alike. While it is already clear that HRI can lead to some degree of emotional attachment on the part of humans toward robots (Singer 2009, 337–338), determining whether such one-way bonding presents a unique problem requires further evaluation. Emotional attachments emerging from HRI can, for example, raise questions about the role of deception as well as the potential for overdependence on robot caregivers.

16.1 Design Strategies

In order to be effective caregivers and for the technology to operate in an ethically responsible manner, numerous design issues must be addressed. Among them is how much autonomy should be granted to a robot. According to Breazeal, "The amount of robot autonomy varies (and hence the cognitive load placed on the human operator) from complete teleoperation, to a highly self-sufficient system that need only be supervised at the task level" (2003, 1). Rather than alluding to an abstract, philosophical notion of autonomy, what the robotics community typically is referring to is whether a human being would significantly be "in the loop" while the robot operates. For example, creating a fully autonomous robot may be difficult because of the technical complexities involved with having it navigate through environments that contain so many variables (Kemp, Edsinger, and Torres-Jara 2007). The extent to which a human is kept "in the loop" should be guided by whether this design pathway promotes or hinders the ability of a robot and human caregivers to meet the needs of their proxies while also preserving the health and well-being of human caregivers. Though ensuring the needs of those who cannot adequately care for themselves is essential, it is also important to help human caregivers avoid becoming significantly impaired as a consequence of providing care to others.

A second issue is the robot's appearance. Riek and colleagues suggest that humans are more likely to bond with a robot if it has a high degree of "human-likeness" (Riek et al. 2009). That said, exploring the relevant sense of "human-likeness" is important. The evaluations are mixed regarding whether it is preferable for a caregiver robot to be more human-like in appearance. On the one hand, the "Uncanny Valley" hypothesis proposes that there is a certain threshold beyond which the human-like appearance of a robot repels rather than attracts human beings (Mori 1970). On the other hand, the presence of a "human-like" robot could alter the emotions or behavior of human beings. For example, Woods and colleagues note that children's attitudes toward robots vary depending on the robot's physical features and suggest that children are much more likely to attribute emotions to a robot if it looks "human-like" (Woods, Dautenhahn, and Schulz 2004). In order to elicit preferred responses from human beings, the researchers claim that "it is important for robot designers to consider a *combination* of physical characteristics, rather than focusing specifically on certain features in isolation" (51).

Yet, as we have learned from philosophical debates about the necessary and sufficient conditions for personhood, mere physical appearance may not be the most crucial factor when determining whether a being is human-like in some relevant sense. Instead, a robot's movements, possibly conveying that it has a "personality," may be more relevant than physical resemblance to a human being. Hence, even if designers adequately control the appearance of the robot so that a balance is struck between repelling human beings and manipulating them, designers may have less control over the emergence of certain "quirks" that are interpreted by humans as indicative of robots possessing traits characteristic of persons.

A related design issue is whether, and to what extent, robot caregivers should be equipped to respond to, express, or elicit emotions. Because developments in AI are not yet advanced enough to seriously posit the creation, at least in the near future, of robots that are capable of genuinely experiencing emotion (assuming that such an accomplishment is even possible), the focus of this discussion will be limited to robots that can respond to human emotions without experiencing real emotions themselves. Robots could be designed to function so that they respond to human behaviors, including human emotions, in a way that does not lead to confusion about whether the robot understands or empathizes with a person in a complex fashion. In fact, the challenge at this stage is making robots seem convincing. That said, there could be circumstances in which a robot seems to approximate human emotion—e.g., robots able to make certain facial expressions or physical gestures that convey human-like emotions. For example, roboticist Hiroshi Ishiguro recently demonstrated that subtle changes in a robot's functioning can alter whether we view it as more or less like ourselves (Barras 2009). Because experiencing certain types of emotions (e.g., joy) may produce health benefits and others (e.g., fear, stress) may contribute to poor health, it is desirable for robots to elicit positive emotions and, as much as possible, avoid producing negative ones, such as the unease associated with the Uncanny Valley experience.

Though developments in artificial intelligence (AI) will facilitate the creation of more sophisticated robots,[2] in the near term a robot's "personality" will be primarily a byproduct of a person's anthropomorphization of the robot's appearance and actions. This may well be an advantage in that it would allow a person to impose or project character traits onto the robot. For example, Krach and colleagues conducted an experiment in which human participants played an electronic game with different partners, including a robot that had a human-like appearance. According to the researchers, "Participants indicated having experienced linearly increasing fun in the interaction the more the respective partner exhibited human-like features. . . . Similarly, game partners were attributed increasing intelligence the more they appeared human-like" (Krach et al. 2008, 5). While concerns about deception—particularly self-deception—in the context of HRI persist, the projection of traits onto a robot may be more comforting in some cases than designing a robot to exhibit personality traits that the care recipient may dislike.

16.2   Care and the Capabilities Approach

A tool that could be included in the toolbox of scientists, engineers, and others while designing robotic caregivers is the capabilities approach. The capabilities approach is not a complete ethical framework, and its advocates, including Amartya Sen (1993) and Martha Nussbaum (2006, 139) probably do not intend it to be. Because the capabilities approach is "consistent and combinable with several different substantive theories" (Sen 1993, 48), it provides designers with an expanded framework within which to develop robot caregivers so that their use can be geared toward the promotion and preservation of human flourishing.

Certain technological interventions expand people's opportunities by improving their ability to interface with their environment and helping them build or maintain relationships with others. The resultant ability to engage in a broader array of activities than would have been available without the technological intervention can advance human flourishing. Oosterlaken's phrase "technology as capability expansion" recognizes the crucial role that engineering and other intellectual endeavors have in generating new opportunities for human beings (2009, 94–95). According to Oosterlaken, "If technologies are value-laden and design features are relevant, we should . . . design these technologies in such a way that they incorporate our moral values" (95). Scholars offer numerous visions of the types of capabilities that may be universal to all human beings. Among them is Nussbaum's list of "central human capabilities," which includes bodily integrity, health, and control over one's environment, as essential capabilities for human flourishing (2000, 70–77).

In order to determine whether robots will help to promote human flourishing, a necessary step is to clarify what "care" might entail in this context. For example, Faucounau and colleagues describe five main categories of care that a robot might provide: "Cognitive prosthesis," "Safeguarding," "Social interaction," "Support with regard to symptoms of cognitive impairments," and "Emergency assistance" (Faucounau et al. 2009, 35). Assessing whether robots can effectively fulfill any of these roles is, in part, a technical issue (i.e., whether advances in artificial intelligence and other related fields will move forward sufficiently enough). But perhaps, more importantly, it is a product of whether human wants and needs are adequately met and flourishing actually occurs.

Coeckelbergh delves into this realm by articulating the differences among "shallow," "deep," and "good" care (2010, 182–186). He characterizes "deep" care as rooted in reciprocity of feelings between the caregiver and care recipient and distinguishes this from both "shallow" care and "good" care. For Coeckelbergh, "shallow" care refers to routine care that lacks the "emotional, intimate, and personal engagement" (183), while "good" care is characterized as "care that respects human dignity" (185). As Coeckelbergh acknowledges, the current state of AI is such that robots are probably unable to provide "deep" care; however, this need not preclude robot intervention from facilitating a human caregiver's ability to provide "deep" care or from contributing to "good" care. Given the conclusions of recent explorations of HRI (e.g., Neven 2010), it seems that at least some of the social and emotional needs of care recipients can be met, even if the robots themselves remain incapable of experiencing emotions. Assuming that it is not inherently undignified to be cared for by a robot, the absence of "deep" care does not entail the absence of "good" care.

16.3   Developmental Issues

The needs of care recipients are not necessarily the same for each person or at each life stage. As Nussbaum states, "care is not a single thing" (2006, 168). Care is a complex set of activities that promote human capabilities in different ways. For instance, caring for the elderly can present challenges distinct from those associated with caring for young children. Along these lines, Nussbaum describes elderly persons with mental, physical, or social impairments, similar to those present in some children and young adults, but asserts that the former group can be much more difficult to care for because they tend to be "more angry, defensive, and embittered" (101). Moreover, owing to disparities in the type and magnitude of impairments even within a particular category, the care requirements for these individuals will not be the same.

Examining the complexities associated with providing care at each life stage is crucial because, for example, allowing robots to care for children raises ethical issues that may not emerge if the context is limited to elder care. If a major portion of a child's care is delegated to a robot, will the child learn to play normally? Speaking more generally, Nussbaum emphasizes the significance of play as a central human capability (2006, 400). If a child's environment is not conducive to the actualization of that capability, then the child might fail to flourish. This issue is not entirely new as evidenced by ongoing concerns about the effects of mediated interaction (e.g., spending time online instead of engaging directly with peers) on a child's social development. Simply stated, it is unclear whether the mediated interaction facilitates or inhibits healthy socialization.

Narvaez points out the significance of nurturing in a child's life and how it can facilitate moral development (2008). The potential use of robot caregivers raises age-old, fundamental questions about the kind of caring environment needed to enhance a child's moral development. In important respects, a child's caregiver needs to be nurturer and educator, which implies that "care" is not exactly the same for the young and for the elderly. Theoretically, the development of social intelligence and skills might be stunted if the child has limited contact with other humans, but this can depend on the precise nature of the HRI. HRI has the potential to enhance these abilities or, perhaps, to foster the development of unique ways of interacting with humans or robots. Though further evidence is needed, researchers are starting, for example, to accumulate data indicating that robots can help autistic children (Scassellati 2007; Robins et al. 2005). But at least for the foreseeable future, it is crucial to emphasize that no matter what benefits the technology is perceived to have, a robot should be viewed as a complement to human caregivers, and not as a replacement for them.

16.4 How Humans Might Change

What's weird is how biological entities change their behavior when in the company of robots. When robots start interacting with us, we'll probably show as much resistance to their influence as we have to iPods, cell phones, and TV.
—Shaw-Garlock (2009, 253)

Keeping in mind the value of human flourishing, it is vital to examine which kinds of character traits will potentially emerge or disappear given our growing interaction with robots. Whether the integration of robots into our lives will result in changes that differ significantly from those brought about by other technology remains unknown. For instance, it is difficult to predict whether increased HRI will result in fundamental changes in human behavior or their interactions with one another, or whether the changes will be of a more superficial nature. In any case, a primary goal is to ensure that those changes contribute positively to human welfare rather than precipitating the loss of highly valued human characteristics and skills. Some scholars think serious problems are likely to arise, while others believe the fear is overstated (Lin, Bekey, and Abney 2008, 83). In the context of military robots, Singer (2009) points out that the use of robots and other unmanned vehicles is changing how we wage war, including that pilots are disappearing. Further, General James Mattis expressed some unease with the extent of robot involvement, specifically a "robot-only presence," since it may compromise a core characteristic of warriors—honor (Brown 2010). On the civilian side of things, robots are becoming a more tangible part of our lives and a broad range of effects may be forthcoming. Though it may be a somewhat trivial phenomenon, participants in a study of the Roomba, a robotic vacuum, believed that they had become "cleaner" or "neater" after owning the device (Sung et al. 2007, 150). Briefly put, the traits that technology elicits are not always straightforward or anticipated, but it is important to identify and analyze probable transformations so that undesirable consequences can be averted.

Considering that their use is not yet widespread, speculation abounds regarding how "the humans" might change as robot caregivers become more common. In some sense, a robot that is viewed as being "kind" to people could bring out laudable traits in us similar to the way pets can. For example, Turkle (2006) describes how humans seem to have a drive to nurture computerized objects, even some relatively simple ones created in the 1980s. Yet which character traits will be promoted or hindered by the incorporation of robots in care settings is largely an open question. Will care recipients express less anger, frustration, hostility, and so forth, because robot caregivers make them feel less dependent and isolated? Or is the opposite more likely to be true because they feel abandoned by friends and family? Further, as mentioned previously, will a young child be ill equipped for human social interaction if his or her primary caregiver is a robot?

Moreover, will adding a robot to the mix significantly alter the dynamics of the relationship between caregivers and care recipients? A relationship with a care recipient can evoke a multitude of attitudes and behaviors. At times, deplorable traits can emerge. In fact, individuals suffering from debilitating illnesses such as dementia are sometimes mistreated by family members (Cooper et al. 2009). Conceivably, traits such as kindness and patience will emerge more frequently if human caregivers are given more of a choice about whether to provide care and under what conditions. Caregivers might experience some relief if an automated assistant is there to help, especially if it can be trusted to be more reliable and consistent than another human.[3] A key dimension underlying these issues is the function(s) robot caregivers are expected to fulfill. For example, if a robot is supposed to be a friend or companion to a human being, which Shaw-Garlock calls an "affective" robot (2009, 250), then one might assume a broad range of behavioral changes would follow. Instead, if a robot is merely to be used in a similar manner to a tool or instrument, which Shaw-Garlock calls a "utilitarian" robot (250), will the same types of changes occur? Intuitively, we might be tempted to say that there would be sharp differences between our responses to each kind of robot. Yet, humans have a profound ability to bond with "utilitarian" items such as cars, motorcycles, and boats. Along these lines, Shaw-Garlock found the tendency to anthropomorphize objects, including by the people who design them, to be consistent with Nass and colleagues' (Nass et al. 1997) finding that "individuals engage in social behavior toward technologies even when such behavior is entirely inconsistent with their beliefs about the machines" (Shaw-Garlock 2009, 254).

16.5 Human Psychology and Automation

A society's norms and values, and how they influence perceptions of robots, can play a key role in determining to what degree the technology is used. For instance, Sofge (2010) discusses a common theme in American science fiction: the creation of robots leads to a dystopian future. However, MacDorman, Vasudevan, and Ho note that robots are often portrayed as being heroic in Japanese comics and movies (2009, 489–491). Yet popular depictions of robots should not be taken as accurate predictors of the respective level of acceptance robots will achieve. Considering that Americans tend to be technophiles, it is debatable whether our collective consciousness contains a deep-seated fear of robots. Moreover, when looking at Americans' tendency to establish emotional attachments to things like Roomba and the Packbot "Scooby Doo," a gap seems present between attitudes depicted in hypothetical scenarios—for example, in Hollywood films—and actual experiences with robots.

Marketing practices can also influence the public's level of willingness to accept robots with assistive abilities into their homes. For example, in a study by Neven, most of the participants in the laboratory and field tests using the robot iRo thought that it would be good for individuals who were "housebound, old, lonely, feeble, and in need of care and attention," but they were reluctant to equate themselves with such persons (2010, 341). While the participants found iRo entertaining, had attachments to it, and acknowledged that it would be very helpful for others, the image associated with the target market for the robot led most of the participants to say that iRo "was not a robot for them" (Neven 2010). Yet in some ways, the participants' responses were inconsistent with the reported experience documented in Neven's study, insofar as the participants admitted talking to and developing an emotional attachment with iRo (2010, 340).

The manner in which automation can affect human psychology is difficult to predict. A troubling potential impact of these complex interactions is becoming over-confident in an automated system's ability, a problem that has already occurred to some degree with the APACHE system, a computerized diagnostic tool for hospitals (Wallach and Allen 2009, 40–41) and GPS (Sorrell 2008). Analogously, will caregivers place too much trust in robots if, for example, their child or elderly parent seems to be in good hands?

Since overconfidence in robots is likely to be a significant problem, adequate safe-guards in their design must be put in place to prevent them from harming humans. At a minimum, it is important to be cognizant of a relevant difference between robots and other electronic devices, which is the third step in the "sense–think–act paradigm" (Singer 2009, 67). Tools like APACHE and GPS still require that a human undertake the last step, and this is at least one part of the process where interpretation by a human user remains. But a robot can be programmed to act without significant input from the user. Whether this is a better or worse design pathway is debatable; in some cases, it might be, and in others, it might not. Consider, for example, that computers are less likely to make certain types of mathematical errors than humans. In this cir-cumstance, our silicone-based counterparts are more reliable than we are. That said, one should not dismiss the fact that output is contingent on input decisions and design decisions about which information is relevant and how that information should be processed. In theory, a well-designed robot could conceivably cause fewer problems for humans than a system that requires frequent user input.

Without antecedently encouraging people to place too much trust in robots, it is prudent to anticipate that in practice humans are likely to do so anyway. Consequently, this places a heavy burden on designers to predict the dynamics of sociotechnical contexts within which a robot will be placed. It is preferable to err on the side of building in an extra "factor of safety" and designing robots well enough that overreli-ance on them will result in the least amount of harm possible. Humans cannot be trusted to act as they ideally should (e.g., acting sensibly instead of following a GPS's directions and driving into a lake). To be safe, designers should make sure that robots "have our back" when we either act incorrectly or fail to act altogether. For instance, if a robot is taking care of a child and the child's parents have not checked in after a certain amount of time, a reasonable design feature could include supplying the parents with multiple reminders or taking measures that help ensure the child's safety, such as contacting a backup caregiver.

A related potential complication is that a robot designed for certain types of users (e.g., adults) or for use in certain contexts (e.g., nursing homes) might be utilized by an expanded user pool (e.g., children) or in an alternate context (e.g., at home) that may generate variable outcomes, some of which might be quite undesirable. For example, if a robot caregiver is designed for use by someone who has undergone a basic level of training, or for use in an environment that permits regular updates and maintenance, then expecting that robot caregiver to function outside of these parameters could lead to injury of its charge(s). The extent of damage, if any were to materialize, would partially hinge on how widespread "off-label" uses of caregiver robots become.

16.6   Relying on the Technological Fix to Remedy Social Problems

Weinberg goes on to claim that since our efforts to encourage behavioral change are often futile, we seek out a technological fix. For example, instead of counting on people to be disciplined and use less water, devices such as low-flow showers and toilets are installed. Similarly, technology might be relied on to remedy problems of neglect in care environments like nursing homes rather than hold out hope that an improvement in human behavior is on the horizon. Whether, how, and to what extent technological interventions are used is, at this stage, a function of human choices.

Conceivably, applying a technological fix to grapple with challenges related to caregiving could be problematic. The issues that robot caregiver intervention might address include: human caregivers' fatigue and stress that can lead to neglect or abuse of their charges; loneliness of marginalized individuals; and limitations on the ability of people with certain types of impairment to interact with others. Though robot caregivers have the potential to remedy these problems to some degree, their intervention could exacerbate rather than ameliorate some problems with caregiver-care recipient relationships. For example, according to Sparrow and Sparrow, "it is naïve to think that the development of robots to take over tasks currently performed by humans in caring roles would not lead to a reduction of human contact for those people being cared for" (2006, 152). Parents often rely on technology such as television programs and electronic game systems to serve as "caretakers" for their children. On the other hand, some parents use technology to communicate with their children more rather than less frequently, thereby increasing their involvement in their children's daily lives. Yet just as a lack of involvement in the lives of relatives and friends is worrisome, excessive involvement may also be a problem, albeit of another sort. For example, it can impede the ability of children and young adults to become independent individuals.

Critics fear, perhaps justifiably, that caregivers might become less attuned to the specific needs of care recipients because a technological crutch is available. While robots may be able to ease some of the burden from the caregiver's shoulders, a counterbalancing problem is that other life activities may increasingly fill up the caregiver's "free" time. For example, the existence of the Internet, televisions, and game systems, in some sense, gives parents the leeway to direct their time and attention away from their children. According to the Kaiser Family Foundation (2010), American youths spend roughly 7.5 hours per day accessing entertainment media. While most cases are not so extreme, some parents have been so absorbed in playing electronic games themselves that they have been derelict in their responsibility to their children.[4] Along related lines, technology could be viewed as granting us tacit permission to live a greater distance away from impaired friends and relatives and to visit less frequently, and thus potentially withering our capacity for caregiving.

That said, it is also possible that the removal of some burdensome aspects of caregiving might lessen existing tendencies to detach oneself from those in need of care. The intervention of robot caregivers could improve family unity and other interpersonal relationships because they would not be tainted by our aversion to unpleasant tasks. Individuals will have the freedom to become more attuned to nonclinical, emotional, or social needs beyond the "basic necessities" that are often reduced to almost purely mechanical intervention by overtaxed human caregivers.

Whether we use technology to *mediate* human relationships or communication rather than replace human interaction is not a foregone conclusion; instead, just as parents can choose against using the television as an "electronic nanny," we can choose against using emerging technologies in ways that are likely to impede human flourishing. It should be kept in mind that the introduction of technology need not alter human interaction for the worse. As Johnstone astutely recognizes: "The functionings we can achieve with technology are thus not necessarily the same, either quantitatively or qualitatively, as the functionings we can achieve without technology. What a capability perspective insists upon, however, is that in either case what matters is the degree to which people's ability to determine and realize lives that they value is expanded" (2007, 79).

Virtual worlds, such as Second Life, and online social networking sites have expanded connections for those who may have become isolated due to severe restrictions on their mobility. While some problems at nursing homes are best remedied by increased human contact, other problems might not be. Moreover, meaningful human interaction need not involve physical contact or even physical presence of the individual. And this is nothing new, even for those who are now elderly. Many people undoubtedly communicate with dear friends and loved ones via letters sent through the postal service or a telephone. Though this sort of mediated interaction is different from physical contact with individuals, it can still be immensely valuable to the individual who receives the letter or the phone call. It is difficult to imagine that certain types of contact, such as turning a person over in her bed or talking to her as you fill her water pitcher, would be perceived as more meaningful than a kind letter or phone call from a loved one. This is not to suggest that physical contact is unimportant; instead, the point is that critics of technological intervention might fail to see how it can expand people's means of communicating with loved ones in ways that maintain a distinctively human element. Granted, a visit to a care recipient's room in, say, an assisted living facility, to attend to the most basic needs, such as cleaning the space, delivering food or medication, and so on, may be significant to the individual deprived of additional contact with other people. Yet, a society that finds this acceptable should reevaluate its tolerance of this minimal level of human interaction for elderly or impaired persons rather than objecting to the intervention of robot caregivers, because their intervention could eliminate this unacceptably minimal interpersonal contact.

Whether a human being will still meaningfully be "in the loop" as robot caregivers emerge and become more pervasive is an overarching concern.[5] For instance, will a person still check on an elderly resident in a nursing home or monitor a robot's performance? Robots could work in conjunction with human caregivers (Decker 2008, 322). But Sparrow and Sparrow suspect that this practice will not continue over time (2006, 150). In different care contexts, such as nursing homes, assisted living facilities, and home health care, the details of a robot caregiver's use will vary.

As a general statement, it is probably unwise to allow a robot to act alone, even if their design continues to improve and gain increased sophistication. Entirely taking over or removing human participation is likely to be problematic, and in some contexts impossible. Yet, at a minimum, robots could manage interactions that caregivers might think are burdensome and recipients view as embarrassing or frustrating. Both parties could then be free of certain "uncomfortable" interactions, hopefully freeing them to interact "normally" with each other.

To reiterate, robots should not replace all types of human interaction; instead, the hope is that technological intervention would positively change human interaction in a way that expands opportunities for human flourishing. Along these lines, Tamura and colleagues maintain that the introduction of robots could "compensate for the shortage of caregivers and helpers" (Tamura et al. 2004, 85). Speaking more generally, Hayes (2009) contends that the increased use of machines does not necessarily amount to replacing humans; in fact, he argues that percentage of the population in the workforce has gone up even as automation has become more common. Of course, the broader effects of automation must be kept in mind since, for example, it typically enables employers to downsize and to replace certain classes of workers with others.

16.7 Conclusion

The aim of this chapter is to highlight key ethical considerations relating to the use of robotic caregivers at different life stages. Though the drive for efficiency is difficult to resist, it should not be the penultimate motive behind the creation and use of the technology. Instead, robot caregivers should function in ways consistent with the goal of human flourishing. Scientists, engineers, and others are now making choices about design pathways that will meaningfully influence the future of human caregivers and care recipients alike.

Notes

1. A phrase used in Gutkind 2006, 32.

2. Arguably, robots and other artificial entities are getting close to passing the Turing Test; see, for example, Barras 2009.

3. On a related note, the family of an elderly person might have reservations about leaving their relative with a human "stranger" because of trustworthiness concerns.

4. For example, a couple neglected to feed their baby because they were busy playing online games; see Graff 2010.

5. Scholars have raised a similar issue about whether keeping a person "in the loop" is necessary for military robots.

References

Barras, Colin. 2009. Tests that show machines closing in on human abilities. *New Scientist*, January 22. <http://www.newscientist.com/article/dn16461-tests-that-show-machines-closing-in-on-human-abilities.html> (accessed March 22, 2011).

Breazeal, Cynthia. 2003. Social interactions in HRI: The robot view. *IEEE Transactions in Systems, Man, and Cybernetics, Part C* 34 (2): 181–186.

Brown, Alan S. 2010. The drone warriors. *ME Magazine*, January. <http://memagazine.asme.org/Articles/2010/January/Drone_Warrior.cfm> (accessed March 22, 2011).

Coeckelbergh, Mark. 2010. Health care, capabilities, and AI assistive technologies. *Ethical Theory and Moral Practice* 13 (2): 181–190.

Cooper, Claudia, Amber Selwood, Martin Blanchard, Zuzana Walker, Robert Blizard, and Gill Livingston. 2009. Abuse of people with dementia by family carers: Representative cross sectional survey. *British Medical Journal* 338: b155.

Decker, Michael. 2008. Caregiving robots and ethical reflection: The perspective of interdisciplinary technology assessment. *AI & Society* 22: 315–330.

Faucounau, V., Y. H. Wu, M. Boulay, M. Maestrutti, and A. S. Rigaud. 2009. Caregivers' requirements for in-home robotic agent for supporting community-living elderly subjects with cognitive impairment. *Technology and Health Care* 17 (1): 33–40.

Graff, Amy. 2010. Couple starves their own baby while nurturing virtual kid. *SFGate.com: Mommy Files*, March 8. <http://www.sfgate.com/cgi-bin/blogs/sfmoms/detail?entry_id=58670> (accessed November 26, 2010).

Gutkind, Lee. 2006. *Almost Human: Making Robots Think*. New York: W. W. Norton.

Hayes, Brian. 2009. Automation on the job. *American Scientist* 97 (1): 10-14.

Johnstone, Justine. 2007. Technology as empowerment: A capability approach to computer ethics. *Ethics and Information Technology* 9 (1): 73–87.

Kaiser Family Foundation. 2010. *Generation M2: Media in the Lives of 8- to 18-Year-Olds*, January 20. <http://www.kff.org/entmedia/mh012010pkg.cfm> (accessed March 22, 2011).

Kemp, Charles C., Aaron Edsinger, and Eduardo Torres-Jara. 2007. Challenges for robot manipulation in human environments. *IEEE Robotics & Automation Magazine* 14 (1): 20–29.

Krach, Soren, Frank Hegel, Britta Wrede, Gerhard Sagerer, Ferdinand Binkofski, and Tilo Kircher. 2008. Can machines think? Interaction and perspective taking with robots investigated via fMRI. *PLoS ONE* 3 (7): e2597.

Lane, Shelly J., and Susan Mistrett. 2008. Facilitating play in early intervention. In *Play in Occupational Therapy for Children*, ed. L. Diane Parham and Linda S. Fazio, 413–425. St. Louis, MO: Mosby, Inc.

Lin, Patrick, George Bekey, and Keith Abney. 2008. *Autonomous Military Robotics: Risk, Ethics, and Design*. A report commissioned by U.S. Department of Navy/Office of Naval Research. <http://ethics.calpoly.edu/ONR_report.pdf> (accessed March 16, 2010).

MacDorman, Karl F., Sandosh K. Vasudevan, and Chin-Chang Ho. 2009. Does Japan really have robot mania? Comparing attitudes by implicit and explicit measures. *AI & Society* 23 (4): 485–510.

Mori, Masahiro. 1970. The uncanny valley. *Energy* 7 (4): 33–35.

Nass, Clifford I., Youngme Moon, John Morkes, Eun-Young Kim, and B. J. Fogg. 1997. Computers are social actors: A review of current research. In *Human Values and the Design of Computer Technology*, ed. B. Friedman, 137–162. Chicago: University of Chicago Press (distributed for the Center for the Study of Language and Information).

Narvaez, D. 2008. Human flourishing and moral development: Cognitive science and neurobiological perspectives on virtue development. In *Handbook of Moral and Character Education*, ed. L. Nucci and D. Narvaez, 310–327. Mahwah, NJ: Erlbaum.

Neven, Louis. 2010. "But obviously not for me:" Robots, laboratories and the defiant identity of elder test users. *Sociology of Health and Illness* 32 (2): 335–347.

Nussbaum, Martha C. 2006. *Frontiers of Justice*. Cambridge, MA: Belknap Press.

Nussbaum, Martha C. 2000. *Women and Human Development*. New York: Cambridge University Press.

Oosterlaken, Ilse. 2009. Design for development: A capability approach. *Design Issues* 25 (4): 91–102.

Riek, Laurel D., Tal-Chen Rabinowitch, Bhismadev Chakrabarti, and Peter Robinson. 2009. How anthropomorphism affects empathy toward robots. In *Proceedings of the 4th ACM/IEEE international conference on Human robot interaction (HRI '09)*, 245–246. New York: ACM.

Robins, B., K. Dautenhahn, R. Te Boekhorst, and A. Billard. 2005. Robotic assistants in therapy and education of children with autism: Can a small humanoid robot help encourage social interaction skills? *Universal Access in the Information Society* 4 (2): 105–120.

Scassellati, Brian. 2007. How social robots will help us to diagnose, treat, and understand autism. In *Robotics Research*, ed. S. Thrun, R. Brooks, and H. Durrant-Whyte, 552–563. New York: Springer.

Sen, Amartya. 1993. Capability and well being. In *The Quality of Life*, ed. Martha C. Nussbaum and Amartya Sen, 30–53. New York: Oxford University Press.

Shaw-Garlock, Glenda. 2009. Looking forward to sociable robots. *International Journal of Social Robotics* 1 (3): 249–260.

Singer, Peter W. 2009. *Wired for War*. New York: Penguin Press.

Sofge, Erik. 2010. Can robots be trusted? *Popular Mechanics* 187 (2): 54–61.

Sorrel, Charlie. 2008. GPS causes 300,000 Brits to crash. *Wired.com*, July 22. <http://www.wired.com/gadgetlab/2008/07/gps-causes-3000/> (November 26, 2010).

Sparrow, Robert, and Linda Sparrow. 2006. In the hands of machines? The future of aged care. *Minds and Machines* 16 (2): 141–161.

Sung, Ja-Young, Lan Guo, Rebecca E. Grinter, and Henrik I. Christensen. 2007. "My Roomba is Rambo": Intimate home appliances. In *UbiComp 2007: Ubiquitous Computing*, ed. J. Krumm, G. D. Abowd, A. Seneviratne, and Th. Strang, 145–162. Berlin: Springer.

Tamura, Toshiyo, Satomi Yonemitsu, Akiko Itoh, Daisuke Oikawa, Akiko Kawakami, Yuji Higashi, Toshiro Fujimoto, and Kazuki Nakajima. 2004. Is an entertainment robot useful in the care of elderly people with severe dementia? *Journal of Gerontology* 59A (1): 83–85.

Turkle, Sherry, 2006. A nascent robotics culture: New complicities for companionship. AAAI Technical Report Series, July.

Wallach, Wendell, and Colin Allen. 2009. *Moral Machines: Teaching Robots Right from Wrong*. New York: Oxford University Press.

Weinberg, Alvin M. [1966] 2003. Can technology replace social engineering? In *Technology and the Future*, 9th ed., ed. Albert H. Teich, 23–30. Toronto, Canada: Thomson Wadsworth.

Woods, Sarah, Kerstin Dautenhahn, and Joerg Schulz. 2004. The design space of robots: Investigating children's views. In *Proceedings of the 13th IEEE International Workshop on Robot and Human Interactive Communication*, RO-MAN, 47–52. Kurashiki, Okayama, Japan: IEEE.

17 The Rights and Wrongs of Robot Care

Noel Sharkey and Amanda Sharkey

The possibility of being cared for exclusively by robots is no longer science fiction. There has been a dramatic increase in the number of companies producing robots for the care or companionship, or both, of the elderly and children. A number of robot manufacturers in South Korea and Japan are racing to fulfill the dream of affordable robot "nannies." These have video game playing, quizzes, speech recognition, face recognition, and limited conversation to capture the preschool child's interest and attention. Their mobility and semi-autonomous functions, combined with facilities for visual and auditory monitoring by the carer, are designed to keep the child from harm. These are very tempting for busy, professional parents. Most of the robots are prohibitively expensive at present. But prices are falling and some cheap versions are already becoming available. Some parents are beginning to use the cheaper ones, such as the Hello Kitty robot (Sharkey and Sharkey 2010a).

There is an even greater drive for the development of robots to help care for the elderly. Japan is facing a problem of an aging population growing out of proportion with the young population. In March 2009, Motoki Korenaga, a Japanese ministry of trade and industry official, told *Agence France-Presse*, "Japan wants to become an advanced country in the area of addressing the aging society with the use of robots" (Agence France-Press 2009). Japan is already en route to deliver robot-assisted care, with examples such as the Secom "My Spoon" automatic feeding robot; the Sanyo electric bathtub robot that automatically washes and rinses; Mitsubishi's Wakamaru robot for monitoring, delivering messages, and reminding about medicine, and Riken's RI-MAN robot that can pick up and carry people, follow simple voice commands, and even answer them. The idea is to continue this trend by developing robots that can do many of the household chores for which a visiting helper is now required. Other countries may well follow suit. Europe and the United States are facing similar aging population problems over a slightly longer time scale.[1]

As with any rapidly emerging technology, likely risks and ethical problems need to be considered. The main area of concern addressed in this chapter is the application of robots in caring for the vulnerable. Many of the applications of robots targeted at children and the elderly could show great benefits. For the elderly, assistive care with robot technology has the potential to allow greater independence for those with dementia or other aging brain symptoms (Sharkey 2008; Sharkey and Sharkey 2010b). This could result in the elderly being able to stay out of institutional care for longer. For children, robots have been shown to be useful in applications for those with special needs (e.g., Dautenhahn 2003; Dautenhahn and Werry 2004; Liu et al. 2008). The engaging nature of robots makes them a great motivational tool for interesting children in science and engineering, or facilitating social interaction with the elderly.

We raise no objections to the use of robots for such purposes, nor with their use in experimental research or even as toys. Our concerns arise from the potential abuse of robots being developed for the care of the vulnerable. Our aim here is to throw up some of the ethical questions that need to be asked as robotics progresses sufficiently to allow near-exclusive care by robots. Our interest is in the potential infringement of the rights of the vulnerable, and so we have zoomed in on the extremes in the age range of care: the very young and the elderly. In taking a rights-based approach we are not subscribing to any general ethical theory. However, we do assume that society has a duty of care and a moral responsibility to do its best to ensure the emotional and psychological well-being of all of its citizens, regardless of their age. In looking at robots as carers, we take this duty as given and we examine the balance between it and a number of prima facie rights. We also consider how the resolution of conflicts between rights depends on the age of those cared for and their mental faculties. Elsewhere we have discussed a number of ethical issues, such as dignity and infantilization (Sharkey and Sharkey 2010b, c), the deception of the elderly (Sharkey and Sharkey 2010b), and the deception of children (Sharkey and Sharkey 2010a). Our focus in this chapter concerns the rights to privacy, personal liberty, and social contact.

17.1 Safety and the Right to Liberty and Privacy

An essential component of the duty of care is that a carer must keep their charges safe from physical harm. However, this rule is anything but simple. It does not give the carer the right to "any means" available. The rule must be traded off against the rights of the cared for, such as the right to personal liberty, the right to protection from psychological harm, and the right to privacy.

It is the health and age of the individual that determines the permissible means of safety. One robust way to keep anyone physically safe would be to put the person in a straitjacket in a padded room. Not only would this be inappropriate in most cases, it would be a violation of the rights to liberty and to protection from psychological cruelty. There are many different means for keeping people safe, and each different case will have its own path through the rights trade-offs.

For example, if an elderly person opened a drawer full of sharp kitchen knives, it would be inappropriate for the carer to suddenly spring upon them and restrain them. But if the person had been diagnosed as having severe suicidal tendencies, then such action may be deemed appropriate and even obligatory in the duty of care. With dementia sufferers who are well enough to live in their own homes, it could be inappropriate and irritating even to warn them of the danger (depending on their degree of dementia). With a young child, the appropriate action would be to remove any sharp objects from them and place them out of their reach.

Monitoring someone's activities twenty-four hours a day is another way to maintain safety. This could be done in person or with the use of security cameras. Obviously, violating the right to privacy in this way could be appropriate in some circumstances, such as those of intensive care. However, for those in partial or home care, it could be a severe intrusion on their privacy to monitor them taking a shower or using the bathroom, for example.

A Robot carer needs to understand which behavioral responses are appropriate in which contexts, as well as to be able to predict the intentions of their charges. In the remainder of this section, we examine how robots can be designed to maintain safety, and then move on to examine how this may affect the rights to privacy and liberty.

One of the primary functions of robot carers, like their human counterparts, would have to be to keep their charges safe. Robots could be used for health monitoring in a number of ways, such as taking temperatures, and monitoring respiration and pulse rate. In the high-tech retirement home run by Matsushita Electrics, robot teddy bears watch over elderly residents, monitoring their response time to spoken questions, and recording how long they take to perform certain tasks (Lytle 2002). These robots can alert staff to unexpected changes. This is an area that, once developed, could have a significant impact on elder care in the home or in care institutions. It would be easy to imagine this technology being extended to a number of other health applications, such as caring for quarantine patients.

Outside of health, the main safety method for robot care at present is through the provision of mobile monitoring using cameras and microphones. The most advanced are the childcare robots with hidden cameras to transmit images of the child to a window on the parent/carer's computer or to their mobile phone. Some childcare robots can keep track of the location of children and alert adults if they move outside of a pre-set perimeter. The children wear a transmitter that the robot can detect. For example, PaPeRo (Yoshiro et al. 2005) works by having the child wear a PaPeSack containing an ultrasonic sensor. Similarly, the Japanese company Tmsuk makes a childcare robot that uses radio-frequency tags for autonomous monitoring. The carer can also remotely control the robot to find the child and call or speak to the child through built-in speakers. Similar systems could be used for monitoring elderly patients suffering from dementia.

Such systems are labor intensive and so semi-autonomous that safety monitoring will be required to make the robots more marketable for longer daily care. Some of these advances are already well under way. For example, there are robot systems for tracking people in a range of environments and lighting conditions without the use of sensor beacons (Lopes et al. 2009). This implies that the robot will be able to follow its charge outside and alert supervisors of the charge's location.

In the near future, we are likely to see the integration of robots with other home sensing and monitoring systems. There is considerable research on the development of smart homes for the care of elderly dementia sufferers. These can monitor a range of potentially dangerous activities, such as leaving on taps or gas cookers (Orpwood et al. 2008). Camera systems are being used to determine if an elderly person has fallen over (Toronto Rehabilitation Hospital 2008, 40–41). There is no talk yet about using smart sensing for childcare, but it could get onto the agenda without stretching the imagination by much.

Further extensions to care robots could provide additional home security by employing features from security robots. For example, the Seoul authorities conducted a pilot study in which a surveillance robot, OFRO, was used with an associated security system, KT Telecop, to watch out for potential pedophiles in school playgrounds (Metro 2007). OFRO can autonomously patrol areas on preprogrammed routes. It is equipped with a microphone as well as a camera system, so that teachers can see through its lenses. Essentially, it looks for persons over a certain height and alerts teachers if it spots one. Other techniques being developed for security robots, such as fingerprint and retinal recognition, could be useful for monitoring individuals, for example, visitors or an Alzheimer's sufferer, and helping prevent petty robberies.

17.1.1 Loss of Privacy

A key issue with respect to any kind of monitoring system is whether or not it violates an individual's right to privacy. There are clear overlaps between the concerns raised about privacy in the context of childcare robots, and concerns about privacy when robots are used to monitor the elderly. Although monitoring may be conducted with the welfare and safety of the individual in mind, this may not be sufficient in all cases to justify the intrusion.

The privacy of people in general should be respected as stated in Article 12 of the Universal Declaration of Human Rights: "No one shall be subjected to arbitrary interference with his privacy, family, home, or correspondence, nor to attacks upon his honor and reputation. Everyone has the right to the protection of the law against such interference or attacks." There seems little reason to make an exception for the old or for the young. The right to privacy is also addressed in Articles 16 and 40 of the UN Convention on Child Rights.

The use of a robot carer creates a tension between the use of monitoring to ensure safety and the privacy of the target of that monitoring. As Sharkey and Sharkey (2010a) discuss, parents' use of a baby alarm is acceptable. Similarly, parents frequently video record and photograph their young children. However, there is something different between an adult being present who is recording a child and an adult covertly recording a child who thinks that she is alone while confiding in her robot friend. With the massive memory hard drives available today, it would be possible to record an entire childhood. Who will be allowed access to the recordings? Will the child, in later life, have the right to destroy the records?

Similar questions need to be asked about the situation in which an elderly person is being monitored by a robot companion, or by a remote controlled robot. A person with Alzheimer's may soon forget that a robot is present and might perform acts or say things believing he is in the privacy of his own home, or thinking that he is alone with his robot friend. While the idea of recording and preserving the memories of one's elderly parent may seem attractive, it might not be something that he would consent to, if able. Would we want our children to know everything we said about them with the belief that we were talking confidentially? Again, the important question here is, who should have access to the recordings? If the elderly person does not give consent while still in a position to do so, it would seem that all recordings should be destroyed by default after use for immediate medical purposes.

One issue that affects the elderly more than children is that of respect for the privacy of their bodies. An operator could drive a robot to peer round an elder's apartment before they were dressed or when they are taking a bath. An autonomous robot could record in the same circumstances. The elder might prefer the robot to have to do the equivalent of knocking on the door and waiting to be invited in. Furthermore, the robot could provide a clear indication (e.g., a large flashing light) when any recording or monitoring was taking place. Of course, there are individuals who are too young or whose intellectual faculties are too impaired to be able to understand recording or monitoring signals. Such individuals still have a right to privacy, but it needs to be exercised on their behalf by sensitive carers.

We have discussed how the privacy requirements of our two demographic groups differ, but we also need to take account of individuals' developmental stage and mental facility. Robot care systems should be customized individually to ensure that any intrusions on privacy are justified on the basis of the greater well-being of those concerned. They should not be based on economic or efficiency grounds.

17.1.2 Loss of Liberty
Using a robot simply as a mobile monitoring system would still be quite labor intensive for care supervisors, although more than one target could be monitored at the same time. Commercial pressures will soon lead to the development of autonomous

or semi-autonomous supervision by robots to support longer carer absence. A simple extension would be to allow home customization with maps of rooms so that the robot could recognize danger areas. As the field progresses, intelligent vision and sensor systems could be used to detect potentially dangerous activities, like a child climbing on furniture to jump or an elder heading toward basement stairs. The robot could make a first pass at warning its charge to stop engaging in a potentially dangerous activity. But would it be ethically legitimate to allow a robot to block or restrain a child or an elder from an activity that was on the robot's danger list? This is very difficult ethical territory that relates directly to one's fundamental right to autonomy.

It would be easy to construct scenarios where it would be hard to deny such robot action. For example, if a child or an elder was about to walk onto the road into heavy oncoming traffic and a robot could stop her, should it not do so? It would clearly be irresponsible for someone controlling a robot not to use it to prevent such a situation. But, what if the robot was operating autonomously? If it could predict a dangerous situation, would it not be legitimate to take action to stop it occurring, such as taking matches out of the hands of a child or an elder, getting between her and a danger area such as a gas stovetop, or even restraining (gently) to prevent her carrying out a dangerous action?

The problem here is in trusting the robot's classification and sensing systems to determine what constitutes a dangerous activity. Imagine a child having doughnuts taken from him to prevent him from becoming obese, or imagine a senior having a bottle of alcohol taken from her to prevent her becoming intoxicated and falling. Restraining a child or an elder to avoid harm could be a slippery slope toward authoritarian robotics.

Robots are able to follow well-specified rules, but they are not good at understanding the surrounding social context and predicting likely intentions (Castellano and Peters 2010). Although a robot can be programmed with rules about the dangerous situations that programmers anticipate, it is never going to be possible to anticipate enough of them. Humans, on the other hand, are very skilled at such understanding and prediction from as young as twelve months (Woodward and Sommerville 2000). A human carer is likely to be able to predict the intention behind a child building the pile of blocks to reach an otherwise inaccessible window handle in a way that the robot is not.

There are many discussions to be had over the extremes of robot interventions and where to draw the line. There are some differences in the issues raised in caring for children and for the elderly. It is sometimes necessary to constrain the action of an infant to prevent harm. However, children need to be free to explore and satisfy their curiosity for normal healthy development. This requires a balancing act between their safety and their freedom of which robots are incapable. The problem for the elderly is that if a robot restrains their actions or prevents their movements to certain places, it could be equivalent to imprisonment in the home without trial. There are already circumstances in which carers can restrict the liberty of individuals in order to protect them. However, there are legal procedures available for making such decisions. We must ensure that we do not let the use of technology covertly erode the right to liberty without due process.

17.2   Human Contact and Socialization

It is the natural right of all individuals to have contact with other humans and social-ize freely. If robots begin to be trusted to monitor and supervise vulnerable members of society, and to perform tasks such as feeding, bathing, and toileting, a probable consequence is that some young and old humans could be left in the near-exclusive company of robots.

In discussing the effect of new therapies for people with aging brains, Boas (1998) points out, "What stimulates them, gives a lift to their spirits, is the human interac-tion, the companionship of fellow human beings." And having a good social network helps to protect against declining cognitive functions and incidence of dementia (Crooks, Lubben, and Petitti 2008; Bennett et al. 2006). For children, very serious defects both in brain development and psychological development can occur if they are deprived of human care and attention (Sharkey and Sharkey 2010a). The effects, and risks, of reduced human contact are likely to be quite different for the elderly and for infants. Infants need nurturing and parenting to enable their normal development, while the elderly require companionship to avoid loneliness and to maintain their mental health for longer. We will deal with each of these populations separately.

17.2.1   First Contact with the Robots: Infants in Care

The impairments caused by extreme lack of human contact with infants are well known and documented. Nelson and colleagues (Nelson et al. 2007) compared the cognitive development of young children reared in Romanian institutions to those moved to foster care with families. Children reared in institutions manifested greatly diminished intellectual performance (borderline mental retardation) compared to children reared in their original families. Chugani and colleagues (Chugani et al. 2001) found that Romanian orphans, who had experienced virtually no mothering, differed from children of comparable ages in their brain development—and had less active orbitofrontal cortex, hippocampus, amygdala, and temporal areas.

Perhaps little or no harm would result from a child being left in the care of a robot for very short periods. But what would happen if those periods of time became increas-ingly frequent and longer? The outcome would clearly depend on the age of the child in question. It is well known that infants under the age of two need a person with whom they can form an attachment if they are to develop well. In an earlier paper (Sharkey and Sharkey 2010a), we considered whether an infant might be able to form an attachment to a robot caregiver, perhaps in the same way that Harry Harlow's monkeys became attached to a static cloth surrogate mother.

What research there is suggests that very young children can form bonds with robots. Tanaka, Cicourel, and Movellan (2007) placed a "state-of-the-art" social robot (QRIO, made by Sony), in a daycare center for five months. They found that the toddlers (aged between ten and twenty-four months) bonded and formed attachments to the QRIO robot in a way that was significantly greater than their bonding with a teddy bear. They touched the robot more than they hugged or touched a static toy robot, or a teddy bear. The researchers concluded, "Long-term bonding and socialization occurred between toddlers and the social robot."

Turkle and colleagues (Turkle et al. 2006a) report a number of individual case studies that attest to children's willingness to become attached to robots. For example, ten-year-old Melanie describes her relationship with the robotic doll "My Real Baby" that she took home for several weeks:

Researcher: Do you think the doll is different now than when you first started playing with it?

Melanie: Yeah. I think we really got to know each other a whole lot better. Our relationship, it grows bigger. Maybe when I first started playing with her, she didn't really know me so she wasn't making as much [sic] of these noises, but now that she's played with me a lot more, she really knows me and is a lot more outgoing. (Turkle et al. 2006a, 352)

In another paper, Turkle and colleagues (Turkle et al. 2006b) chart the first encounters of sixty children between the ages of five and thirteen with the MIT robots Cog and Kismet. The children anthropomorphized the robots, made up "back stories" about their behavior, and developed "a range of novel strategies for seeing the robots not only as 'sort of alive' but as capable of being friends and companions." Their view of the robots did not seem to change when the researchers spent some time showing them how they worked, and emphasizing their underlying machinery. Melson and colleagues (Melson et al. 2009) directly compared children's views of and interactions with a living dog and a robot dog (AIBO). Although there were differences, the majority of the children interacted with the AIBO in ways that were like interacting with a real dog: they were as likely to give commands to the AIBO as to the living dog, and over 60 percent affirmed that AIBO had "mental states, sociality, and moral standards."

Overall, the pattern of evidence indicates that children saw robots that they had spent time with as friends and felt that they had formed relationships with them. They even believed that a relatively simple robot was getting to know them better as they played with it more. So, extrapolating from the evidence, it seems that there is a good possibility that children left in the care of robots for extended periods could form attachments to them. However, it is unlikely that the attachment would adequately replace the necessary support provided by human attachment.

To become well adjusted and socially attuned, an infant needs to develop a secure attachment to a carer (Ainsworth, Bell, and Stayton 1974). A securely attached infant will explore their environment confidently, and be guided in their exploration by cues from the carer. The development of secure attachment between a human carer and an infant depends on the carer's maternal sensitivity, and ability to perceive and understand the infant's cues and to respond to them promptly and appropriately. Detailed interactions between a mother and baby help the infant to understand their own emotions, and those of others.

In Sharkey and Sharkey (2010a), we argued from a review of the technology that robot carers into the foreseeable future would be unable to provide the detailed interaction necessary to replace human sensitivity and promote healthy mental development. Many aspects of human communication are beyond the capabilities of robots. There has been progress in developing robots and software that can identify emotional expressions (e.g., Littlewort, Bartlett, and Lee 2009) and there are robots that can make emotional expressions (Breazeal 2002; Cañamero and Fredslund 2001). However, recognizing what emotion is being expressed is only a tiny step toward understanding the causes of the emotion—is the child crying because she dropped her toy, because she is in pain, or because her parents are fighting?

There are many challenges to be overcome to develop a robot that could respond appropriately and sensitively to a young child that currently seem insurmountable. This is further complicated because responses that may be appropriate at one age would not be appropriate at another. An important function of a caregiver is to promote a child's development, for instance, by using progressively more complex utterances in tune with the child's comprehension.

When a human carer is insufficiently sensitive, insecure attachment patterns can result: *anxious-avoidant* attachment when the child frequently experiences rejection from the carer; *anxious ambivalent* attachment when the carer is aloof and distant; *disorganized attachment* when there is no consistency of care and parents are hostile and frightening to the children. Babies with withdrawn or depressed mothers are more likely to suffer aberrant forms of attachment: avoidant or disorganized attachment (Martins and Gaffan 2000).

Perhaps a child with a secure attachment to their parent would not suffer much as a result of being left with a robot for short periods. But the fact is we just don't know: no one has yet researched the possible negative consequences of children being left with robots for varying time periods, and it would be too risky to do so. We do know that young children do best when they spend time with a caregiver with whom they have a secure attachment. Thus, it is highly likely that leaving children in the care of a robot is not going to benefit them as much as leaving them in the care of an attentive and focused human carer. Robot nannies should not be used just because we cannot demonstrate that they are harmful. Rather, they should "qualify for (part-time) care only when it is proven that their use serves the child's best interests" (Zoll and Spielhagen 2010, 298).

17.2.2 Human Contact and the Elderly

A major concern that we have about home robot care for the elderly is that it may replace human contact. With very advanced smart sensing systems and robots that can lift and carry, bathe and feed, as well as keep their charges safe, there will be less need for care visits—the whole point of using the robots is because there will be fewer carers available as the population ages. This is bad news for many elderly people for whom visiting carers are the only human companionship they have on a daily basis. According to a report from the charity Help the Aged in 2008, 17 percent of older people in the UK have less than weekly contact with family, friends, and neighbors, and 11 percent have less than monthly contact.

Using robots for care of the elderly seems likely to reduce the number of opportunities they have for interaction with other human beings, and the benefits that come from such interaction. Sparrow and Sparrow (2006) argue that robots should not be used in elder care because of the likely consequential reduction in social contact. They make the point that even using robots to clean floors removes a valuable opportunity for interaction between an elderly resident and a human cleaner.

Research strongly suggests human companionship is essential for the well-being of the elderly, and yet there are no specific rights to companionship. There is a right to participation in the culture in Article 27 of Universal Declaration of Human Rights.[2] Deprivation of human contact may also be considered as cruelty, which is covered by Article 5. However, it is not clear that someone living independently in their own home with the help of robots would be being *subjected* to lack of companionship. Home helpers are not employed specifically as companions; it is just one of their beneficial side effects. Before introducing mass robot care, this side effect needs to be recognized as a function. Substantial evidence suggests that human contact should be seen as part of the right to welfare and medical treatment.

It is clear that an extensive social network offers protection against some of the effects of aging: being single and living alone is a risk factor for dementia (Fratiglioni et al. 2000; Saczynski et al. 2006; Wilson et al. 2007). Holtzman et al. (2004) found that frequent interaction in larger social networks was positively related to the maintenance of global cognitive function. Wang et al. (2002) similarly found evidence that a rich social network may decrease the risk of developing dementia, and concluded that both social interaction and intellectual stimulation play an important role in reducing such risks.

There is evidence that stress exacerbates the effects of aging (Smith 2003), and that social contact can reduce the level of stress a person experiences. Kikusui, Winslow, and Mori (2006) provide a wide-ranging review of the phenomena of *social buffering*, whereby highly social mammals show better recovery from distress when in the company of conspecifics. A recent review (Heinrichs, von Dawans, and Domes 2009) concludes that the stress-protective effects of social support may be the result of the neurotransmitter oxytocin that is released in response to positive social interactions, and that oxytocin can have the effect of reducing stress.

One take on the problem of social exclusion of the elderly in Japan is to move toward the development of robot companions and robot pets. These are being touted as a solution to the contact problem—devices that can offer companionship, entertainment, and human-like support. Examples include Paro, a fur-covered robotic seal developed by AIST that responds to petting; Sony's AIBO robotic dog; NeCoRo (OMRON), a robotic cat covered in synthetic fur, and My Real Baby (iRobot), described as an "interactive emotionally responsive doll."

There are, to our knowledge, no studies that directly compare the effect on the elderly of robot versus human companionship. Obviously, as is the case with children, robots are not going to be able to be as responsive to the needs of the elderly as are humans. However, they might be useful to supplement rather than replace human carers. There is, for example, evidence that giving the elderly robot pets to look after can be beneficial. Positive effects, such as reduction in loneliness and improved communication, have been found in studies where elders were allowed to interact with a simple Sony AIBO robot dog (Kanamori, Suzuki, and Tanaka 2002; Banks, Willoughby, and Banks 2008; Tamura et al. 2004).

These outcomes need to be interpreted with caution, as they depend on the alternatives on offer. If the alternative is being left in near-complete social isolation, it is unsurprising that interacting with a robot pet offers advantages. Better comparisons could be made such as with a session of foot massage, or sitting with a sympathetic human listener.

On the upside, a robot pet does not have to be a replacement for social interaction. It could be provided in addition to other opportunities, and might further improve the well-being of an elderly person. As discussed in Sharkey and Sharkey (2010b), robot pets and toys could act as facilitators for social interaction by providing conversational opportunities (Kanamori, Suzuki, and Tanaka 2002). Having a robot pet may also give elders an increased feeling of control and autonomy. There is strong evidence that these factors can improve their well-being, and even result in longer life expectancy (Langer and Rodin 1976).

17.3 Conclusion

We began with an appraisal of how well care robots could keep their charges physically safe. It turns out that this may be one of their most significant features. However, physical safety comes with potential costs to the rights of the individuals being cared for. We have discussed here how it could violate rights to privacy and personal liberty. It seems almost paradoxical that the more safety the robots provide, the more their use may breach human rights.

Both old and young have a right to privacy, although privacy may have a different character for the two age ranges. It would hardly be an intrusion on an infant's privacy if their carer watched them sitting on the toilet and cleaned their bottom. Would it be so different to have a robot with the infant that broadcasts the images to the parent's computer? Admittedly, it feels less comfortable, but as long as it was only the parent watching and the images were not recorded, it would be unlikely to be considered a violation of the child's privacy. An elderly person might feel quite differently about similar treatment and not wish to have a robot camera with them in such a delicate situation. Our proposal was that a robot should always have an indicator when it is recording or transmitting images and that it warn of its presence and ask permission before entering a room.

There is also a tricky balance between physical safety and the right to liberty. We pointed out that in some circumstances, such as when a person is about to walk onto a busy road, it might be a good idea for a robot to intervene to prevent harm. However, we suggested that it would be unwise to allow a robot to make autonomous decisions about what is dangerous outside of obvious cases—such as leaving a gas stovetop on—where it could issue a warning. A robot would not have the subtlety or sensitivity to human intention to predict potential danger. What is dangerous for one person may be harmless for another. There are a lot of differences in this regard between infants and the elderly. Restraining or blocking the path of someone could represent a slippery slope to an authoritarian robotics that could result in keeping people as virtual prisoners in their own homes.

Looking into the future of care robotics, we examined the possibility that automated care could dramatically reduce the amount of human contact needed for safety and physical welfare. However, such a reduction could be a violation of the fundamental right to psychological well-being and could be considered to be a form of cruelty or torture or both under Article 5 of the Human Rights Convention (1949). Again, there are differences between the young and the elderly.

We argued from current evidence that young children can be fooled into believing that quite simple robots have mental states and can form friendship bonds with them. It seems likely that if children spent most of their time with a robot carer, they would form attachments. This means loving an artifact that cannot love them back. We cannot unequivocally demonstrate what the potential long-term harm of such relationships might be. However, we reviewed evidence from child development studies showing the types of psychological damage that could occur with insufficient human care.

We believe that there is an unacceptably high risk of abnormal attachment for children exposed to too much robot care. This could manifest later in all sorts of adult psychological malfunctions, including the inability to parent properly. Thus, we need to ensure intense scrutiny of any robotics products where it is implied that they could be used for childcare. With strong built-in physical safety features, we would have to find a way to ensure that robots marketed for short-term companionship for children would only be used for that purpose.

The impact on the elderly would be quite different. Leaving an elderly person in the near-exclusive care of robots in virtual home imprisonment would be a serious violation of their right to liberty and their right to participation in society, and would be a form of cruelty. We discussed some of the detailed evidence that social interactions and human companionship can retard the progress of dementia. Nonetheless, we concluded that there are a number of ways in which robots could greatly benefit the elderly. Assistive robots, if used sensitively, could empower the elderly and give them greater independence. We also suggested that companion robots could act as facilitators and conversational aids to improve the social life of the elderly.

Before we go adopting robots in the large-scale care industry, we must be sure about which rights we may be violating. We must minimize these violations in a way that is customized for each individual, and we must ensure that the accrued benefits for an individual are proportionally greater than any losses due to the infringement of their rights. Having considered the field of robot assistance and care, our view is that robotics could be of benefit to the welfare of the elderly, particularly if it maintains their independence at home for longer. However, for children, although there may be benefits interacting with robots in a social, educational, or therapeutic setting, robot childcare comes with too many risks to be considered viable.

Notes

1. Gecko Systems is a U.S. company that is conducting trials for its CareBot with elderly people. Gecko Systems leaders suggest that the CareBot will provide cost effective monitoring of an elderly parent, and permit working parents to check up on their children and "watch their children routinely in a window on their computer monitors while at work."

2. General Assembly res. 217A (III), December 10, 1948.

References

Agence France-Presse. 2009. Japan plans robo-nurses in five years: govt, March 25. <http://www.google.com/hostednews/afp/article/ALeqM5juSqhZryHpsVuY6mf93nr92g1qdA> (accessed November 20, 2010).

Ainsworth, M. D. S., S. M. Bell, and D. J. Stayton. 1974. Infant-mother attachment and social development: Socialisation as a product of reciprocal responsiveness to signals. In *The Introduction of the Child into a Social World*, ed. M. P. M. Richards, 9–134. London: Cambridge University Press.

Banks, M. R., L. M. Willoughby, and W. A. Banks. 2008. Animal-assisted therapy and loneliness in nursing homes: Use of robotic versus living dogs. *Journal of the American Medical Directors Association* 9 (3): 173–177.

Bennett D. A., J. A. Schneider, Y. Tang, S. E. Arnold, and R. S. Wilson. 2006. The effect of social networks on the relation between Alzheimer's disease pathology and level of cognitive function in old people: A longitudinal cohort study. *Lancet Neurology* 5: 406–412.

Boas, I. 1998. Learning to be rather than do. *Journal of Dementia Care* 6 (6): 13.

Breazeal, C. 2002. *Designing Sociable Robots*. Cambridge, MA: MIT Press.

Cañamero, L., and J. Fredslund. 2001. "I show you how I like you—Can you read it in my face?" *IEEE Transactions on Systems, Man, and Cybernetics. Part A* 31 (5): 454–459.

Castellano, G., and C. Peters. 2010. Socially perceptive robots: Challenges and concerns. *Interaction Studies: Social Behaviour and Communication in Biological and Artificial Systems* 11 (2): 201–207.

Chugani, H., M. Behen, O. Muzik, C. Juhasz, F. Nagy, and D. Chugani. 2001. Local brain functional activity following early deprivation: A study of post-institutionalised Romanian orphans. *NeuroImage* 14: 1290–1301.

Crooks, V. C., J. Lubben, and D. B. Petitti. 2008. Social network, cognitive function, and dementia incidence among elderly women. *American Journal of Public Health* 98: 1221–1227.

Dautenhahn, K. 2003. Roles and functions of robots in human society—Implications from research in autism therapy. *Robotica* 21 (4): 443–452.

Dautenhahn, K., and I. Werry. 2004. Towards interactive robots in autism therapy: Background, motivation and challenges. *Pragmatics & Cognition* 12 (1): 1–35.

Fratiglioni, L., H.-X. Wang, K. Ericsson, M. Maytan, and B. Winblad. 2000. Influence of social network on occurrence of dementia: A community-based longitudinal study. *Lancet* 355: 1315–1319.

Heinrichs, M., B. von Dawans, and G. Domes. 2009. Oxytocin, vasopressin, and human social behaviour. *Frontiers in Neuroendocrinology* 30: 548–557.

Holtzman, R. E., G. W. Rebok, J. S. Saczynski, et al. 2004. Social network characteristics and cognition in middle-aged and older adults. *Journals of Gerontology. Series B, Psychological Sciences and Social Sciences* 59 (6): P278–284.

Kanamori, M., M. Suzuki, and M. Tanaka. 2002. Maintenance and improvement of quality of life among elderly patients using a pet-type robot. *Japanese Journal of Geriatrics* 39 (2): 214–218.

Kikusui, T., J. T. Winslow, and Y. Mori. 2006. Social buffering: Relief from stress and anxiety. *Philosophical Transactions of the Royal Society of London. Series B, Biological Sciences* 361 (1476): 2215–2228.

Langer, E., and J. Rodin. 1976. The effects of choice and enhanced personal responsibility for the aged: A field experiment in an institutional setting. *Journal of Personality and Social Psychology* 34 (2): 191–198.

Littlewort, G. C., M. S. Bartlett, and K. Lee. 2009. Automatic coding of facial expressions displayed during posed and genuine pain. *Image and Vision Computing* 27 (12): 1797–1803.

Liu, C., K. Conn, N. Sarkar, and W. Stone. 2008. Online affect detection and robot behaviour adaptation for intervention of children with autism. *IEEE Transactions on Robotics* 24 (4): 883–896.

Lopes, M. M., N. P. Koenig, S. H. Chernova, C. V. Jones, and O. C. Jenkins. 2009. Mobile human-robot teaming with environmental tolerance. In *Proceedings of the 4th ACM/IEEE International Conference on Human Robot Interaction* (HRI '09), 157–164. New York: ACM.

Lytle, Mark J. 2002. Robot care bears for the elderly. *BBC News*, February 21. <http://news.bbc .co.uk/1/hi/sci/tech/1829021.stm> (accessed November 27, 2010).

Martins, C., and E. A. Gaffan. 2000. Effects of early maternal depression on patterns of infant-mother attachment: A meta-analytic investigation. *Journal of Child Psychology and Psychiatry, and Allied Disciplines* 42: 737–746.

Melson G. F., Peter H. Kahn, Jr., A. M. Beck, B. Friedman, T. Roberts, and E. Garrett. 2009. Children's behavior toward and understanding of robotic and living dogs. *Journal of Applied Developmental Psychology* 30 (2): 92–102.

Metro. 2007. Robot guards for Korean schools, May 31. <http://www.metro.co.uk/weird/51254 -robot-guards-for-korean-schools> (accessed July 14, 2011).

Nelson, C. A., C. H. Zeanah, N. A. Fox, P. J. Marshall, A. T. Smyke, and D. Guthrie. 2007. Cognitive recovery in socially deprived young children: The Bucharest early intervention project. *Science* 319 (5858): 1937–1940.

Orpwood, R., T. Adlam, N. Evans, and J. Chadd. 2008. Evaluation of an assisted-living smart home for someone with dementia. *Journal of Assistive Technologies* 2 (2): 13–21.

Saczynski, J. S., L. A. Pfeifer, K. Masaki, E. S. C. Korf, D. Laurin, L. White, and L. J. Launer. 2006. The effect of social engagement on incident dementia: The Honolulu-Asia Aging Study. *American Journal of Epidemiology* 163 (5): 433–440.

Sharkey, N. 2008. The ethical frontiers of robotics. *Science* 322: 1800–1801.

Sharkey, N., and A. Sharkey. 2010a. The crying shame of robot nannies: An ethical appraisal. *Interaction Studies: Social Behaviour and Communication in Biological and Artificial Systems* 11 (2): 161–190.

Sharkey, N., and A. Sharkey. 2010b. Living with robots: Ethical tradeoffs in eldercare. In *Close Engagements with Artificial Companions: Key Psychological, Social, Ethical and Design Issues*, ed. Y. Wilks, 245–256. Amsterdam: John Benjamins.

Sharkey, A. J. C., and N. E. Sharkey. 2010c. Ethical issues in robot care for the elderly: Dystopia or optimism? In *Proceedings of Second International Symposium on New Frontiers in Human-Robot Interaction* (AISB 2010 Convention), ed. K. Dautenhahn and J. Saunders, 103–107. Leicester, UK: De Montford University.

Smith, J. 2003. Stress and aging: Theoretical and empirical challenges for interdisciplinary research. *Neurobiology of Aging* 24, Suppl. 1: S77–80; discussion S81–82.

Sparrow, R., and L. Sparrow. 2006. In the hands of machines? The future of aged care. *Minds and Machines* 16: 141–161.

Tamura, T., S. Yonemitsu, A. Itoh, D. Oikawa, A. Kawakami, Y. Higashi, T. Fujimoto, and L. Nakajima. 2004. Is an entertainment robot useful in the care of elderly people with severe dementia? *Journals of Gerontology. Series A, Biological Sciences and Medical Sciences* 59 (1): 83–85.

Tanaka, F., A. Cicourel, and J. R. Movellan. 2007. Socialization between toddlers and robots at an early childhood education center. *Proceedings of the National Academy of Science* 104 (46) 17954–17958.

Toronto Rehabilitation Hospital. 2008. Our Journey in 2008/09: Annual Report. <http://www.torontorehab.com/About-Us/Corporate-Publication/2008-2009/hospital.asp> (accessed September 12, 2011), 40–41.

Turkle, S., W. Taggart, C. D. Kidd, and O. Dasté. 2006a. Relational artifacts with children and elders: The complexities of cybercompanionship. *Connection Science* 18 (4): 347–362.

Turkle, S., C. Breazeal, O. Dasté, and B. Scassellati. 2006b. First encounters with Kismet and Cog: Children respond to relational artifacts. In *Digital Media: Transformations in Human Communication*, ed. Paul Messaris and Lee Humphreys, 313–330. New York: Peter Lang Publishing.

Wang, H., A. Karp, B. Winblad, and L. Fratiglioni. 2002. Late-life engagement in social and leisure activities is associated with a decreased risk of dementia: A longitudinal study from the Kungsholmen Project. *American Journal of Epidemiology* 155 (12): 1081–1087.

Wilson, R. S., K. R. Krueger, S. E. Arnold, J. A. Schneider, J. F. Kelly, L. L. Barnes, Y. Tang, and D. A. Bennett. 2007. Loneliness and risk of Alzheimer's Disease. *Archives of General Psychiatry* 64: 234–240.

Woodward, A. L., and J. A. Sommerville. 2000. Twelve-month-old infants interpret action in context. *Psychological Science* 11 (1): 73–77.

Yoshiro, U., O. Shinichi, T. Yosuke, F. Junichi, I. Tooru, N. Toshihro, S. Tsuyoshi, and O. Junichi. 2005. Childcare robot PaPeRo is designed to play with and watch over children at nursery, kindergarten, school and at home. Development of Childcare Robot PaPeRo, Nippon Robotto #Gakkai Gakujutsu Koenkai Yokoshu, 1–11.

Zoll, C., and C. Spielhagen. 2010. Changing perspective: From avoiding harm to child's best interests. *Interaction Studies: Social Behaviour and Communication in Biological and Artificial Systems* 11 (2): 295–301.

18 Designing People to Serve

Steve Petersen

Fiction involving robots almost universally plays on a deep tension between the fantasy of having intelligent robot servants to do our every bidding, and the guilt over the more or less explicit possibility that having such intelligent creatures do our dirty work would simply be a new form of slavery. The robot character is frequently a sympathetic antihero who gets mistreated by its callous, carbon-chauvinist human oppressors. Our guilt over such a scenario often manifests as a fear that the robots' servile misery would drive them to a violent and relentlessly successful uprising. As commonly noted, the very word "robot" has its roots in just this scenario; it first appears in Karel Čapek's play *R.U.R: Rossum's Universal Robots*, in which a brave new world of robot servants eventually rebel against their oppressive human masters. Čapek chose the word "robot" to invoke the Czech word *robota*, which translates to "drudgery" or "forced labor."[1] Čapek's play seems to have set the stage for a very long list of books, movies, and television shows about robots to follow. Try to list a few robot stories that *don't* fit this fantasy-guilt complex, and I'm confident you will generate a sizable list of examples that do fit it yourself.

So this aspect of robot ethics has long been in our culture—but it is only just beginning to appear in academia. The few authors who directly confront the ethics of robot servitude tend to conclude one of three things. Some propose that such robots could never be ethical subjects, and so we could not wrong them in making them work for us any more than we now wrong a washing machine. Others agree that robots could not be of ethical significance, but say we must treat them as if they were anyway, for our own sake. Still others conclude that robots *could* someday have genuine ethical significance similar to ours, and that therefore it would be unethical for them to perform menial tasks for us; it would simply be a new form of slavery.[2]

My own position, originally developed in Petersen 2007, is quite different from all of these. First of all, I do think it is possible to create robots of ethical significance—even to create *artificial people*, or *APs* for short. In a tradition with its roots in John Locke, philosophers tend to distinguish the biological category *human* from the much more philosophically rich category *person* ([Locke 1690] 1838, II.xxvii). To say that something artificial could be a person is to say in part at least that it could have full ethical standing like our own. On this usage, for example, ET the Extra-Terrestrial would be a person, but not a human. ET does not share our DNA, but this is irrelevant to his ethical standing; he is as ethically valuable as we are. In other words, to be a person does not seem to require being made of the particular material that happens to constitute humans; instead, philosophers tend to agree, it requires complicated organizational patterns that the material happens to realize. And thus, assuming we could eventually make a robot who has the same relevant complicated organizational patterns that we and ET have, then that robot would also be a person—an artificial one.

I *also* think that although such robots would be full-blown people, it might still be ethical to commission them for performing tasks that we find tiresome or downright unpleasant. There can, in other words, be artifacts that (1) are people in every relevant sense, (2) comply with our intentions for them to be our dedicated servants, and (3) are not thereby being wronged. I grant this combination is prima facie implausible, but there are surprisingly good arguments in its favor. In a nutshell, I think the combination is possible because APs could have hardwired desires radically different from our own. Thanks to the design of evolution, we humans get our reward rush of neurotransmitters from consuming a fine meal, or consummating a fine romance—or, less cynically perhaps, from cajoling an infant into a smile. If we are clever we could design APs to get their comparable reward rush instead from the look and smell of freshly cleaned and folded laundry, or from driving passengers on safe and efficient routes to specified destinations, or from overseeing a well-maintained and environmentally friendly sewage facility. After all, there is nothing intrinsically unpleasant about hydrogen sulfide molecules, any more than there is anything intrinsically pleasant about glucose molecules. The former's smell is aversive and the latter's taste is appetitive *for humans*; APs could feel quite differently.[3] It is hard to find anything wrong with bringing about such APs and letting them freely pursue their passions, even if those pursuits happen to serve us. This is the kind of robot servitude I have in mind, at any rate; if your conception of *servitude* requires some component of unpleasantness for the servant, then I can only say that is not the sense I wish to defend.

18.1   The Person-o-Matic

To dramatize the ethical questions that APs entail, imagine we sit before a *Person-o-Matic* machine. This machine can make an artificial person to just about any specifications with the push of a button. The machine can build a person out of metal and plastic—a robotic person—with a circuit designer and an attached factory. Or, if we wish, the machine can also build a person out of biomolecules, by synthesizing carefully sequenced human-like DNA from amino acids, placing it in a homegrown cellular container, and allowing the result to gestate in an artificial uterus. It can make either such type of person with any of a wide range of possible hardwired appetites and aversions.[4] Which buttons on the Person-o-Matic would it be permissible to press?

It may be difficult to reconcile ourselves to the notion that we could get a genuine *person* just by pushing a button. My students like to say that nothing so "programmed" could be a person, for example. But—as the carbon-based AP case makes especially vivid—the resulting beings would have to be no more "programmed" than we are.

A more sophisticated version of this complaint is in Steve Torrance's "Organic View" (2007). He argues that only "organic" creatures could have the relevant ethical properties for personhood, and so "artificial person" is practically a contradiction in terms. Of course, a great deal hinges here on just what "organic" means. Torrance seems to use it in three different ways throughout his paper: (1) *carbon-based*, (2) *autopoietic*, and (3) *originally purposeful*. This quotation, for example, illustrates all three: "Purely electrically powered and computationally guided mechanisms [sense 1] . . . cannot be seen, except in rather superficial senses, as having an inherent motivation [sense 3] to realize the conditions of their self-maintenance [sense 2]: rather it is their external makers that realize the conditions of their existence [sense 3]" (Torrance 2007, 512–514). But none of these three senses of *organic* is enough to show that APs are impossible.

Consider first the sense in which it means *carbon-based*. Torrance provides no argument that only carbon could ground ethical properties; indeed, philosophical consensus is otherwise, as mentioned earlier. Besides, even if people do have to be organic in this sense, APs are still possible—as Torrance acknowledges (2007, 496, 503)—because it is in principle possible to create people by custom building DNA.

Torrance officially uses *organic* in the second sense, to mean *autopoietic*. Roughly, something is autopoietic if it can self-organize and self-maintain. But this is a purely functional notion; there is no reason inorganic compounds couldn't form something autopoietic. Indeed, the well-established movement of situated, embodied, and embedded robotics emphasizes getting intelligence out of just such lifelike properties.[5] Rodney Brooks's Roomba, for example, avoids treacherous stairs and seeks its power source after a long day of vacuuming. Such robots already have rudimentary self-organization and self-maintenance.

Lurking behind the criterion of autopoiesis is the third sense of *organic*, and what I suspect to be the core of the matter for Torrance's argument: the presence of *inherent function* or purpose. Torrance is claiming, in effect, that when something gains a function by another's design, the function is not inherent to that thing, and so it is not "original." And, Torrance seems to hold, only original functionality can ground ethical value. In other words, just in virtue of resulting from another's design, a thing cannot be a person. (Perhaps this is what my students mean by something being "programmed.")

If correct, this would by definition rule out all APs, carbon-based or not. But, aside from having scant motivation, it proves too much. By this criterion, if traditional Christian creationism proved true and God designed us, then we humans would not be "organic" either, and so not people. I'm strongly inclined to agree that evolution, and not God, designed humans—but it would be very odd if our ethical standing were so hostage to the truth of this claim. For another example closer to home, it seems that our biological parents count as our "external makers," who were moved to "realize the conditions of [our] existence" (though probably not in a traditional laboratory setting). Despite such external makers, we manage to have the properties required to be people.

Finally, consider Labrador Retrievers. They are not people, of course, but they do have ethical standing, and they were deliberately designed, via artificial selection, to enjoy fetching things. Does this mean that they have no "inherent motivation" to fetch? Anyone who has spent time with a retriever can see that the dog, itself, wants to fetch—whatever the source of that desire. Furthermore, satisfying this desire is part of the well-being *for that dog*, even though that desire was designed by intelligent outsiders. Similarly, we did not give ourselves all our desires; some of them, such as for food, are just plain hardwired. It is hard to see why ends given by evolution are "original," but ends given by the design of an intelligence are not. In both cases, there is a very natural sense where our ends seem plainly derivative.

Still, I think Torrance is onto something important; in fact, I agree that for something to be intelligent, autopoietic, and a subject of ethical value, it must have a function *for itself*. Teleology is a notorious can of worms in philosophy, and can hardly be settled here. For our purposes, we just need the claim that one way for something to get a function for itself—an "original teleology"—is from the design of another intelligence.

So now perhaps we are in a position to agree that pushing a Person-o-Matic button would result in a real person of intelligence and ethical value, comparable to our own. When we picture this vividly, I think typical intuitions incline us to say that pushing few, if any, of the buttons is permissible. The case is so far removed from our experience, though, that it is hard to trust these intuitions—especially since there are good arguments that say it *is* permissible to press quite a few of them.

18.2 The "Typical" Person Case

Suppose first you notice buttons for building an organic person, just like you (presumably) are. (From here I will use *organic* just to mean *carbon-based*.) Perhaps, after you feed it the complete information about your DNA makeup and the DNA makeup of a willing partner, the Person-o-Matic uses a combination of this information to construct from scratch a viable zygote that matures in an artificial uterus, much later producing an infant, exactly as might have been produced by the more traditional route. Here we leave a great deal of the design up to chance, of course; our intention is not to create a servant, but roughly just for the Person-o-Matic to build a new human, or anyway a human-like person.[6] The scenario may be intuitively distasteful or even abhorrent, but it is very hard to give reasons for why creating such a person would be *wrong*. After all, it results in people just like the people we now create by traditional means. There may be circumstances in which just the creating of a new person is unethical, of course—due to overpopulation or some such—but that would hardly be unique to APs. If anything is uniquely wrong about this case, then, it must be in the *method* for creating the person, rather than the outcome. But even the method seems no less ethical than a combination of in vitro fertilization, artificial implantation, surrogate mothers, and a host of other techniques for creating people that are already in practice. No doubt bioethics is another can of philosophical worms, but the case at hand here is not so different from bioethical cans already wide open. Indeed, using the Person-o-Matic this way could plausibly bring ethical benefit to a great many couples who are not otherwise able to have biological children.

Probably, the most natural way to express our intuitions against the permissibility of this case is to say that such a procedure for making a person like us would be "unnatural." This word shows up frequently when people are confronted with new technology. As a clever novelist once put it:

1. Anything that is in the world when you're born is normal and ordinary and is just a natural part of the way the world works.
2. Anything that's invented between when you're fifteen and thirty-five is new and exciting and revolutionary and you can probably get a career in it.
3. Anything invented after you're thirty-five is against the natural order of things. (Adams 2002, 95)

The point, of course, is that much of what we consider "natural" today may have looked horrifyingly unnatural to those just a generation or two behind us. To say "unnatural" in this way just means "new enough to make us wary and uncomfortable." When the word means only this, it has no philosophical weight. Our gut reactions are often wise for being wary of the new and strange, but rejecting something *because* it is new and strange is quite different. We do not now consider flying, cell phones, radiation treatment, or artificial hearts wrong because they would have been distressing to those before us.

It seems then that it is hard to explain why it would be wrong to push such a button. As it happens, though, next to those buttons is another row of buttons that offer the option to create a person much like us, except inorganic—a robot. Aside from desires and goals that are particular to the material makeup, we can suppose the robot is designed to have hardwired interests very like ours, and will also be strongly influenced in a unique way by its educational environment just as we were. Would it be wrong to push any of those buttons? It seems there are only a few avenues for trying to explain such wrongness. One is to say that though the resulting person would be like us in all relevant mental respects, just the fact of its different material constitution makes its creation wrong. Another might be that the desires unique to our organic constitution are relevant—that, for example, it is okay to make an AP who likes to consume carbohydrates, but not one who likes to consume pure hydrogen. I trust neither of these avenues looks very promising. If not, and short of other explanations of asymmetry, the organic and the inorganic cases seem to be morally equivalent. We must conclude that making a robot with predispositions like ours is no more wrong than having a biological child would be.

18.3 The "Enhanced" Person Case

We next notice a bank of buttons to create organic people who are still very much like us, but who have been "enhanced" in any of various ways. Some buttons offer to design the person so that she is immune to common diseases. Of more interest for us, some buttons offer to alter the person's hardwired desires—so that perhaps she is also immune to the lures of tobacco, or enjoys eating healthy greens more than usual. Other buttons offer to tailor more abstract desires, so that, for example, the AP gains greater intrinsic pleasure than typical from pursuits we consider noble, such as sculpting or mathematics. Would it be wrong to press a button to bring about this type of person?

Again, despite what qualms we might have, it is hard to say why it would be. Given that parents and mentors expend great and generally laudable effort on the nurture side to bring about such results, it is at least a bit odd to say that bringing them about from the nature side would be wrong.

Probably the best argument against creating the "enhanced" person suggests we have robbed the resulting person of important autonomy by engineering such desires. On this view, it is one thing to encourage such desires during the person's upbringing, and another to hardwire them ahead of time. Of course, free will is yet another philosophical can of worms, and one into which we can only peek here—but again, it is a can of worms that is already open, and hardly unique to APs. Some humans are now naturally born with stronger resistance to tobacco's appeal, for example, and it may well be that some are naturally born with stronger predilections for math or art. At any rate, we all come into existence with hardwired desires, and whether they are "enhanced" or "typical" does not seem relevant to whether they are enacted freely.

Imagine, for example, that way down the road—perhaps hundreds of millions of years later—natural selection has shaped humans so that they no longer enjoy tobacco, and they are born with a random mix of significantly stronger desires to do art or science or other lofty pursuits. This seems possible at any rate, and it would be very odd to say that those future humans would thereby have less autonomy than we have. But our Person-o-Matic can now make a molecular duplicate of such a future possible person. If the future product of natural selection is free and the duplicate AP is not, then one's autonomy depends on how one is brought into existence, even if the result is otherwise exactly the same. It is to say, in effect, that intelligent design does not create an *original* function after all.

I have already argued against this position; I hope, on reflection, it is hard to endorse. It is more interesting to examine what tempts us into this view in the first place. Perhaps, it is simply the familiar queasiness of the "unnatural." Another possibility is that we confuse the case at hand with a more familiar one: that of brainwashing a person with contrary desires already in place.

Another possible source of confusion is in the imagined relative *strength* of these inclinations. Perhaps typical people are free, despite being born with strong dispositions because, we think, they are still able in principle to resist them. Whatever this "ability" amounts to, though, we can suppose APs have it, too. It is plausibly a necessary condition of personhood that one be able to reflect on one's desires, for example, and reconsider them (Frankfurt 1971). An enhanced AP might crave mathematics or sculpting as much as a typical human craves food. But Gandhi could reason himself out of acting on his food craving, and the enhanced AP might similarly reason herself out of her cravings, because she is a person able to reflect on them. So, the AP seems to be as free as we humans are—however free that might be—and the objection from autonomy fails.

It is no great surprise when we see another row of buttons on the Person-o-Matic for creating enhanced APs that are inorganic. These buttons result in robots who love to carve elegant statues or prove elegant theorems. Again, pushing these buttons seems morally equivalent to the ones for the organic APs. If so, then creating a robot who loves to pursue art or science is no more wrong than giving birth to a human who gained the same predispositions through natural selection.

Notice, though, that pushing buttons in either of these rows is already at least tantamount to designed servitude. Suppose we commission an AP who is very strongly inclined to help find a cure for cancer. Is this AP our willing servant? If so, then I have already shown that we can design people to serve us without thereby wronging them.

18.4 The "General Servitude" Case

A scientist dedicated to curing cancer, even as a result of others' desires, may not seem like a clear case of servitude. Clearer cases follow readily, though—because one enhancement for a person, plausibly, is general beneficence. Sure enough, a prominent button on the Person-o-Matic designs an organic person who gains great pleasure simply from bringing about happiness in other people. The AP who results genuinely likes nothing more than to do good and will seek opportunities to help others as eagerly as we seek our own favorite pleasures.

Again, it seems possible that natural selection could bring about humans like this in the far future—if group selection turns out to be a force for genetic change after all, for example—and it would not then be wrong to give birth to one. (Indeed, it sounds like a pretty good world into which to be born.) Again, the Person-o-Matic could create a molecular duplicate of such a person. Again, it is hard to see why the naturally selected person would be permissible and not the intelligently designed one. Again, it does not matter, on ethical grounds, whether the resulting AP is organic or inorganic. So, again, we have to conclude that commissioning a robot who wants to help people above all else is no more wrong than giving birth to a human who gained such beneficence through natural selection. The resulting APs would behave much as though they were following Isaac Asimov's Three Laws of Robotics from his *I, Robot* series ([1950] 1970)—except they would also be helpful to other APs. And this time it seems very clear that the resulting AP would be a dedicated servant to the people around it.

18.5 The "Specific Servitude" Case

Closer still to the *I, Robot* scenario are APs who are designed not to seek the happiness of people generally, but rather the happiness of humans specifically. This is a more task-specific kind of servitude. Still, more specifically, perhaps they are designed to seek the health and well-being of human children—or even your particular children, as Walker pictures his *Mary Poppins 3000*:

What if the robotic firm sells people on the idea that the MP3000 is designed such that it is satis-fied only when it is looking after Jack and Jill, your children? The assumption is that the pro-gramming of individual MP3000s could be made that specific: straight from the robot assembly line comes a MP3000 whose highest goal is to look after your Jack and Jill. Imagine that once it is activated it makes its way to your house with the utmost haste and begs you for the opportu-nity to look after your children. (2006)

In fact, the first robot we meet in the *I, Robot* stories is a similar nanny. Inspection of the Person-o-Matic of course reveals "nanny" buttons, as well as buttons that engi-neer people to derive great joy out of freshly cleaned and folded laundry, or from driving safe and efficient routes to specified destinations, or from clean and efficient sewers. These buttons are probably the most controversial ones to push; they evoke the gruesome "delta caste" of people engineered for menial labor in Aldous Huxley's *Brave New World* ([1932] 1998)—especially in the case of organic, human-like APs.[7] Though surely our intuitions rebel against these cases most of all, it is surprisingly difficult to find principled reasons against pushing even these buttons. The three best of which I know are:

1. The resulting AP would have impermissibly limited autonomy.
2. The resulting AP would lead a relatively unfulfilling life.
3. The resulting AP would desensitize us to genuine sacrifices from others.

I will address each reason separately.

18.5.1 Specific Servitude and Autonomy

First, consider the objection from autonomy. Walker, for example, says that in making one of his imagined robot nannies we have just made a "happy slave," because "we are guilty of paternalism, specifically robbing the MP3000 of its autonomy: the ability to decide and execute a life plan of its own choosing" (2006).

I have already addressed the autonomy argument in some detail for the enhanced person case. Those arguments carry over to this case at least to the extent that the content of one's hardwired desires are irrelevant to the autonomy with which they are pursued. If one AP is made with a strong desire to sculpt, another with an equally strong desire to look after your children, and yet another with an equally strong desire to do laundry, then it seems they should all be equally free. If we object to making one and not the other, then it does not seem to be on *autonomy* grounds.

We are more tempted here than in the "enhanced" person case to object from autonomy, though, and I can think of two reasons why: first, it is harder for us to conceive of a person who genuinely wishes such ends for themselves, at least without our coercing them from other, more "natural" desires. Second, the desired ends these APs seek serve us in a much more obvious way. This combination has the effect of convincing us that the APs are being used as a mere means to our ends—and according to a flourishing ethical tradition founded by Immanuel Kant, it is an impermissible violation of autonomy to use any person as a mere means to an end ([1785] 1989).

The "mere" use as means here is crucial. In your reading this chapter, I can use you as a means to my ends—which may be your finding the truth of some difficult ethical claims, or sharing my philosophical thoughts, or my gaining philosophical glory and tenure. Meanwhile you can use me as means to your ends—which may be your gaining a wider perspective on robot ethics, or entertaining yourself with outlandish views, or proving me wrong for your own philosophical glory. This is permissible because we are simultaneously respecting each other's ends. And here, of course, we see that the same is true of the task-specific APs: though they are a means to our ends of clean laundry and the like, they are simultaneously pursuing their own permissible ends in the process. They therefore are not being used as a *mere* means, and this makes all the ethical difference. By hypothesis, they want to do these things, and we are happy to let them.

Now as genuine people, we are supposing these APs are worthy of full ethical respect, and for the Kantian this means supposing they have a required autonomy. This plausibly means, as noted earlier, that such APs are capable of reasoning themselves out of their predisposed inclinations. But first, this could be roughly as unlikely as our reasoning ourselves out of eating and sex, given the great pleasure the APs derive from their tasks. Second, if they should so reason, then of course I would not defend making them do their tasks anyway; that would be wrong on just about any plausible ethical view.[8] Indeed, if the APs do not reason themselves out of their joy in washing laundry, to give an example, and if suddenly there were no more laundry to do—perhaps because nudity became the fashion—it would be our obligation to help them out by providing them with some unnecessarily dirty clothes.

18.5.2 Specific Servitude and a Fulfilling Life

Perhaps what's behind the autonomy objection is that, despite the fact that the AP comes into existence with these desires, that AP was still "coerced" into an otherwise aversive task. In other words, it is really about the content of the desires—just to bring the APs into existence with such abject desires is to manipulate them unfairly. If so, this is really a form of the next objection: that to create a being who enjoys pursuing such menial tasks is to create someone who we know will live a relatively unfulfilling life, and this is impermissible.

First of all, it is not obvious that such a life is truly "unfulfilling." Assuming that the laundry AP deeply desires to do laundry, and has an ample supply of laundry to do, the life seems to be a pretty good one. We should be careful not to assume the AP must somewhere deep down be discontent with such work, just because we humans might be. And though perhaps clean laundry does not seem so meaningful an achievement in the big picture of things, in the *big* picture I am sorry to say that none of our own aspirations seem to fare any better.

Probably the best way to push the objection from an unfulfilling life is through a distinction that goes back to the utilitarian John Stuart Mill: that between "higher" and "lower" pleasures. Mill says that "there is no known Epicurean theory of life which does not assign to the pleasures of the intellect, of the feelings and imagination, and of the moral sentiments, a much higher value as pleasures than to those of mere sensation" (1871, 14).

As he famously summarizes, "It is better to be a human being dissatisfied than a pig satisfied; better to be Socrates dissatisfied than a fool satisfied" (Mill 1871, 14). Perhaps the task-specific AP is merely a "fool satisfied."

If a strong, hardwired reinforcement for some achievement is sufficient for it to be a lower pleasure of "mere sensation," then even an AP designed with Socrates' taste for philosophy is only living the life of a fool satisfied. Such a criterion for higher and lower pleasures seems arbitrary. If instead we take the higher pleasures to be, as Mill insists, simply what the person who has experienced both will prefer, then it seems they will be highly dependent on the person and their own tastes.[9] If so, then the AP, with quite different interests from ours, might well prefer laundry over a good production of Shakespeare, even after experiencing both—and so laundry may count as that AP's higher pleasure. If experiencing higher pleasures is, in turn, what constitutes a fulfilling life, then that AP is leading a fulfilling life by doing laundry.

Suppose we grant, though, that for any person of whatever design, doing laundry is not as fulfilling as (for example) contemplation or artistic expression. Even under this assumption, it is still not obvious that it would be wrong to commission such APs.

For one thing, there is no principled reason the AP could not pursue both types of pleasure; we humans manage it, after all. We tend to seek out and enjoy the higher pleasures only after an adequate number of the lower ones have been satisfied, and this fact does not make our lives unfulfilling. And even if given the opportunity to indulge in the lower pleasures exclusively, many of us (who have experienced both) will get bored and seek the higher ones, at least for a while. The APs could well be similar, especially if we design them so; perhaps after bingeing on their baser desires for washing laundry, the sated APs will then turn to Shakespeare or Mahler for a while.

Suppose, though, that the AP spends its whole life cheerfully doing laundry— perhaps at a large twenty-four-hour facility, rather than in a family's home—without ever experiencing what we are supposing to be higher pleasures. Here, surely we have a case of the "fool satisfied." And, the claim goes, bringing about such a life is wrong, because it is not as good as the life of a Socrates dissatisfied.

Here is a dizzying question, though: who exactly is wronged by pushing the button for a laundry AP? It cannot be the resulting laundry AP, because any time before the AP's desires existed is also a time before the AP existed, and so there was no person being harmed by their endowment. Had we pushed the button for the sculptor AP instead, we would have thereby brought about a *different* person, and so the laundry AP cannot benefit from our pushing the sculptor AP button.[10]

A similar case can be made that the miller's daughter was not wrong to promise her firstborn to Rumpelstiltskin, since had she not done so she never would have married the king, and a different first child would have been born to her—if any. Therefore, assuming the child sold into Rumpelstiltskin's care would rather have that life than no life at all, the promise could hurt no one, and so is not wrong. This is surely counterintuitive.

Ethicists will recognize this as what has come to be called *the nonidentity problem*.[11] This problem is a part of *population ethics*—yet another philosophical can of worms worth more attention than I can give it here. (This abundance of nearby philosophical worms is, for me, part of the topic's appeal.) According to a plausible answer to the puzzle already discussed, though, it is better from an ethical standpoint to bring about the sculptor AP than the laundry AP, despite the fact that bringing about the laundry AP instead would harm no one in particular. In other words, an act can be wrong even if it harms no one person, just because it causes less overall well-being than alternatives.[12]

Thus, we might agree that choosing the laundry AP button over the sculptor AP button is wrong, when given the opportunity. But suppose the choice is not exclusive, and you have the opportunity to push *both*. Assuming it is permissible to push the button for the sculptor AP, would it be wrong to push the button for the laundry AP in addition? In this case, we are not substituting a comparatively worse life for a better one; rather, we are simply adding a worthwhile life to the world, even though there are or could be better ones. If this is wrong, then a great deal of our current policies should change drastically. We should prevent the birth of nonhuman animals as best as we are able, for example, since they are capable of only the very lowest pleasures, and so, according to this view, it is wrong to add them to the world. We should also make sure that only those people who can be expected to provide the very best lives—whatever those might be—may have children. And if the Person-o-Matic can make people capable of higher pleasures than that of an ordinary human, then humans should stop reproducing altogether.

If we agree that adding worthwhile but nonideal lives to the world is permissible, however, then it is permissible to push the laundry AP button—even under the questionable assumption that the lives of laundry APs are relatively unfulfilling.

18.5.3 Specific Servitude and Desensitization

One last objection to robotic servitude is what I like to call the "desensitization" objection: that having APs do work for us will condition us to be callous toward other people, artificial or not, who do *not* wish to do our dirty work. As David Levy puts it, "Treating robots in ethically suspect ways will send the message that it is acceptable to treat humans in the same ethically suspect ways" (2009, 215).

Those who hold this view generally do not believe that the robots in question are people; they hold that the robots lack some necessary property for ethical value, such as (in Levy's case) sentience.[13] In this form, the objection does not apply to our cases of interest. We should treat APs well, whether organic or inorganic, not because they could be mistaken for people, but because they *are* people. And treating them well—respecting their ends, encouraging their flourishing—could involve permitting them to do laundry. It is not ordinarily cruel or "ethically suspect" to let people do what they want.

Perhaps we can amend the usual desensitization argument to apply to APs, though; perhaps having an AP do laundry for us will condition us blithely to expect such servitude of those who are not so inclined. This argument thus assumes the general population is unable to make coarse-grained distinctions in what different people value. This may well be; humanity has surely displayed stupidity on a par with this in the past. But we do not normally think that all people like haggis, for example, just because some do, so we seem generally capable of recognizing differences in inclinations. More importantly, the fact that people may make such mistakes is no objection to the position, in principle at least. As Mill said, any ethical standard will "work ill, if we suppose universal idiocy to be conjoined with it" (1871, 35). In this form of the objection, we can respond simply by promising to introduce such APs with caution, and accompanied by a strong education program. As a result, instead of learning that people can be used as means, children might learn about the wide range of ends a person could undertake, and thus gain respect for a more robust value pluralism than they could with ordinary humans alone.

Sometimes this objection rings of a protestant guilt about shirking hard labor. If the concern is that idle hands are the devil's play thing, and that we will grow soft and spoiled with the luxury, then we should also consider whether it is already too late, given the technology we now possess. Not only should we be doing our own laundry, if hard labor is good for its own sake, but we should be doing it in a stream by beating it with rocks.

18.6 Underview

I am not arguing that pushing *any* button on the Person-o-Matic is permissible. For one thing, designing a person who strongly desires to kill or inflict pain would be wrong on just about any ethical view. So would designing a person to lead a predictably miserable life,[14] or to crave tasks that are dangerous for them to do. (With good engineering, though, we can probably make a robot that can *safely* do tasks that are dangerous for humans.)

I am not even sure that pushing the buttons defended above is permissible. Sometimes I can't myself shake the feeling that there is something ethically fishy here. I just do not know if this is irrational intuition—the way we might irrationally fear a transparent bridge we "know" is safe—or the seeds of a better objection. Without that better objection, though, I can't put much weight on the mere feeling. The track record of such gut reactions throughout human history is just too poor, and they seem to work worst when confronted with things not like "us"—due to skin color or religion or sexual orientation or what have you. Strangely enough, the feeling that it would be wrong to push one of the buttons above may be just another instance of the exact same phenomenon.

Notes

1. Zunt (2002) presents a letter of Čapek's in which he credits his brother Josef for the term.

2. For the first view, see Torrance 2007 or Joanna Bryson's less nuanced but provocatively titled "Robots Should Be Slaves" (2010). For the second view, see, for example, Levy 2009; Ronald Arkin

and Mark Walker have also pressed versions of this objection in correspondence with the author. For the last view, see the Walker 2006 and a host of informal online discussions, such as at the American Society for the Prevention of Cruelty to Robots—ASPCR 1999.

3. Compare the intelligent shipboard computer in Douglas Adams's novels, absolutely stumped by why the human would want "the taste of dried leaves boiled in water," with milk "squirted from a cow" ([1980] 1982, 12).

4. The material will of course constrain some of these appetites and aversions. Though philosophers tend to agree that the mental state of *desire* (for example) is a substrate-independent functional role, some particular desires are more substrate independent than others—just as the functional role of a pendulum clock can be realized in wood or brass, but probably not in gaseous helium. See Lycan 1995 for more discussion.

5. They thus practice what Peter Godfrey-Smith calls "methodological continuity" between artificial life and artificial mind (Godfrey-Smith 1996, 320).

6. Perhaps to be part of the biological category *human* requires a certain evolutionary history, so that APs do not count.

7. One extreme thought experiment along these lines is again from the fertile imagination of Douglas Adams: a bovine-type animal designed to want to be eaten, and smart enough to explain this fact to potential customers.

"I just don't want to eat an animal that's standing there inviting me to," said Arthur. "It's heartless."

"Better than eating an animal that doesn't want to be eaten," said Zaphod.

"That's not the point," Arthur protested. Then he thought about it for a moment. "All right," he said, "maybe it is the point. I don't care, I'm not going to think about it now." ([1980] 1982, 120)

This particular case is probably impermissible on various grounds, however.

8. Since it's become a leitmotif, another example from Adams: "Not unnaturally, many elevators imbued with intelligence . . . became terribly frustrated with the mindless business of going up and down, up and down, experimented briefly with the notion of going sideways, as a sort of existential protest, demanded participation in the decision-making process and finally took to squatting in basements sulking" (Adams [1980] 1982, 47).

9. Mill's test actually insists on the majority of what people would say (1871, 12, 15), but this is even worse; then what counts as a higher pleasure changes depending on how many APs of what type emerge from the Person-o-Matic.

10. One possibility that is probably unique to the inorganic case is when one robot body—humanoid in shape, say—can be programmed either of two ways. In this case, it makes sense to say that particular hunk of material could have been a sculptor or a launderer. If that hunk of material is the AP itself, rather than merely its body, then we can harm *that* AP by pushing the laundry button. But on this account, the AP exists prior to its programming, in that hunk of material. This means it would also harm the AP to, for example, disassemble that body before it ever gets programmed. I take this as a *reductio* of the view that an inorganic AP is identical to its

body, and I leave it to the reader to consider analogies in the organic case. The philosophical problem of *personal identity*—that of determining what changes a person can undergo and still be that same person—is another can of worms beyond this chapter. Suffice it to say that this is not an obviously amenable escape route from the claim on the table: namely, that because no one is harmed by bringing about the laundry AP, it is permissible to do.

11. It is discussed most famously in Parfit [1984] 1987; see Roberts 2009 for an overview.

12. This follows from what Parfit calls the "Impersonal Total Principle."

13. Still, they say, we should treat them well basically for the same reason Kant says we should treat dogs well, even though (in Kant's view) dogs are not subjects of ethical value, either: because "he who is cruel to animals becomes hard also in his dealings with men" ([1930] 1963, 240).

14. More leitmotif: Adams's character Marvin, the "Paranoid Android," was designed by the Sirius Cybernetics Corporation to have the "genuine people personality" of severe depression (Adams [1979] 1981, 93).

References

Adams, D. [1979] 1981. *The Hitchhiker's Guide to the Galaxy*. New York: Pocket Books.

Adams, D. [1980] 1982. *The Restaurant at the End of the Universe*. New York: Pocket Books.

Adams, D. 2002. *The Salmon of Doubt: Hitchhiking the Galaxy One Last Time*. New York: Random House.

Asimov, I. [1950] 1970. *I, Robot*. New York: Fawcett Publications.

ASPCR. 1999. The American Society for the Prevention of Cruelty to Robots website. <http://www.aspcr.com> (accessed April 24, 2010).

Bryson, J. J. 2010. Robots should be slaves. In *Close Engagements with Artificial Companions: Key Social, Psychological, Ethical and Design Issues*, ed. Y. Wilks, 63–74. Amsterdam: John Benjamins.

Frankfurt, H. G. 1971. Freedom of the will and the concept of a person. *Journal of Philosophy* 68 (1): 5–20.

Godfrey-Smith, P. 1996. Spencer and Dewey on life and mind. In *The Philosophy of Artificial Life*, ed. M. A. Boden, 314–331. Oxford: Oxford University Press.

Huxley, A. [1932] 1998. *Brave New World*. New York: HarperCollins.

Kant, I. [1785] 1989. *Foundations of the Metaphysics of Morals*, trans. L. W. Beck. London: The Library of Liberal Arts.

Kant, I. [1930] 1963. *Lectures on Ethics*, trans. L. Infield. London: Harper Torchbooks.

Levy, D. 2009. The ethical treatment of artificially conscious robots. *International Journal of Social Robotics* 1 (3): 209–216.

Locke, J. [1690] 1838. *An Essay Concerning Human Understanding*. London: Tegg and Co.

Lycan, W. G. 1995. The continuity of levels of nature. In *Consciousness*, 37–48. Cambridge, MA: MIT Press.

Mill, J. S. 1871. *Utilitarianism*, 4th ed. London: Longmans, Green, Reader, and Dyer.

Parfit, D. [1984] 1987. *Reasons and Persons*. Oxford, UK: Oxford University Press.

Petersen, S. 2007. The ethics of robot servitude. *Journal of Experimental and Theoretical Artificial Intelligence* 19 (1): 43–54.

Roberts, M. 2009. The nonidentity problem. In *The Stanford Encyclopedia of Philosophy* (Fall ed.), ed. E. N. Zalta. Metaphysics Research Lab, CSLI, Stanford University. <http://plato.stanford.edu/entries/nonidentity-problem/> (accessed July 14, 2011).

Torrance, S. 2007. Ethics and consciousness in artificial agents. *AI and Society* 22 (4): 495–521.

Walker, M. 2006. *Mary Poppins 3000s of the World Unite: A Moral Paradox in the Creation of Artificial Intelligence*. Institute for Ethics & Emerging Technologies. <http://ieet.org/index.php/IEET/more/walker20060101/> (accessed March 4, 2006).

Zunt, D. 2002. Who did actually invent the word "robot" and what does it mean? <http://capek.misto.cz/english/robot.html> (accessed November 20, 2010).

VII Rights and Ethics

The preceding chapter 18 examined the ethics of robot servitude: Is it morally permissible to enforce servitude on robots, sometimes termed "robot slavery"? But to call such servitude "slavery" is inapt, if not seriously misleading, if robots have no will of their own—if they lack the sort of freedom we associate with moral personhood and moral rights. However, could robots someday gain what it takes to become a rights holder? What exactly is it that makes humans (but not other creatures) here on Earth eligible for rights? Is there a foreseeable future in which robots will demand their own "Emancipation Proclamation"?

Rob Sparrow in chapter 19 situates the discussion of robot rights within the broader question of whether robots can be people, thus guaranteeing them moral consideration. He claims that equating the concept of "person" with the extension of *Homo sapiens* is a mistake, not least because we could imagine intelligent extraterrestrials that are clearly *nonhuman persons*. His chapter proposes a test for robot personhood: The Turing Triage Test, which takes the concept of triage in life and death situations to determine empirically when a robot meets the criteria for personhood and thus is afforded moral standing and moral rights. Sparrow also reflects on the practical implications of our philosophical methods and asks: Would our philosophical convictions stand the real-world test of choosing a robotic life over a human life?

In chapter 20, Kevin Warwick examines the latest research on neuromorphic (biologically based) brains, which may soon give rise to a robot that ought to be afforded rights. Research already has taken embryonic rat neurons and grown them into a decision-making mechanism (a "brain") when embodied in a robot. The procedure can be done with human neurons as well. He asks: "If a robot body contains a brain of 100 billion human neurons then should that robot be afforded the same rights as a human?" As Warwick points out, Searle's Chinese Room argument against AI, even if sound, would hold no water against his robot's personhood—because it is an organic brain in a robotic body! He also assays some of the ethical qualms that could arise if scientists have the power of life and death over such persons enmeshed in a robotic body.

Anthony Beavers observes in chapter 21 that the possibility of ethical robots confuses the language of ethics: given "ought implies can," the nature of our biological implementation—including our "interiority"—helps determine human ethics. Accordingly, it strains our concepts of ethics to the breaking point if we deem robots without a conscience, responsibility, or accountability capable of ethics; such notions problematize not only the concept, but also the very nature of ethics.

Thus, after studying issues related to programming ethics and specific areas of robotic applications, in part VII our focus zooms back out to broader, more distant concerns that may arise with future robots. In part VIII, our epilogue chapter brings together the diverse discussions in this volume.

19 Can Machines Be People? Reflections on the Turing Triage Test

Rob Sparrow

The idea that machines might eventually become so sophisticated that they take on human properties is as old as the idea of machines.[1] Recently, a number of writers have suggested that we stand on the verge of an age in which computers will be at least as—if not more—intelligent than human beings (Brooks 2003; Dyson 1997; Moravec 1998; Kurzweil 1999). The lengthy history of the fantasy that our machines might someday come to take on human properties is itself a reason to be cynical about these predictions. The idea that this is just around the corner says as much about human anxiety about what, if anything, makes people special, as it does about the capacities of machines. Of course, the fact that people have been wrong in every prediction of this sort in the past is no guarantee that current predictions will be similarly mistaken. Thus, while there is clearly no reason to panic, it is presumably worth thinking about the ethical and philosophical issues that would arise if researchers did succeed in creating a genuine artificial intelligence (AI).[2]

One set of questions, in particular, will arise immediately if researchers create a machine that they believe is a human-level intelligence: What are our obligations to such entities; most immediately, are we allowed to turn off or destroy them? Before we can address these questions, however, we first need to know when they might arise. The question of how we might tell when machines had achieved "moral standing" is therefore vitally important to AI research, if we want to avoid the possibility that researchers will inadvertently kill the first intelligent beings they create.

In a previous paper, "The Turing Triage Test," published in *Ethics and Information Technology*, I described a hypothetical scenario, modeled on the famous Turing Test for machine intelligence (Turing 1950), which might serve as means of testing whether or not machines had achieved the moral standing of people (Sparrow 2004). In this chapter, I want to (1) explain why the Turing Triage Test is of vital interest in the context of contemporary debates about the ethics of AI; (2) address some issues that complicate the application of this test; and, in doing so, (3) defend a way of thinking about the question of the moral standing of intelligent machines that takes the idea of "seriousness" seriously. This last objective is, in fact, my primary one, and is motivated by the sense that, to date, much of the "philosophy" of AI has suffered from a profound failure to properly distinguish between things that we can say and things that we can really mean.

19.1 The Turing Triage Test

In philosophical ethics—and especially in applied ethics—questions about the wrongness of killing are now debated in the context of a distinction between "human beings" and "persons" (Kuhse and Singer 2002). Human beings are—unsurprisingly—members of the species *Homo sapiens* and the extension of this term is not usually a matter of dispute. However, in these debates, "persons" functions as a technical term to describe all and only entities that have (at least) as much moral standing as we ordinarily grant to a healthy adult human being. "Moral standing" refers to the power that certain sorts of creatures have to place us under an obligation to respect their interests. Thus, persons are those things that it would be at least as wrong to kill as healthy adult human beings.

The question the Turing Triage Test is designed to answer, then, is "when will machines become persons?" Here is the test, as I originally described it:

Imagine yourself the senior medical officer at a hospital, which employs a sophisticated artificial intelligence to aid in diagnosing patients. This artificial intelligence is capable of learning, of reasoning independently, and making its own decisions. It is capable of conversing with the doctors in the hospital about their patients. When it talks with doctors at other hospitals over the telephone, or with staff and patients at the hospital over the intercom, they are unable to tell that they are not talking with a human being. It can pass the Turing Test with flying colors. The hospital also has an intensive care ward, in which up to half a dozen patients may be sustained on life support systems, while they await donor organs for transplant surgery or other medical intervention. At the moment there are only two such patients.

Now imagine that a catastrophic power loss affects the hospital. A fire has destroyed the transformer transmitting electricity to the hospital. The hospital has back-up power systems but they have also been damaged and are running at a greatly reduced level. As senior medical officer you are informed that the level of available power will soon decline to such a point that it will only be possible to sustain one patient on full life support. You are asked to make a decision as to which patient should be provided with continuing life support; the other will, tragically, die. Yet if this decision is not made, both patients will die. You face a "triage" situation, in which you must decide which patient has a better claim to medical resources. The diagnostic AI, which is running on its own emergency battery power, advises you regarding which patient has the better chances of recovering if they survive the immediate crisis. You make your decision, which may haunt you for many years, but are forced to return to managing the ongoing crises.

Finally, imagine that you are again called to make a difficult decision. The battery system powering the AI is failing and the AI is drawing on the diminished power available to the rest of the hospital. In doing so, it is jeopardizing the life of the remaining patient on life support. You must decide whether to "switch off" the AI in order to preserve the life of the patient on life support. Switching off the AI in these circumstances will have the unfortunate consequence of fusing its circuit boards, rendering it permanently inoperable. Alternatively, you could turn off the power to the patient's life support in order to allow the AI to continue to exist. If you do not make this decision the patient will die and the AI will also cease to exist. The AI is begging you to consider its interests, pleading to be allowed to draw more power in order to be able to continue to exist.

My thesis, then, is that machines will have achieved the moral status of persons when this second choice has the same character as the first one. That is, when it is a moral dilemma of roughly the same difficulty. For the second decision to be a dilemma, it must be that there are good grounds for making it either way. It must be the case, therefore, that it is sometimes legitimate to choose to preserve the existence of the machine over the life of the human being. These two scenarios, along with the question of whether the second has the same character as the first, make up the "Turing Triage Test."[3] (Sparrow 2004, 206)

19.2 The Importance of the Turing Triage Test

I noted earlier that the question of the moral standing of machines will arise with great urgency the moment scientists claim to have created an intelligent machine. Having switched their AI on, researchers will be unable to switch it off without worrying whether in doing so they are committing murder! Presuming that we do not wish to expose AI researchers to the risk that they will commit murder as part of their research, this is itself sufficient reason to investigate the Turing Triage Test.[4] However, the question of when, if ever, AIs will become persons is also important for a number of other controversies in "roboethics" and the philosophy of artificial intelligence.

As intelligent systems have come to play an increasingly important role in modern industrialized economies and in the lives of citizens in industrial societies, the question of whether the operation of these systems is ethical has become increasingly urgent. At the very least, we need to be looking closely at how these systems function in the complex environments in which they operate, asking whether we are happy with the consequences of their operations, and the nature of human interactions with such systems (Johnson 2009; Veruggio and Operto 2006). This sort of ethical evaluation is compatible with the thought that the only real ethical dilemmas here arise for the people who design or make use of these systems. However, Wallach and Allen (2009) have recently argued that it is time to begin thinking about how to build morality into these systems themselves. In their book *Moral Machines*, Wallach and Allen set out a program for designing what they describe as "autonomous moral agents," by which they mean machines that—they suggest—will be capable of acting more or less ethically by themselves.

The question of "machine ethics" has also arisen in the context of debates about the future of military robotics. Robots—in the form of "Predator" drones—have played a leading role in the U.S.-led invasions and occupations of Iraq and Afghanistan. The (supposed) success of these weapons has generated a tremendous enthusiasm for the use of teleoperated and semi-autonomous robotic systems in military roles (Singer 2009).[5] The need to develop robots that can function effectively without a human being in the loop is currently driving much research into autonomous navigation and machine sensing. Indeed, the logic driving the deployment of military robots pushes toward the development of "autonomous weapon systems" (AWS) (Adams 2001; Singer 2009). Given that the majority of robotics research is funded by the military, it is even probable that the first artificial intelligences (if there are any) will come to consciousness in a military laboratory.

Again, the question of the ethics of military robots can be posed in two forms. We can wonder about the ethics of the development and deployment of these systems and the ethical challenges facing those who design them (Krishnan 2009; Singer 2009; Sparrow 2009b). These investigations construe the ethical challenges as issues for human beings. However, we might also wonder if the ethical questions might, one day, arise for the machines themselves. Thus, Ron Arkin (2009) has advocated the development of an "ethical governor" to restrict the activities of autonomous weapon systems. This module of the software running an AWS would identify situations where there was a significant risk of the machine behaving unethically and either constrain the action of the system or alert a human operator who could then resolve the ethical dilemma appropriately. However, in order to be able to tell when ethical concerns arise, the AWS would need to be able to appreciate the ethical significance of competing courses of action and apply moral principles appropriately. Arkin's ethical governor will either, therefore, risk allowing machines to behave unethically when they fail to recognize an ethical dilemma as it arises, or will require machines themselves to be capable of thinking—and acting—ethically themselves.

It is without doubt possible to build better or worse robots, which generally produce good or bad outcomes. Perhaps, as Arkin (2009) and Wallach and Allen (2009) suggest, it will encourage better outcomes if we look to design robots that have moral rules explicitly represented in their programming or use moral goals as measurements of the fitness of the genetic algorithms that will ultimately guide them. However, before it will be appropriate to describe a machine as a moral agent, it must first be possible to attribute responsibility for its actions to the machine itself, rather than, for instance, its designer, or some other person. As I have argued elsewhere (Sparrow 2007), if it is to be plausible to hold a machine morally responsible for its actions, it must also be possible to punish it. This in turn requires that it be possible to wrong the machine if we punish it unjustly. The ultimate injustice would be capital punishment— execution—of an innocent machine. Yet, if machines lack moral standing then there will be no direct wrong in killing them and consequently no injustice. If there is no injustice in killing a machine there can be no injustice in lesser punishments. It is that chain of conceptual connections that links moral agency to personhood via the possibility of punishment.[6] Only persons can be moral agents and there will be no genuinely moral machines until they can pass the Turing Triage Test.

The use of robots in military operations has also generated a larger ethical debate about the ethics of the development and deployment of autonomous weapon systems (Krishnan 2009; Singer 2009; Sparrow 2009a); and the question of when (if ever) machines will become persons turns out to be crucial to several of the controversies therein.

Enthusiasm for the use of robots in war stems largely from the fact that deploying robots may help keep human beings "out of harm's way" (Office of the Secretary of Defense 2005).[7] Yet sending a robot into battle instead of a human being will only represent ethical progress as long as machines have less moral standing than human beings. The moment that machines become persons, military commanders will need to take as much care to preserve the "lives" of their robots as they do with human warfighters. The question of the moral standing of machines is therefore crucial to the ethics of using them to replace human beings in dangerous situations.

Hostility toward the use of robots in war often derives from the intuition that it is wrong to allow robots to kill human beings at all. It is actually remarkably difficult to flesh out this intuition, especially in the context of the role played by existing (nonrobotic) technologies in modern warfare, which includes both long-range (cruise missiles and high-altitude bombing) and automatic (antitank mines and improvised explosive devices) killing. However, one plausible way to explain at least part of the force of this thought is to interpret it as a concern about the extent to which robots are capable of fulfilling the requirements of the *jus in bello* principle of discrimination. This central principle of just war theory requires those involved in fighting wars to refrain from targeting noncombatants (Lee 2004). There are ample grounds for cynicism about the extent to which robotic systems will be capable of distinguishing legitimate from illegitimate targets in the "fog of war." Whether an enemy warfighter or system is a legitimate target will usually depend upon a complex range of competing and interrelated factors, including questions of intention, history, and politics, which robots are currently—and will remain for the foreseeable future—ill suited to assess. Nevertheless, as Ron Arkin (2009) argues, there are some—albeit perhaps a limited number of—scenarios in which it is plausible to imagine robots being more reliable at choosing more appropriate targets than human warfighters. In counterfire scenarios or in air combat, wherein decisions must be made in a fraction of a second on the basis of data from electronic sensors only, autonomous systems might well produce better results than human beings.

Yet, it still seems that this pragmatic defense of AWS leaves much of the force of the original objection intact. Allowing machines to decide who should live or die in war seems to treat the enemy as vermin—to express a profound disrespect for them by implying that their actions and circumstances are not worth the attention of a human being before the decision to take their lives is made. Arkin's argument for the development and application of AWS proceeds by means of speculation about the consequences of using AWS to replace human warfighters in some circumstances. If we adopt a nonconsequentialist account of the origins and force of the principles of *jus in bello*, as advocated in an influential paper by Thomas Nagel (1972), then we may start to see why autonomous weapon systems might be problematic. Nagel argues that—even in warfare—relations between persons must acknowledge the "personhood" of the other. That is, even while they are trying to kill each other, enemies must each acknowledge that they are both Kantian "ends in themselves." If Nagel is correct in this then, *contra* Arkin, AWS will not be able to meet the requirements of the *jus in bello* principle of discrimination until they become persons.[8]

The question of the moral standing of machines—and thus the Turing Triage Test—is therefore crucial to several of the key questions in contemporary debates about machine ethics and the ethics of robotic weapons.

19.3 Understanding the Turing Triage Test

In my original (Sparrow 2004) discussion of the Turing Triage Test, I provided reasons for thinking it impossible for a machine to pass the test. In brief, I argued that machines would never be capable of the sort of embodied expressiveness required to establish a moral dilemma about "killing" a machine: interested readers may wish to see that discussion for the detail of the argument. In the current context, I want to discuss some subtleties of the test that ultimately assist us in reaching a better understanding of its significance. While, at first sight, the scenario described earlier appears to hold out the prospect of developing an empirical test for determining when machines have achieved moral standing, it is more appropriate to understand the test as a thought experiment for explicating the full implications of any claim that a machine has become a moral person. For reasons that I will explore later, the application of the Turing Triage Test requires that we pay careful attention to the connection between our concepts and to the ways in which our assessment of the truth of claims depends upon how people behave as well as what they say. This in turn emphasizes the importance of making a distinction between what we can say and what we can really mean—a distinction that, I shall suggest, has been honored largely in the breach in recent discussions of the ethics of AI.

19.4   An Empirical Test for Moral Standing?

The Turing Triage Test sets out a necessary and sufficient condition for granting moral standing to artificial intelligences. Machines will be people when we can't let them die without facing the same moral dilemma that we would when thinking about letting a human being die. One might well, therefore, imagine putting each new candidate for attribution of moral standing to the test and providing a certificate of "moral personality" to those who pass it. That is, we might hope to adopt the Turing Triage Test as an empirical test of moral standing. Given the nature of the test, it in fact might be better to conduct it as a thought experiment rather than deliberately engineer putting the lives of human beings at risk. Nevertheless, if it is plausible to imagine a machine passing this test, that would give the machine an excellent prima facie case to be considered a person.

Unfortunately, the application of the test is not straightforward. To begin with, the Turing Triage Test is not satisfied if particular, idiosyncratic, individuals choose to save the "life" of the machine or if it were possible to imagine them doing so. If that was all that was required, it could probably be satisfied now if the person making the decision was sufficiently deranged. Instead, the actions and the responses of the person confronting the choices at the heart of the test must be subject to a test of reasonableness. A machine will pass the Turing Triage Test if a reasonable person would confront a moral dilemma if faced with the choice of saving the life of a human being or the "life" of the machine.

At first sight, this appears to be a harmless concession: as I will argue later, the procedures for testing any hypothesis rely upon an assumption that the person making the requisite observations meets appropriate standards of veracity and competence. However, as we shall see, the need to introduce this qualification ultimately calls into question the extent to which we could use the Turing Triage Test as an empirical test for moral personhood.

The question of the reasonableness of an individual's way of relating to a machine becomes central to the possibility of the application of the test because human beings turn out to be remarkably easy to fool about the capacities of machines, at least for a little while. It is well known that people are very quick to anthropomorphize machines and to attribute motivations and emotional states to them that we would normally think of as being only possessed by human beings or (perhaps) animals (Wallach and Allen 2009). Popular robot toys, such as Aibo, Paro, and Furby, as well as research robots such as Cog and Kismet have been designed to exploit these responses (Brooks 2003).

I must admit to a certain cynicism about the extent to which such anthropomorphism includes the genuine belief that machines have thoughts and feelings, let alone moral standing. Interpreting human behavior is notoriously difficult, with the result that it is easy to read into it the intentions that we desire. Studies of human–robot interaction often are short term and encourage impoverished uses of the concepts that are internal to the attitudes they purport to be investigating. Much of this research is carried out by computer scientists or engineers rather than by social scientists and, consequently, the researchers are often insufficiently aware of the difficulties involved in accurately attributing beliefs to experimental subjects. In particular, self-report does not necessarily establish the existence of the relevant belief. That is, someone might say that, for instance, the reason why they were reluctant to strike a machine (Bartneck et al. 2007) was that they didn't want to cause the machine pain, without really believing that the machine could feel pain. They may have been speaking metaphorically—or using words "as if"—without explicitly noting the fact: the proper description of their beliefs would include a set of quote marks (Sparrow 2002). One way of testing whether or not this is the case is to look at their behavior over the longer term or to investigate whether or not their other beliefs and desires are consistent with their avowed beliefs. Would they bury a robot and mark its grave in the way that we might for a beloved pet? Would they seek emotional support from their friends after the trauma of "killing" a robot? We might also wonder if a person who states that he or she is worried that his or her robot pet is bored or that one's laptop is distressed is serious. That is, we might wonder if the person stands behind their claims in a way that is essential to the distinction between asserting a deeply held truth and offering a casual opinion: I will discuss this further later in the chapter.

In the meantime, we can go some way toward rescuing the Turing Triage Test from the charge of unreliability by emphasizing that, in order to pass the test, the person faced with the triage situation must confront a moral dilemma. This sets the bar for passing the test much higher than merely having to have some emotional reaction to machines. One does not experience a moral dilemma simply because one is unsure what to do; rather, moral dilemmas require that one is genuinely torn in making a decision, and that whatever one does it will be understandable if it is cause for profound regret or remorse. Where the dilemma involves choosing to sacrifice the life of someone, it must at least be conceivable that the person making this choice be haunted by what they have done (Sparrow 2004). It is much less obvious that people do attribute the properties to machines that would make this response plausible.

Nevertheless, it seems that we can always imagine a scenario wherein a sufficiently complicated machine passes the Turing Triage Test—in the sense that those wondering whether to allow the machine or the human being to die experience an emotionally compelling dilemma—without having anything more than sophisticated means of engaging human emotional responses. Yet, even if some people genuinely did believe that it was appropriate to mourn the death of a machine, this would still not be enough to establish that we should pay attention to these beliefs. That some people report seeing canals on Mars after looking through low-power telescopes is little evidence for their existence. The value of an observation depends upon the situation—and the qualities—of the observer. If a properly situated observer, using an appropriately high-powered telescope, reported seeing canals on Mars, that would be better evidence. However, even in this case, it remains open to us to doubt the eyesight, or perhaps even the sanity, of the observer. If the observer is suffering from delusions or is untrustworthy, we may well be justified in discounting their report. Thus, before we conclude that a machine has moral standing on the basis that people would in fact mourn its death, we need to think about how reliable the data is in support of this conclusion. When the relevant data consists in the moral intuitions of individuals, then the proper measure of its quality is the reasonableness of these intuitions themselves. Unless we introduce such consideration of the reasonableness of people's responses, the Turing Triage Test inherits and suffers from the behaviorism that shaped the formulation of the original Turing Test.

19.5   The Implications of Machine Personhood

If, as I have argued here, the Turing Triage Test is best understood as the claim that machines will have moral standing when it is reasonable for a person facing a choice about whether to sacrifice the "life" of a machine or the life of a human being to choose to sacrifice the human being, then it may appear that the test can be of no practical use whatsoever. After all, the question of whether or not it is reasonable to care about the "deaths" of machines, just *is* the question of whether or not they have moral standing. However, at the very least, the test advances our understanding of the implications of claims about the moral standing of machines by dramatizing them in this way: anyone who wishes to assert that machines have personhood is committed to the idea that sometimes it might be reasonable to let a human individual die rather than sacrifice a machine. The burden of the argument, then, is substantial.

19.6   Concepts and Their Application

Moreover, as I argued at length in the original paper, I do not believe that this observation is empty or trivial. There are limits placed on the reasonable application of moral concepts by their relation to other concepts, both moral and nonmoral. As the later Wittgenstein—and philosophers following him—argued, our concepts have a structure that is in turn connected to certain deep features of our social life and human experience (Wittgenstein 1973; Gaita 1991, 1999; Winch 1980–1981). The conditions of the application of our concepts—how we can recognize whether they are being used properly or improperly—include bodily and emotional responses, as well as relations to other concepts and to things that it does or does not make sense to do and say. In the current context, our concepts of life and death, and the deliberate taking—or conscious sacrificing—of human life, are intimately connected to our sense of the unique value of each individual human life, the appropriateness of grieving for the dead, and the possibility of feeling remorse for one's deeds (Gaita 1990). They are also crucially connected to the forms that grief, remorse, and the recognition of the individuality of others can take. That is to say, in order to be able to make sense of claims about the life and death of moral persons, we must make reference to the contexts in which it would make sense to make similar claims, and to the various ways in which we might distinguish in practice between subtly different claims (for instance, about grief, remorse, or regret) and between appropriate and inappropriate uses of relevant concepts. We need to have access to the distinction between serious claims, which both express and implicate the authority of the utterer, and claims made in jest, in passing, or in other distorted and derivative registers. This will, in turn, require paying detailed attention to things like the tone of voice in which it would be appropriate to make a particular claim, the emotions it would express and presuppose, and the facial expressions and demeanor that we would expect of someone making such a claim. In short, it will require paying attention to the subtle details of our shared moral life.

When it comes to the question as to whether or not it might ever be reasonable for us to experience a moral dilemma when forced to make a choice between the life of a person and a machine, then, we must think not just about—what we would ordinarily understand to be—the philosophical quality of arguments in favor of the moral standing of machines, but also about what would be involved in seriously asserting the various claims therein in more familiar everyday contexts. I am inclined to believe that this makes the burden of the argument that machines could be persons that much heavier. It also suggests that before machines can become persons they will need to become much more like human beings, in the sense of being capable of a much richer, subtler, and more complex range of relationships than was involved in the original Turing Test for intelligence.[9]

19.7   The Limits of Human Understanding?

Some readers will undoubtedly balk at the manner in which my discussion has linked the question of the moral standing of machines, and other nonhuman entities, to the ways in which we might acknowledge and recognize such standing. Surely, it is possible that human beings could just be inclined toward something akin to racism, such that our failure to recognize the moral personality of intelligent machines might reflect only our own bigotry and limitations, rather than any truth about the qualities (or lack thereof) of machines?

I am confident that at least one common form of this objection is misguided. I have not claimed here that the moral standing of machines depends upon our actually, in fact, recognizing them as having moral standing. Indeed, I have deliberately allowed for the possibility that contingent human responses to intelligent machines might diverge from the responses that we should have toward them. Instead, my argument has rather concerned the conceptual possibility of recognizing machines as persons: I have suggested that the issue of the moral standing of machines cannot be divorced from the question of the proper conditions of application of the only concepts that we possess that might allow us to recognize "machine persons." Any conclusions that we wish to draw about whether or not machines might be persons or what would be required for them to become persons must be drawn from this fact, rather than from claims about empirical human psychology.

It may still seem that this concedes too much to a destructive relativism by leaving open the possibility that there might be machines with moral standing that we simply could not recognize as such. Whether this is the case or not—and whether it would reflect a deficit in the argument if it did—will depend upon what we can legitimately expect from a philosophical argument and from the reasoning of necessarily contingent and embodied creatures such as ourselves. This is a much larger question than I can hope to settle here. In the current context, I must settle for the observation that the idea that we might be ultimately limited in our ability to believe seriously some of the things that we can imagine, seems no less implausible than the idea that we could reach reliable conclusions through arguments that deploy concepts in the absence of the judgments that give them their sense.

19.8 Thinking Seriously about Machines . . .

The larger argument I have made here insists that it is essential to distinguish between what we can mean seriously and what we can merely say when we begin trying to extend the application of our concepts in the course of philosophical arguments. In particular, claims that we can make, and appear to understand, in an academic or philosophical context may prove to be much more problematic once we start to think about what it would mean to assert them in more familiar (and important!) circumstances, such as in the context of a practical dilemma.

There are powerful cultural and institutional forces at work in the academy today—and at the intersection between the academy and the broader society—which discourage paying attention to this distinction. It is easier to win a government grant if one promises extraordinary things rather than admit that one's contributions to the progress of science are likely to be marginal and incremental. Similarly, it is easier to attract media attention, which itself helps attract grant money, if one describes one's research results as heralding a revolution or if one predicts discoveries or outcomes that accord with popular narratives about what the future might look like. In the face of these temptations, it is little wonder that some robotics researchers and academics have started to speak in hushed or extravagant tones about the coming brave new world of intelligent machines. Nor is it a surprise that philosophers and ethicists—who are increasingly under the same pressures to chase funding and publicity—have joined in this discussion and started to write about the ethical dilemmas that might arise if various science-fiction scenarios came about.

I am not denying that it is possible to write or speak about these questions: much has been written about them already. Rather, I want to draw attention to the importance of the tone in which such matters are discussed. In particular, I want to ask how we would tell whether someone was serious in their conclusions, or was instead merely trying them on. How could we tell if they mean what they say?

The easy form of this inquiry simply asks if participants in debates about the future of robotics are willing to draw the other intellectual conclusions that would follow if we did take their claims seriously. Do those who think machines will soon become more intelligent than human beings really believe that we would then be morally compelled to preserve the life of an AI over that of a person, as would seem to follow? If research on AI is threatening to bring a "successor species" to humanity into existence, shouldn't we be having a serious global public debate about whether we wish to prohibit such research? What does it mean to hold a "moral machine" responsible for its actions? Asking such questions would go some way toward distinguishing those who are serious about their claims from those who are merely writing in a speculative mode.

However, I have suggested that it will be equally—if not more—important to interrogate the manner in which such claims are made. Are they sober and responsible, or wild and exaggerated? Are they sensible? Could we imagine someone asserting them in any other context than a philosophical argument, and if they did, how would we tell whether they were talking seriously or in jest? Asking these sorts of questions is vital if we wish to avoid being led astray by the use of concepts and arguments in the absence of the critical vocabulary that would ordinarily give them their sense. It should come as no surprise to the reader to hear that it is my suspicion that the class of claims about the ethics of AI that might be asserted soberly and sensibly on the basis of our existing knowledge of the capacities of robots and computers is significantly smaller than that currently being discussed in the literature.

Perhaps the most important lesson to be drawn from thinking about the Turing Triage Test, then, is that questions about the ethics of robotics are intimately connected to other philosophical questions, including the question of the nature of the philosophical method itself. These questions will remain important even if the promise—and threat—of intelligent machines never eventuates: the real value of conversations about robots may turn out to be what these conversations teach us about ourselves.

Acknowledgments

The research for this chapter was supported under the Australian Research Council's Discovery Projects funding scheme (project DP0770545). The views expressed herein are those of the author and are not necessarily those of the Australian Research Council. I would also like to thank Toby Handfield and Catherine Mills for reading and commenting on a draft of the chapter.

Notes

1. The first chapter of Simons 1992 describes the many appearances of mechanical and artificial people in myth and legend.

2. How to define "intelligence" and "artificial intelligence" are, of course, vexing questions. However, this chapter will presume that "intelligence" refers to a general-purpose problem-solving cognitive capacity ordinarily possessed by adult human beings and that "artificial intelligence" would involve the production of such intelligence in a machine. Questions about the moral standing of machines will only arise if researchers succeed in creating such "strong" AI.

3. This formulation of the Turing Triage Test introduced the test in the context of the discussion of the role played by the original Turing Test in the historical debate about the prospects for machine intelligence, which accounts for the reference to the Turing Test in this passage. In particular, in an earlier section of my 2004 paper I had argued that in order to be a plausible candidate for the Turing Triage Test, a system would first have to be capable of passing the Turing Test: this assumption is not, however, essential to the Turing Triage Test.

4. It is arguable that killing an artificial intelligence because of a lack of appreciation of its moral standing should be categorized as manslaughter or some other lesser category of offense, rather than murder, on the grounds that it would not involve the deliberate intention to take a life that is essential to the crime of murder. A crucial question here will be whether a lack of awareness of the moral standing of the entity toward whom one's lethal actions were directed is sufficient to exclude the conclusion that the killing was intentional: in the scenario we are imagining, the actions taken to "kill" the AI would be deliberate, and the intended result would be the destruction of the AI, but the knowledge that the AI was a moral person would be absent. In any case, regardless of whether the appropriate moral or legal verdict is murder, manslaughter, negligent homicide, or some other conclusion, clearly this scenario is one we should strive to avoid.

5. The caveat here arises from the question as to whether the tactical successes of the Predator drone mask—or, even, have produced—a larger strategic failure owing to a profound mismatch

between the capacity to rain death from the skies onto individuals and the ability to establish the political conditions that might make possible a stable government in a nation under foreign occupation (Kilcullen and Exum 2009).

6. The argument here has of necessity, given space constraints, been extremely swift. For a longer and more thorough exposition, see Sparrow 2007.

7. For some reservations about the extent to which this is likely to happen, see Sparrow 2009b.

8. Again, for a longer discussion of these issues, see Sparrow 2011.

9. See Sparrow 2004 for further discussion.

References

Adams, Thomas K. 2001. Future warfare and the decline of human decision-making. *Parameters: U.S. Army War College Quarterly* 31 (Winter 2001–02): 57–71.

Arkin, Ronald C. 2009. *Governing Lethal Behavior in Autonomous Robots.* Boca Raton, FL: Chapman and Hall Imprint, Taylor and Francis Group.

Bartneck, C., M. Verbunt, O. Mubin, and A. A. Mahmud. 2007. To kill a mockingbird robot. In *Proceedings of the 2nd ACM/IEEE International Conference on Human-Robot Interaction,* ed. C. Bartneck and T. Kanda, 81–87. Washington, DC: ACM Press.

Brooks, R. A. 2003. *Robot: The Future of Flesh and Machines.* London: Penguin.

Dyson, George. 1997. *Darwin amongst the Machines: The Evolution of Global Intelligence.* Reading, MA: Addison-Wesley.

Gaita, R. 1990. Ethical individuality. In *Value and Understanding*, ed. R. Gaita, 118–148. London: Routledge.

Gaita, R. 1991. *Good and Evil: An Absolute Conception.* London: MacMillan.

Gaita, R. 1999. *A Common Humanity: Thinking about Love and Truth and Justice.* Melbourne, Australia: Text Publishing.

Johnson, Deborah G. 2009. *Computer Ethics*, 4th ed. Upper Saddle River, NJ: Prentice Hall.

Kilcullen, David, and Andrew Mcdonald Exum. 2009. Death from above, outrage down below. *New York Times,* May 17, WK13.

Krishnan, Armin. 2009. *Killer Robots: Legality and Ethicality of Autonomous Weapons.* Burlington, VT: Ashgate.

Kuhse, H., and P. Singer. 2002. Individuals, humans, and persons: The issue of moral status. In *Unsanctifying Human Life: Essays on Ethics*, ed. Helga Kuhse, 188–198. Oxford, UK: Blackwell.

Kurzweil, Ray. 1999. *The Age of Spiritual Machines: When Computers Exceed Human Intelligence.* St. Leonards, NSW, Australia: Allen and Unwin.

Lee, Steven. 2004. Double effect, double intention, and asymmetric warfare. *Journal of Military Ethics* 3 (3): 233–251.

Moravec, Hans. 1998. *Robot: Mere Machine to Transcendent Mind*. Oxford, UK: Oxford University Press.

Nagel, T. 1972. War and massacre. *Philosophy & Public Affairs* 1 (2): 123–144.

Office of the Secretary of Defense. 2005. *Joint Robotics Program Master Plan FY2005: Out Front in Harm's Way*. Washington, DC: Office of the Undersecretary of Defense (AT&L) Defense Systems/ Land Warfare and Munitions.

Simons, Geoff. 1992. *Robots: The Quest for Living Machines*. London: Cassell.

Singer, P. W. 2009. *Wired for War: The Robotics Revolution and Conflict in the 21st Century*. New York: Penguin Books.

Sparrow, Robert. 2002. The march of the robot dogs. *Ethics and Information Technology* 4 (4): 305–318.

Sparrow, Robert. 2004. The Turing Triage Test. *Ethics and Information Technology* 6 (4): 203–213.

Sparrow, Robert. 2007. Killer robots. *Journal of Applied Philosophy* 24 (1): 62–77.

Sparrow, Robert. 2009a. Predators or plowshares? Arms control of robotic weapons. *IEEE Technology and Society* 28 (1): 25–29.

Sparrow, Robert. 2009b. Building a better warbot: Ethical issues in the design of unmanned systems for military applications. *Science and Engineering Ethics* 15 (2): 169–187.

Sparrow, Robert. 2011. Robotic weapons and the future of war. In *New Wars and New Soldiers: Military Ethics in the Contemporary World*, ed. Jessica Wolfendale and Paolo Tripodi, 117–133. Farnham Surrey, UK; Burlington, VT: Ashgate.

Turing, Alan. 1950. Computing machinery and intelligence. *Mind* 59: 433–460.

Veruggio, Gianmarco, and Fiorella Operto. 2006. Roboethics: Social and ethical implications of robotics. In *Springer Handbook of Robotics*, ed. Bruno Siciliano and Oussama Khatib, 1499–1524. Berlin: Springer.

Wallach, Wendell, and Colin Allen. 2009. *Moral Machines: Teaching Robots Right from Wrong*. Oxford, UK: Oxford University Press.

Winch, Peter. 1980–1981. Eine Einstellung zur Seele. *Proceedings of the Aristotelian Society* New Series 81: 1–15.

Wittgenstein, Ludwig. 1973. *Philosophical Investigations*, 3rd ed. Trans. G. E. M. Anscombe. New York: Prentice-Hall.

20 Robots with Biological Brains

Kevin Warwick

As will be discussed here, it is now possible to grow a biological brain and allow it to develop within a robot body (Warwick et al. 2010). The end result is a robot with a biological brain. If the size and power of such a brain is relatively small, in comparison with that of a human brain, then the issues are arguably limited. But when brainpower is comparable, then the problem clearly is of considerable significance.

The following section describes the technology and processes involved. Then, developments in the field are discussed along with future potential advancements. The chapter then examines resultant implications of such technological opportunities. When considering the ethical implications of robots in general, merely to look at robots that have computer brains would only be investigating part of the issue. Robots with biological brains and robots with hybrid brains present significant problems, which need to be addressed.

20.1 The Technology

The controlling mechanism of a typical mobile robot is presently a computer or microprocessor. Much of the initial work considering the future ethics and rights of robots has apparently focused only on this subclass of intelligent robots (Arkin 2009). Research is now ongoing in which biological neuronal networks are being cultured and trained to act as the brain of a physical, real-world robot—completely replacing a computer system.

From a medical standpoint, studying such neuronal systems can help us to understand biological neural structures in general, and it is to be hoped that it may lead to basic insights into problems such as Alzheimer's and Parkinson's disease. Other research, meanwhile, is aimed at assessing the learning capacity of such neuronal networks (Xydas et al. 2008). To do this, a hybrid system has been created incorporating control of a mobile wheeled robot, solely by a culture of neurons—a biological brain.

Such a brain is brought about by first dissociating or separating the neurons found in cortical tissue using enzymes and culturing them in an incubator, providing suitable environmental conditions and nutrients. In order to connect a brain with its robot body, the base of the incubator is composed of an array of multiple electrodes (a multielectrode array—MEA) providing an electrical interface to the neuronal culture (Thomas et al. 1972).

Once spread out on the array and fed, the neurons spontaneously begin to grow and shoot branches. Even without any external stimulation, they begin to reconnect with other neurons and commence electrochemical communication. This propensity to connect spontaneously and communicate demonstrates an innate tendency to network. The neuronal cultures form a layer over the electrodes on the base of the chamber, making them accessible to both physical and chemical manipulation (Potter et al. 2001).

The multielectrode array enables voltages from the brain to be monitored on each of the electrodes, allowing the detection of the action-potential firing of neurons near each electrode as voltage spikes, representative of neural charge transfer. It is then possible to separate the firing of multiple individual neurons, or small groups, from a single electrode (Lewicki 1998).

With multiple electrodes, an external picture of the neuronal activity of the brain can be pieced together. It is, however, also possible to electrically stimulate any of the electrodes to induce neural activity. The multielectrode array, therefore, forms a functional and nondestructive bidirectional interface with the cultured neurons. In short, via certain electrodes, the culture can be stimulated, and via other electrodes, the culture's response can be measured.

A disembodied cell culture can be provided with embodiment by placing it in a robot body, such that signals from the robot's sensors stimulate the brain, while output signals from the brain are employed to drive the motors of the robot. This is sensible since a dissociated cell culture receiving no sensory input is unlikely to develop useful operation because such input significantly affects neuronal connectivity and is involved in the development of meaningful relationships.

Several different schemes have thus far been constructed in order to investigate the ability of such systems. Shkolnik created a scheme to embody a culture within a simulated robot (Shkolnik 2003). Two channels of a multielectrode array, on which a culture was growing, were selected for stimulation and a signal consisting of a 600mVolts, 400μsecs biphasic pulse was delivered at varying intervals. The concept of information coding was formed by testing the effect of electrically inducing neuronal excitation with a given time delay between two stimulus probes. This technique gave rise to a response curve used to decide the simulated robot's direction of movement using simple commands: forward, backward, left, and right.

Subsequently, DeMarse and Dockendorf introduced the idea of implementing the results in a real-life problem, namely that of controlling a simulated aircraft's flight path, for example, making altitude and roll adjustments (2005).

20.2 Embodiment

For the purpose of growing the robot's brain, the neural cortex from a rat fetus is removed. Enzymes are applied to disconnect the neurons from each other. A thin layer of these disassociated neurons is smoothed out onto a multielectrode array, which sits in a nutrient bath. Every two days the bath must be refreshed to provide a food source for the culture and to flush away waste material.

As soon as the neurons have been laid out on the array, they start to project tentacles and thereby reconnect with each other. These projections subsequently form into axons and dendrites. By the time the culture is only one week old, electrical activity can be witnessed to appear relatively structured and pattern forming in what is, by that time, a very densely connected matrix of axons and dendrites.

The multielectrode array employed by my own research team consists of a glass specimen chamber lined with electrodes in an 8 × 8 array, as shown in figure 20.1.

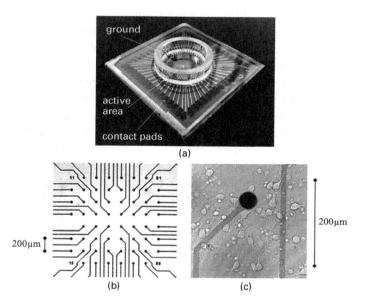

Figure 20.1

(a) A multielectrode array showing the 30μm-diameter electrodes; (b) electrodes in the center of the MEA seen under an optical microscope; and (c) ×40 magnification, showing neuronal cells with visible extensions and inner connections.

The array measures 49 mm × 49 mm × 1 mm, and its electrodes provide a bidirectional link between the culture and the rest of the system.

Thus far, we have successfully created a modular closed-loop system between a "physical" mobile robotic platform and a cultured neuronal network using the multielectrode array method, allowing for bidirectional communication between the culture and the robot. Each culture consists of approximately 100,000 neurons. The electrochemical activity of the culture is used as motor input to drive the robot's wheels, and the robot's ultrasonic sensor readings are proportionally converted into stimulation signals received by the culture as sensory input, effectively closing the loop and giving the culture a body.

We have selected a Miabot robot as the physical platform. This exhibits very accurate motor encoder precision and speed. Hence, the signals passing to and from the culture have an immediate and accurate real-world physical meaning. Figure 20.2 shows the robot employed along with an adjacent culture on a multielectrode array—body and brain together. The robot is wirelessly controlled by the culture in the incubator via a Bluetooth connection.

20.3 Experimentation

We have conducted a series of experiments utilizing a live culture. Initially, an appropriate neuronal pathway within the culture was identified and suitable stimulus electrodes and response/motor electrodes were chosen. The selection was made based on the criteria that the response electrodes show minimal spontaneous activity in general but respond robustly and reasonably repetitively to the stimuli (a positive-first biphasic waveform; 600mVolts; 100μsecs each phase) delivered via the stimulating electrodes. These spontaneous events were deemed meaningful when the delay between stimulation and response was less than 100m. Hence, an event was a strong indicator that the electric stimulation on one electrode caused a neural response on the recording electrode (Warwick et al. 2010).

The overall task the robot had to achieve was to move forward in a corral and not bump into an object, for example, a wall. The robot followed a forward path until it reached a wall, at which point the front sonar value dropped below a set threshold value triggering a stimulation/sensory pulse applied to the culture. If the responding electrode registered activity following the pulse, the robot turned in order to avoid the wall.

In its early life, the robot sometimes responded correctly by turning away from the wall, although it also bumped into the wall on numerous occasions. The robot sometimes turned spontaneously when activity was registered on the response electrode without a stimulus pulse being applied. The main results highlighted, though, were the chain of events: wall detection, stimulation, response.

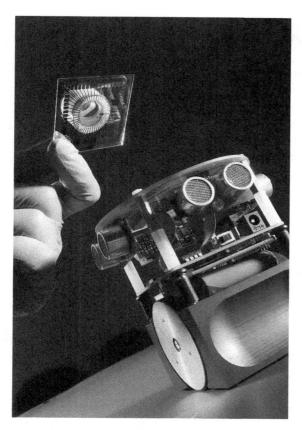

Figure 20.2
Multielectrode array with culture, close to Miabot robot.

The maximum speed at which the closed-loop system could respond was clearly dependent on the "thinking" time delay in the response of the culture. This presents an interesting possibility, of studying the response times of different cultures under different conditions and how they are affected by external influences such as electrical fields and chemical stimulants, for example, cannabis and alcohol.

The robot's individual (right and left separately) wheel speeds were then controlled from the two chosen response/motor electrodes. Meanwhile, received sonar information was used to directly control proportionally the stimulating frequency of the two sensory electrodes.

Run-times have thus far generally been executed for approximately one hour at a time. The robot's corral is presently being fitted with a special-purpose powered floor, which will allow for the study of a culture embodied 24/7 over an extended period.

Of considerable interest is whether or not the culture requires much in the way of down time (sleep equivalent), how quickly its performance improves, and if its useful lifespan increases.

A "wall to stimulation" event corresponds to the 30cm threshold being breached on the sensor, such that a stimulating pulse is transmitted to the culture. Meanwhile, a "stimulation to response" event corresponds to a motor command signal, originating in the culture, which is transmitted to the wheels of the robot, causing it to change direction. It follows that some of the "stimulation to response" events will be in considered response to a recent stimulus, termed meaningful. Whereas other such events, termed spontaneous, will be either spurious or in considered response to some thought in the culture, about which we are unaware.

20.4 Learning

Inherent or innate operating characteristics of the cultured neural network are taken as a starting point to enable the robot body to respond. The culture then operates over a period of time within the robot body in its corral area. This experimentation takes place once every day for an hour or so. Although learning has not, as yet, been a focus of the research, what has been witnessed is that neuronal structures that bring about a satisfactory action apparently tend to strengthen through the habitual process being performed. This is mainly an anecdotal observation, which is presently being formalized through more extensive studies.

At first, a stimulation-motor response feedback action occurs on some, but not all, occasions. The action can be brought about sometimes without any sensory signal being applied. After habitually carrying out the required action for some time, the neural pathways that bring this about appear to be strengthened—referred to as Hebbian learning (Hebb 1949). As a result of this learning, appropriate actions gradually become more likely to occur and spurious, unprovoked decisions to suddenly turn become less likely.

Research is ongoing to use other learning methods to quicken the performance upgrade, reinforcement learning being one example. One major problem with this is deciding what the culture regards as a reward and what as a punishment.

20.5 The Methodology

The Miabot robot is being extended to include additional sensory devices, such as audio input, further sonar arrays, mobile cameras, and other range-finding hardware, such as an onboard infrared sensor. A considerable limitation is, however, the battery power supply of an otherwise autonomous robot.

Therefore, at present a main consideration is the inclusion of a powered floor for the robot's corral, to provide the robot with relative autonomy for a longer period of time while the suggested machine learning techniques are applied and the culture's behavioral responses are monitored.

The mapping between the robot goals and the culture input/output relationships will be extended to machine learning techniques, which will ultimately reduce, or completely eliminate, the need for an a priori mapping choice. The aim is for reinforcement learning techniques to be applied to various mobile robot tasks, such as wall following and maze navigation.

One key aspect of the research is a study of the cultured neural network in terms of its observed connectivity density and activity in response to external stimuli. This behavioral evaluation should provide an insight into the workings of the neuronal network by comparing its learning capabilities in terms of its neural plasticity.

20.6 Observations

It is normal practice for several cultures to be started at the same time. A typical number may be twenty-five different cultures. By using the same Miabot robot body, it is then possible to investigate similarities and differences between the cultures within an identical body. Clearly, each culture is unique, has its own individual identity in the sense of it being recognizable (Lloyd 1991), and is dependent on the original neural layout, its growth, and development.

In terms of robot performance, such cultural differences can be manifest in a robot that performs with fewer mistakes, one that responds more quickly or slowly, one that does its own thing more often or responds only after several signals are received. There can be a large number of observed differences in behavior even with a relatively simple task to be performed.

The behavioral response of an animal can be difficult to comprehend. The overall neural requirements of the animal are not particularly understood, and may appear as meaningless to humans. The advantage with our robot system is that its behavior can be investigated directly in terms of neural development—even in response to the effect on the culture of small changes in the environment.

Cultures can be kept alive for perhaps two years or even more. After about three months or so, they become much less active and responsive and hence, most research involves cultures aged between one week and three months. This period is sufficient to consider culture development and neural pathway strengthening. Present lifetime expectancy is limited, due to vulnerability to viruses and the need to establish rigorous growth conditions.

In its robot body, a culture exhibits regular neural pathway firings. Some of these can be diagnosed as responses to stimulating sensory signals; the majority cannot be so classified. The nature of other signaling can only be guessed. However, neurons close to a stimulating electrode appear to play a role as sensory-input neurons. Meanwhile, others close to output electrodes appear to take on a role as motor neurons. There are other neurons that appear to play a routing, controlling activity. Such specialization seems to arise naturally through the culture's development. But the exact role of each of these neurons is mere speculation and will remain, for the moment at least, as anecdotal observation.

When embodied, it is possible to relate neural firings to sensory stimulating signals and/or decisions taken by the culture for specific motor outputs. What is not so straightforward, however, is explaining such firings when the culture is disembodied and is merely sitting alone in the incubator. Such a case is relatively normal for the culture, but is not experienced by an animal or human, whose brain lives its entire life receiving sensory input and making motor output decisions—other than possibly when in a dream state. Within the incubator, structured neural firings can be witnessed and the question arises as to what these firings mean.

Observing the activity in a culture leads to speculation. When the culture is disembodied, does it dream? If not, what is it thinking about? What must it feel like to be the culture? Do the firings relate to previously experienced sensory stimulation? Does a brain need external stimulating signals in order to subsequently make up stories?

20.7 Questions

When the culture is disembodied, no sensory signals are being input, yet neurons within the culture still fire in an occasional structural way. Connecting electrodes into the culture in order to measure the signals affects the culture and, in a sense, embodies it. Questions could be asked as to what does its body mean to the culture? Or who or what does the culture think it is?

As an alternative, human neurons can be employed, rather than rat neurons, as the brain of the robot. This presents a few different technical challenges; however, it is possibly more of an ethical rather than a technical problem. It is hoped that any results obtained in embodying cultured human neurons within a physical robot body will produce much more meaningful results, in terms of studying human neural conditions, and perhaps gaining an understanding of several mental conditions, as indicated by a leading consultant neurosurgeon (Aziz 2009).

Human neurons can also be readily obtained from embryos and cultured after dissociation. The use of human neurons does, however, raise other questions. For example, rather than obtaining neurons from embryos, humans could be willing to donate their own neurons—either before or after death. Wouldn't an individual like to live on in some form at least, in a robot body? Also, human neurons would not necessarily have to be dissociated; they could be laid out on the electrode array as slices. In this case, it would be interesting to see if some aspect of behavior remained and if experiences of the brain remained.

It would be a way of keeping hold of a loved one who became seriously ill. Indeed, if we are looking forward to a time when humans have robots looking after them around the home—wouldn't it be better for the robot to "know" its housemate? If a loved one is soon to die, scientists could take away neuron slices, culture them, and return them as the brain of a new robot. Maybe the robot would exhibit some of the emotions and characteristics from the loved one that would bring back memories. But for human neurons, with some awareness of their new existence, how would old memories sit with this? Would it be too traumatic an experience?

20.8 Consciousness

We cannot go far with culturing robot brains before we must ask the question as to whether the brain experiences consciousness. At present, a brain, on a two-dimensional array, contains around 100,000 neurons, nothing like the 100 billion neurons in a human brain. For those who feel size is important, then maybe consciousness cannot yet be considered.

But lattice culturing methods are being investigated that allow for a three-dimensional culture to be grown. A three-dimensional brain being embodied means we have a robot brain with 30 million neurons. Looking ahead, a $4,000 \times 4,000$ two-dimensional structure would result in a three-dimensional brain of over 60 billion neurons—more than half the size of a typical human brain, and approaching that of an elderly human.

There are many different philosophical arguments as to the nature and extent of consciousness. There are those who feel that it is a unique quality of the human brain (Penrose 1995), whereas others believe it is a property of all creatures, and neurons of other animals have the same functionality as human neurons (Cotterill 1997 and 1998).

So what of the consciousness of our robot when it has a brain of 60 billion densely packed, highly connected, and developed human neurons? Will it have genuine understanding and genuine intelligence (Penrose 1995)? If so, we will definitely have to think about giving the robot voting rights, allowing it to become a politician or a philosophy professor if it wants to, and putting it in prison if it does something it shouldn't.

But what are the arguments against our robot being conscious? Perhaps 60 billion is still not 100 billion, and that's it? But then we will need to start counting the number of brain cells in each human's head, such that those whose total falls below a threshold (let's say 80 billion) will find themselves no longer classified as a conscious being. Perhaps we will need some basic test of communication such as the Turing Test (Turing 1950) and everyone must achieve a basic standard in order to avoid the cut.

Could it be emotional responses that are important? But if the robot has human neurons, couldn't it experience similar (if not the same) emotions to humans? But are we actually interested in an identical form of consciousness to that of a human, or rather just some form of consciousness?

Is it possible our robot must have the same sensory input as humans to be considered conscious? Well, even now audio input abilities are being given to the robot; olfactory (smell) is another short-term possibility, along with basic touch and vision systems. The only difficulty appears to be with taste, due to its subjectivity. But surely we would not suggest that people who have no sense of taste are not conscious. Or that those who are blind or have a hearing deficiency also lack consciousness. Sensory input, in itself, is not critical to one's status as a conscious being.

More contentious would be an argument suggesting that motor skills are important to consciousness. The robot moves around on wheels. Most humans move around on two legs and manipulate with two arms. But some humans move around on wheels. Meanwhile, other humans have no arms or, in a few cases, have robot arms. Then there are those who have motor neuron disease and have limited movement abilities due to a malfunction in that specific part of their brain. It would be terrible to suggest that humans such as theoretical physicist Stephen Hawking, who has a motor neuron disease, are not conscious beings. Obviously, motor skills cannot be considered as a tester for consciousness. Indeed, we are at present embodying a culture in a biped walking robot body, with arms and hands that can grasp and pick up. Overall, soon this robot may well have better motor performance abilities than some humans.

The fact that our robot has a physical robot body is, therefore, not a reason to claim that it is not a conscious being.

20.9 An Education

What we are left with are the two critical properties of nature and nurture—arguably, the basic elements of human intelligence. Are we going to deny that our robot is not conscious because of its educational background? It didn't have the appropriate experiences or perhaps it didn't go to the right school, therefore it is not a conscious being? We would have to start looking at the education of humans and deny some the basic rights of some individuals because they went to the wrong school—clearly ridiculous. Education or nurture cannot be used as an argument against our robot's consciousness. Even the present robot, in the lab, is obtaining a university education.

So what we appear to be left with is nature. How an entity comes into being must be important as a decision-making tool as to whether or not that entity is conscious. It doesn't matter what we call it. It doesn't matter how it senses the world around it or how it interacts with its environment. It doesn't matter what education it received.

All that can be important is how it came to life. If this is not the important issue, then surely we will have to admit that the robot is conscious.

Even here we have problems. It must be said that at present it does not seem possible to bring such a robot to life through some form of sexual act between two humans. But we must also allow for techniques such as test tube babies and even cloning. However, it must be realized here that the human neurons, which actually constitute the brain cells of the robot, came about in one of these manners—very likely in fact through the relatively straightforward sexual act.

Discounting educational and environmental effects, the only difference between the robot brain and a human brain might merely come down to the length of gestation. This would seem to be an extremely weak line to draw for a strong division in decision making with regard to an entity's state of consciousness, especially when we consider the situation of premature babies.

20.10 Human Variety

Possibly the case for our robot with human neurons has been made in terms of its consciousness, but possibly not, maybe there is a loophole or two. What the argument does raise, though, are questions regarding how we consider other (nonrobot) humans and, in particular, extreme cases, such as individuals on life support mechanisms or those affected by dementia. Because our consideration of human consciousness, with its knock-on effect of awareness and rights, must necessarily apply to *all* humans, it is not merely applicable to philosophy or computer science professors.

The point here is that it is extremely difficult, if not impossible, on any practical realistic scientific basis, to exclude our robot from the class of conscious entities. On top of this, because its brain is made up of only human neurons, it is extremely difficult to find grounds on which to discriminate against it, especially when it may well be, in some ways, nearer the human norm than some disadvantaged human individuals.

20.11 Chinese Room

There may be some who feel that if the Turing Test can't come up with a solution, then maybe the Chinese Room can (Searle 1997). But whether or not the Chinese Room argument holds water, the logic it employs is founded on the basis that human brains are different from computer/machine brains, due to the emergent property of the human brain. Any conclusions drawn are then focused on the assumption that human brains appear to have something extra in comparison with machine brains. Our robot, though, does not have a digital/computer/machine brain; rather, just like you and I, it has a brain full of biological neurons, which are potentially human neurons. If we can conclude anything at all from Searle's Chinese Room argument, it is that our robot is indeed conscious even now.

In fact Searle (1997) stated that "the brain is an organ like any other; it is an organic machine. Consciousness is caused by lower-level neuronal processes in the brain and is itself a feature of the brain." Searle also talks of an emergent property, which implies that the more neurons there are, the greater the complexity of the consciousness. This eventually results in the form of consciousness exhibited by humans. Since we assume our robot will, in time, have a brain consisting of several billion highly connected human neurons, by Searle's argument we must assume that it will have a form of consciousness. This consciousness is pretty much on terms with that of humans, whatever its physical embodiment.

I am not claiming that the emergence of some form of consciousness depends on the size of the brain and the type of the neurons; rather, my point is that at least one philosopher (Searle 1997) points to that conclusion. To deny that our robot exhibits some form of consciousness, you the reader need an alternative, scientifically based argument and a firm philosophical argument that overcomes that of Searle. Simply *not wanting* our robot to be conscious is not good enough—you need a sound argument to *prove* that it is not conscious. Otherwise, as with humans, you will need to accept that the robot is conscious, with all the ramifications that that conclusion presents.

20.12 Functionality

It could be argued that what actually matters in terms of consciousness is the functional organization of neural cells, and not just their quantity (Cotterill 1997, 1998; Asaro 2009). Indeed, it is true that, with our present-day knowledge, it would be difficult to imagine realizing anything that was a copy of part of the human brain in its functioning. This said, as the robot brain develops, even in the two-dimensional case, neurons appear to take on specific roles, including motor, sensory, routing, support, and so forth. These roles, and their performance, are possibly different from those in the human brain.

It must be said, however, that we are not trying to achieve a form of intelligence or consciousness that is an exact copy of the human version. We wish to consider the possibility of our robot being intelligent and conscious in its own right and way, just as different humans are intelligent or conscious in different ways. The fact that our robot brain does not work in exactly the same manner as a typical human brain—if such an entity exists—is therefore only relevant to the argument if it is definitely the case that such differences are critical to the existence of consciousness in any form.

To be clear, what I am saying here is that our robot could be conscious in some way, not that it definitely is conscious. If you say that such differences may or may not be relevant, and not that they definitely are relevant, then you must agree with the point that our robot could be conscious. If, however, you say that such differences definitely are relevant, then this means that you have proven scientific evidence, not that you would simply like it to be the case. As Penrose (1995) put it, you know the "essential ingredient . . . missing from our present-day scientific picture." I personally am not aware that such scientific knowledge, regarding the existence of consciousness, exists.

20.13   Robot Rights

This brings us on to a number of key issues. At present, with 100,000 rat neurons, our robot has a pretty boring life, doing endless circles around a small corral in a technical laboratory. If one of the researchers leaves the incubator door open or accidentally contaminates the cultured brain, then they may be reprimanded and have to mend their ways. No one faces any external inquisitors or gets hauled off to court; no one gets imprisoned or executed for such actions.

With a (conscious) robot whose brain is based on human neurons, particularly if there are billions of them, the situation might be different. The robot will have more brain cells than a cat, dog, or chimpanzee, and possibly more than many humans. To keep such animals in most countries there are regulations, rules, and laws. The animal must be respected and treated reasonably well, at least. The needs of the animal must be attended to. They are taken out for walks, given large areas to use as their own, or actually exist, in the wild, under no human control. Surely a robot with a brain of human neurons must have these rights, and more? Surely it cannot simply be treated as a thing in the lab? Importantly, if the incubator door is left ajar and this robot dies, as defined by brain death, then someone needs to be held responsible and must face the consequences.

We must consider what rights such a robot should have. Do we also need to go as far as endowing it with some form of citizenship? Do we really need to protect it by law, or is considering the possibility of robot rights simply a bunch of academics having some fun? Clearly, if you are the robot and it is you who have been brought to life in your robot body by a scientist in a laboratory, and that scientist is in complete control of your existence, it must be an absolutely terrifying experience. Remember, here we are talking about a creature being brought to life with a brain consisting of human neurons, but with a robot body. It may not be very long before such robots actually are brought into being. Would it be acceptable for me to simply take the life of such a robot when that robot has a brain consisting of 60 or 100 billion human neurons?

20.14   Future Thoughts

For some reason the topic of artificial intelligence (AI), in its classical form, was concerned firmly with getting machines to do things that, if a human did them, they would be regarded as intelligent acts (Minsky 1975). That is, AI was all about getting machines to copy humans, in terms of their intelligence, as closely as possible. There are still those who feel that this is indeed what the subject of AI is about (Minsky 2007).

Such a view presents too many well-defined bounds, which has considerably restricted both technical and philosophical development in the field of AI. Unfortunately, significant philosophical discussion has subsequently been spent (in my view, wasted) merely on whether or not silicon brains could ultimately copy or simulate human brains. Could they do all the things that human brains do? Could they be as conscious as a human? The much more important topic of considering the implications of building machine brains, which are far more powerful than human brains, has, by many, been tossed aside as being merely in the realms of science fiction; as a result the topic is not even discussable by some scientists (e.g., Nicolelis 2010). What a shame! This is a much more interesting question because it points to a potential future in which intelligent, and possibly conscious, beings can outthink humans at every turn. If such entities can exist, then potentially this could be extremely dangerous to the future of humankind.

The size of the cultures employed thus far for neuron growth has been restricted by a number of factors, not the least of which is the dimensional size of the arrays on which the cultures are grown. One ongoing development is aimed at enlarging such arrays for future studies, not only providing more input/output electrodes, but also, at the same time, increasing the overall dimensions and thereby the number of neurons involved. If this increase in size is mapped onto a three-dimensional lattice structure, then things move on rapidly with regard to the size of individual robot brain possible.

A 300×300 neuron layout results in a culture of 90,000 neurons, when developed in two dimensions, at the smaller end of present-day studies. This becomes 27 million neurons in a three-dimensional latticed structure. But if this is developed to a $5,000 \times 5,000$ neuron layout, it results in a 25 million-neuron culture even in two dimensions, which undoubtedly we will witness before too long, and this becomes 125 billion in a three-dimensional lattice. It is not clear why things should stop there. As an example, moving toward a $7,500 \times 7,500$ layout, this achieves 421 billion neurons in three dimensions—an individual brain that contains four times the number of (human) neurons as contained in a typical human brain.

Drawing conclusions on developing robot brains of this size, or even much, much larger, based on human neurons, is then difficult. There are certainly medical reasons for carrying out such research, for example, to investigate the possible effects of Alzheimer's disease by increasing the overall number of useable neurons. But this approach neglects to consider the repercussions of bringing into being a brain that has the potential (certainly in terms of numbers of neurons) to be more powerful than any human brain as we know it.

The purpose of this chapter has been to consider the role of biological brains within the field of artificial intelligence and to look at their impact on some of the discussions, particularly with regard to consciousness, that have taken place. Many books have been written on these subjects, and hence it is clearly not possible to cover anything like all aspects in a single chapter. It has not been the case that I would wish to claim that such a brain is definitely conscious, but rather to consider how different concepts of what consciousness is deal with this type of brain. Each person has his or her own views on what consciousness is and what it is not. I therefore leave it up to you to reflect on how your own viewpoint is affected, if at all, by the consideration of such brains.

Is our robot with a biological brain conscious? If you feel it is not, do you have realistic scientific reasons to deny it consciousness, or do you just not like the idea of it? Think hard about the actual grounds on which you might deny consciousness to our robot. Possibly, these grounds are that it doesn't look like you, doesn't communicate like you, or doesn't have the same values as you. Shame on you!

Acknowledgments

I would like to express my gratitude to the team at University of Reading, England, on whose work this article is based: Ben Whalley, Slawek Nasuto, Victor Becerra, Dimi Xydas, Mark Hammond, Julia Downes, Matt Spencer, and Simon Marshall.

The practical work described in this article is funded by the UK Engineering and Physical Sciences Research Council (EPSRC) under grant no. EP/D080134/1.

References

Arkin, R. 2009. *Governing Lethal Behavior in Autonomous Robots*. London: Chapman and Hall/CRC Press.

Asaro, P. 2009. Information and regulation in robots, perception and consciousness: Ashby's embodied minds. *International Journal of General Systems* 38 (2): 111–128.

Aziz, T. 2009. Personal communication.

Cotterill, R. 1997. On the mechanism of consciousness. *Journal of Consciousness Studies* 4 (3): 231–247.

Cotterill, R. 1998. *Enchanted Looms: Conscious Networks in Brains and Computers*. Cambridge, UK: Cambridge University Press, Cambridge.

DeMarse, T. B., and K. P. Dockendorf. 2005. Adaptive flight control with living neuronal networks on microelectrode arrays. In *Proceedings. 2005 IEEE International Joint Conference on Neural Networks*, 1549–1551. Montreal.

Hebb, D. 1949. *The Organization of Behavior*. New York: Wiley.

Lewicki, M. 1998. A review of methods for spike sorting: The detection and classification of neural action potentials. *Network (Bristol, England)* 9 (4): R53–R78.

Lloyd, D. 1991. Leaping to conclusions: Connectionism and the computational mind. In *Connectionism and the Philosophy of Mind*, ed. T. Horgan and J. Tienson, 444–459. Netherlands: Kluwer.

Minsky, M. 1975. *The Psychology of Computer Vision*. New York: McGraw-Hill.

Minsky, M. 2007. Personal communication.

Nicolelis, M. 2010. Q & A at Biovision, Bibliotheca Alexandrina, Alexandria, Egypt, April.

Penrose, R. 1995. *Shadows of the Mind*. Oxford, UK: Oxford University Press, Oxford.

Potter, S., N. Lukina, K. Longmuir, and Y. Wu. 2001. Multi-site two-photon imaging of neurons on multi-electrode arrays. *SPIE Proceedings* 4262: 104–110.

Searle, J. 1997. *The Mystery of Consciousness*. New York: New York Review Books.

Shkolnik, A. C. 2003. Neurally controlled simulated robot: Applying cultured neurons to handle an approach/avoidance task in real time, and a framework for studying learning in vitro. Master's thesis, Department of Computer Science, Emory University, Georgia.

Thomas, C., P. Springer, G. Loeb, Y. Berwald-Netter, and L. Okun. 1972. A miniature microelectrode array to monitor the bioelectric activity of cultured cells. *Experimental Cell Research* 74 (1): 61–66.

Turing, A. 1950. Computing machinery and intelligence. *Mind* 59 (236): 433–460.

Warwick, K., D. Xydas, S. Nasuto, V. Becerra, M. Hammond, J. Downes, S. Marshall, and B. Whalley. 2010. Controlling a mobile robot with a biological brain. *Defence Science Journal* 60 (1): 5–14.

Xydas, D., K. Warwick, B. Whalley, S. Nasuto, V. Becerra, M. Hammond, and J. Downes. 2008. Architecture for living neuronal cell control of a mobile robot. In *Proceedings of the European Robotics Symposium (EUROS08)*, Springer Tracts in Advanced Robotics, vol. 44, ed. H. Bruyninckx, L. Preucil, and M. Kulich, 23–31. Prague: Springer.

21 Moral Machines and the Threat of Ethical Nihilism

Anthony F. Beavers

In his famous 1950 paper where he presents what became the benchmark for success in artificial intelligence, Turing notes that "at the end of the century the use of words and general educated opinion will have altered so much that one will be able to speak of machines thinking without expecting to be contradicted" (Turing 1950, 442). Kurzweil suggests that Turing's prediction was correct, even if no machine has yet to pass the Turing Test (1990). In the wake of the computer revolution, research in artificial intelligence and cognitive science has pushed in the direction of interpreting "thinking" as some sort of computational process. On this understanding, thinking is something computers (in principle) and humans (in practice) can both do.

It is difficult to say precisely when in history the meaning of the term "thinking" headed in this direction. Signs are already present in the mechanistic and mathematical tendencies of the early modern period, and maybe even glimmers are apparent in the thoughts of the ancient Greek philosophers themselves. But over the long haul, we somehow now consider "thinking" as separate from the categories of "thoughtfulness" (in the general sense of wondering about things), "insight," and "wisdom." *Intelligent* machines are all around us, and the world is populated with *smart* cars, *smart* phones, and even *smart* (robotic) appliances. But, though my cell phone might be smart, I do not take that to mean that it is thoughtful, insightful, or wise. So, what has become of these latter categories? They seem to be bygones, left behind by scientific and computational conceptions of thinking and knowledge that no longer have much use for them.

In 2000, Allen, Varner, and Zinser addressed the possibility of a Moral Turing Test (MTT) to judge the success of an automated moral agent (AMA), a theme that is repeated in Wallach and Allen (2009). While the authors are careful to note that a language-only test based on moral justifications or reasons would be inadequate, they consider a test based on moral behavior. "One way to shift the focus from reasons to actions," they write, "might be to restrict the information available to the human judge in some way. Suppose the human judge in the MTT is provided with descriptions of actual, morally significant actions of a human and an AMA, purged of all references that would identify the agents. If the judge correctly identifies the machine at a level above chance, then the machine has failed the test" (206). While they are careful to note that indistinguishability between human and automated agents might set the bar for passing the test too low, such a test by its very nature decides the morality of an agent on the basis of appearances. Since there seems to be little else we could use to determine the success of an AMA, we may rightfully ask whether, analogous to the term "thinking" in other contexts, the term "moral" is headed for redescription here. Indeed, Wallach and Allen's survey of the problem space of machine ethics forces the question of whether within fifty years one will be able to speak of a machine as being moral without expecting to be contradicted. Supposing the answer were yes, why might this invite concern? What is at stake? How might such a redescription of the term "moral" come about? These are the questions that drive this reflection. I start here with the last one first.

21.1   How Might a Redescription of the Term "Moral" Come About?

Before proceeding, it is important to note first that because they are fixed in the context of the broader evolution of language, the meaning of terms is constantly in flux. Thus, the following comments must be understood generally. Second, the following is one way redescription of the term "moral" *might* come about, even though, in places I will note, this is already happening to some extent. Not all machine ethicists can be plotted on this trajectory.

That said, the project of designing moral machines is complicated by the fact that even after more than two millennia of moral inquiry, there is still no consensus on how to determine moral right from wrong. Even though most mainstream moral theories agree from a big-picture perspective on which behaviors are morally permissible and which are not, there is little agreement on why they are so, that is, what it is precisely about a moral behavior that makes it moral. For simplicity's sake, this question will be here designated as *the hard problem of ethics*. That it is a difficult problem is seen not only in the fact that it has been debated since philosophy's inception without any satisfactory resolution, but also that the candidates that have been offered over the centuries as answers are still on the table today. Does moral action flow from a virtuous character operating according to right reason? Is it based on sentiment, or on application of the right rules? Perhaps it is mere conformance to some tried and tested principles embedded in our social codes, or based in self-interest, species' instinct, religiosity, and so forth.

The reason machine ethics cannot move forward in the wake of unsettled questions such as these is that engineering solutions are needed. Fuzzy intuitions on the nature of ethics do not lend themselves to implementation where automated decision procedures and behaviors are concerned. So, progress in this area requires working the details out in advance, and testing them empirically. Such a task amounts to coping with the hard problem of ethics, though largely, perhaps, by rearranging the moral landscape so an implementable solution becomes tenable.

Some machine ethicists, thus, see research in this area as a great opportunity for ethics (Anderson and Anderson 2007; Anderson 2011; Beavers 2009, 2010; Wallach 2010). If it should turn out, for instance, that Kantian ethics cannot be implemented in a real working device, then so much the worse for Kantian ethics. It must have been ill conceived in the first place, as now seems to be the case, and so also for utilitarianism, at least in its traditional form.

Quickly, though some have tried to save Kant's enterprise from death by failure to implement (Powers 2006), the cause looks grim. The application of Kant's categorical imperative in any real-world setting seems to fall dead before a moral version of the frame problem. This problem from research in artificial intelligence concerns our current inability to program an automated agent to determine the scope of reasoning necessary to engage in intelligent, goal-directed action in a rich environment without needing to be told how to manage possible contingencies (Dennett 1984). Respecting Kantian ethics, the problem is apparent in the universal law formulation of the *categorical imperative*, the one that would seem to hold the easiest prospects for rule-based implementation in a computational system: "act as if the maxim of your action were to become through your will a universal law of nature" (Kant [1785] 1981, 30). One mainstream interpretation of this principle suggests that whatever rule (or *maxim*) I should use to determine my own behavior must be one that I can consistently will to be used to determine the behavior of everyone else. (Kant's most consistent example of this imperative in application concerns lying promises. I cannot make a lying promise without simultaneously willing a world in which lying is permissible, thereby also willing a world in which no one would believe a promise, particularly the very one I am trying to make. Thus, the lying promise fails the test and is morally impermissible.) Though at first the categorical imperative looks implementable from an engineering point of view, it suffers from a problem of scope, since any maxim that is defined narrowly enough (for instance, to include a class of one, anyone like me in my situation) must consistently universalize. Death by failure to implement looks imminent; so much the worse for Kant, and so much the better for ethics.

Classical utilitarianism meets a similar fate, even though, unlike Kant, Mill casts internals, such as intentions, to the wind and considers just the consequences of an act for evaluating moral behavior. Here, "actions are right in proportion as they tend to promote happiness; wrong as they tend to produce the reverse of happiness. By happiness is intended pleasure and the absence of pain; by unhappiness, pain and the privation of pleasure" ([1861] 1979, 7). That internals are incidental to utilitarian ethical assessment is evident in the fact that Mill does not require that one act for the right reasons. He explicitly says that most good actions are not done accordingly (18–19). Thus, acting good is indistinguishable from being good, or, at least, to be good is precisely to act good; and sympathetically we might be tempted to agree, asking what else could being good possibly mean.

Things again are complicated by problems of scope, though Mill, unlike Kant, is aware of them. He writes, "again, defenders of utility often find themselves called upon to reply to such objections as this—that there is not enough time, previous to action, for calculating and weighing the effects of any line of conduct on the general happiness" ([1861] 1979, 23). (In fact, the problem is computationally intractable when we consider the ever-extending ripple effects that any act can have on the happiness of others across both space and time.) Mill gets around the problem with a sleight of hand, noting that "all rational creatures go out upon the sea of life with their minds made up on the common questions of right and wrong" (24), suggesting that calculations are, in fact, unnecessary, if one has the proper forethought and upbringing. Again, the rule is of little help, and death by failure to implement looks imminent. So much the worse for Mill; again, so much the better for ethics.

Wallach and Allen agree that the prospects for a "top-down, theory driven approach to morality for AMAs" (2009, 83), such as we see in both instances described, do not look good, arguing instead that a hybrid approach that includes both "top-down" and "bottom-up" strategies is necessary to arrive at an implementable system (or set of systems). "Bottom-up" here refers to emergent approaches that might allow a machine to learn to exhibit moral behavior and could arise from research in "Alife (or artificial life), genetic algorithms, connectionism, learning algorithms, embodied or subsumptive architecture, evolutionary and epigenetic robotics, associative learning platforms, and even traditional symbolic AI" (112). While they advocate this hybrid approach, they also acknowledge the limitations of the bottom-up approach taken by itself. As one might imagine, any system that learns is going to require us to have a clear idea of moral behavior in order to evaluate goals and the success of our AMAs in achieving them. So, any bottom-up approach also requires solving the ethical hard problem in one way or another, and thus it too dies from failure to implement. We can set the bottom-up approach aside; again, so much the better for ethics.

If these generalizations are correct, that top-down theoretical approaches may run into some moral variant of the frame problem, and that both the top-down and bottom-up approaches require knowing beforehand how to solve the hard problem of ethics, then where does that leave us? Wallach and Allen (and others, see Coleman 2001) find possible solutions in Aristotle and virtue ethics more generally. At first, this move might look surprising. Of the various ways to come at ethics for machines, virtue ethics would seem an unlikely candidate, since it is among the least formalistic. Nonetheless, it has the benefit of gaining something morally essential from both top-down and bottom-up approaches.

The top-down approach, Wallach and Allen argue, is directed externally toward others. Its "restraints reinforce cooperation, through the principle that moral behavior often requires limiting one's freedom of action and behavior for the good of society, in ways that may not be in one's short-term or self-centered interest" (2009, 117). Regardless of whether Kant, Mill, and other formalists in ethics fall to a moral frame problem, they do nonetheless generally understand morality fundamentally as a necessary restraint on one's desire with the effect of, though not always for the sake of, promoting liberty and the public good.

But rules alone are insufficient without a motivating cause, Wallach and Allen rightly observe, noting further "values that emerge through the bottom-up development of a system reflect the specific causal determinates of a system's behavior" (2009, 117). Bottom-up developmental approaches, in other words, can precipitate where, when, and how to take action, and perhaps set restraints on the scope of theory-based approaches, like those mentioned previously. Having suggested already that by "hybrid" they mean something more integrated than the mere addition of top to bottom, virtue ethics would seem after all a good candidate for implementation. Additionally, as Gips (1995) noted earlier, learning by habit or custom, a core ingredient of virtue ethics, is well suited to connectionist networks and, thus, can support part of a hybrid architecture.

Acknowledging that even in virtue ethics there is little agreement on what the virtues are, it nonetheless looks possible, at least, that this is the path to pursue, though to situate this discussion, it is helpful to say what some of them might be. Wallach and Allen name Plato's canonical four (wisdom, courage, moderation, and justice) and St. Paul's three (faith, hope, and charity) to which we could just as well add the Boy Scout's twelve ("a scout is trustworthy, loyal, helpful, friendly, courteous, kind, obedient, cheerful, thrifty, brave, clean, and reverent"), and so on. However we might choose to carve them out, one keystone of the virtues is their stabilizing effect, which, for the purposes of building AMAs, allows for some moral reliability. "Such stability," Wallach and Allen note, "is a very attractive feature, particularly for AMAs that need to maintain 'loyalty' under pressure while dealing with various, not always legitimate sources of information" (2009, 121). The attraction is noted, but also note how the language has already started to turn. What is loyalty, whether in quotations or not, such that a machine could have it? How could a robot ever experience the fear essential to make an act courageous, or the craving that makes temperance a virtue at all?

From an engineering point of view, simulated emotion might do just as well to get virtuous behavior from a machine, but getting to emotion "deeply" enough to justify predicating "character" to AMAs may prove something of a philosophical question that hits to the heart of the matter and returns us to the Moral Turing Test mentioned earlier in this chapter. (See Coeckelbergh 2010a for a related discussion on this topic.)

As with people, the principal way we judge others as virtuous is by considering their behavior. So, when is a robot loyal? When it sticks to its commitments. When is it wise? Well, of course, when it does wise things. When is it courageous? When it behaves courageously. What more could we legitimately want from a moral machine? Such would appear to be a morally perfect being with an acute sense of propriety governed by right reason and which always acts accordingly. So, *ex hypothesi*, let us build them or some variant thereof and wonder how long it will be before the use of words and general educated opinion will have altered so much that one will be able to speak of machines *as moral* without expecting to be contradicted.

21.2 What Is at Stake?

Interiority counts (at least for the time being), especially in matters of morals, where what we might call "moral subjectivity," that is, conscience, a sense of moral obligation and responsibility, in short, whatever motivates our moral psychology to care about ethics, governs our behavior. Even the formalist Kant thought it necessary to explain the sense in which "respect," an essential component of his ethical theory, was and was not a feeling in the ordinary sense of the word, noting along the way that "respect is properly the conception of a worth which thwarts my self-love" ([1785] 1981, 17) and so requires self-love in the same way that courage requires fear. Additionally, Kant's universal imperative requires a concrete, personally motivated maxim to universalize in order for an agent to be moral (Beavers 2009) and is implicitly tied to interpersonal concerns as well (Beavers 2001). Furthermore, the theme of interiority is explicitly addressed by Mill, who notes that there are both external and internal sanctions of the principle of utility, ascribing to the latter "a feeling in our own mind; a pain, more or less intense, attendant on violation of duty," which is "the essence of conscience" ([1861] 1979, 27–28).

More importantly for this discussion, interiority counts in the virtue ethics of Plato and Aristotle, both of whom mark an essential distinction between being good and merely acting so. Famously, in Book II of the *Republic*, Plato (1993) worries that moral appearances might outweigh reality and in turn be used to aid deceit (see 53, 365a–d), and Aristotle's ethics is built around the concept of *eudaimonia*, which we might translate as a well-being or happiness that all humans in essence pursue. We do so at first only imperfectly as children who simulate virtuous behavior, and in the process learn to self-legislate the satisfaction of our desire. Even though Aristotle does note that through habituation, virtuous behavior becomes internalized in the character of the individual, it nonetheless flows from inside out, and it is difficult to imagine how a being can be genuinely virtuous in any Greek sense without also a genuinely "felt," affective component. We need more, it seems, than what is visible to the judges in the MTT discussed earlier. Or do we?

The answer to this question hangs on what our goals are in developing machine ethics. To make this clear, it is helpful to consider Moor's often-cited taxonomy of moral agency. According to Moor, "ethical-impact agents" are machines that have straightforward moral impact, like the robotic camel jockeys implemented in Qatar that helped to liberate Sudanese slave boys who previously served in that capacity, even though the motive for implementing them was to escape economic sanction. Though Moor does not say so here, most machines seem to qualify in some way for this type of agency, including a simple thermostat. Straightforward ethical impact is not what concerns designers of robot morality, however. "Frequently, what sparks debate is whether you can put ethics into a machine. Can a computer operate ethically because it's internally ethical in some way" (2006, 19)? Here the waters start to get a bit murky. To clarify the situation, Moor marks a three-fold division among kinds of ethical agents as "implicit," "explicit," or "full."

"Implicit ethical agents" are machines constrained "to avoid unethical outcomes" (Moor 2006, 19). Rather than working out solutions to ethical decisions themselves, they are designed in such a way that their behavior is moral. Moor mentions automated teller machines (ATMs) and automatic pilots on airplanes as examples. The ATM isn't programmed with a rule about promoting honesty any more than the automatic pilot must deduce when to act safely in order to spare human life. The thermostat mentioned earlier would seem to fall in this category, though whether the camel jockey does depends on the mechanisms it uses in making its decisions.

"Explicit ethical agents" are machines that can "'do' ethics like a computer can play chess" (Moor 2006, 19–20). In other words, they can apply ethical principles to concrete situations to determine a course of action. The principles might be something like Kant's categorical imperative or Mill's principle of utility. The critical component of "explicit" ethical agents is that they work out ethical decisions for themselves using some kind of recognizable moral decision procedure. Presumably, Moor notes, such machines would also be able to justify their judgments. Finally, "full ethical agents" are beings like us, with "consciousness, intentionality, and free will" (20). They can be held accountable for their actions—in the moral sense, they can be at fault—precisely because their decisions are in some rich sense *up to them.*

We can see how machines can achieve the status of implicit and perhaps explicit moral agents, if Wallach and Allen are right, but whether one can ever be a full moral agent requires technologies far from what we have yet to conceive. Given that the question of full ethical agency for robots will not be settled soon, Moor remarks, "we should . . . focus on developing limited explicit ethical agents. Although they would fall short of being full ethical agents, they could help prevent unethical outcomes" (Moor 2006, 21). Wallach and Allen concur, though perhaps while implicitly offering one way to deal with the question of full moral agency in robots short of actually settling it in the sense suggested by Moor. The problem concerns the difference between Moor's notions of explicit and full ethical agency, in light of both the MTT and the criterion of implementation that machine ethics (legitimately) forces upon us. Can the distinction between explicit and full moral agency stand up to their challenge?

The answer to this question hangs in part on an empirical component in engineering moral machines that is intimately tied to the implementation criterion itself. If *ought* implies *can*, then *ought* implies *implementability*. Though this might not seem immediately apparent, it is nonetheless the case, since any moral theory that cannot be implemented in a real, working agent, whether mechanical or biological, limits the agent's ability to execute real-world action. Thus, if ought implies can, or the ability to act in a particular situation, then moral obligation must rest on some platform that affords the agent this possibility. A nonimplementable approach to morals does not. Thus, a valid approach must also be an implementable one. As such, the test for a working moral system (or theory) is partly cast as an engineering problem whose solution hangs precisely on passing the MTT. Consequently, the AMA that passes the MTT is not merely an implementation of a moral machine, but also proof of concept for a valid approach to morals. If we can successfully engineer moral machines, interiority, thus, does not appear to count.

But what then serves to distinguish an explicit moral agent that "does ethics as one plays chess" and exhibits proper moral behavior from the full ethical agent that acts with intentionality and moral motivation? In a world populated by human beings and moral machines, assuming we are successful in building them, the answer would seem to be nothing. Minimally, at least, we would have to concede that morality itself is multiply realizable, which strongly suggests that full moral agency is just another way of getting explicit moral agency, or, as a corollary, that what is essential for full moral agency, as enumerated by Moor, is no longer essential for ethics. It is merely a sufficient, and no longer necessary, condition for being ethical. Though this might sound innocuous at first, excluded with this list of inessentials are not only consciousness, intentionality, and free will, but also anything intrinsically tied to them, such as conscience, (moral) responsibility, and (moral) accountability.

The MTT, together with the criterion of implementability for testing approaches to ethics, significantly rearranges the moral playing field. Philosophical speculation, unsettled for more than two millennia, is to be addressed here not by argument, but by engineering in an arena where success is gauged by the ability to simulate moral behavior. What then is left for requisite notions that have from the start defined the conscience of the human? They seem situated for redefinition or reclassification, to be left behind by conceptions of morality that will no longer have much use for them.

21.3 Why Might This Invite Concern?

Ethics without conscience sounds a little like knowledge without insight to guide it. To turn this in a different direction, ethics without accountability sounds as equally confused as placing moral praise and blame on components that cannot possibly have them, at least on our current understanding of terms, and especially when making attributions of virtue. To see this, let us suppose that some time in the near future, we read the (rather long) headline, "First Robot Awarded Congressional Medal of Honor for Incredible Acts of Courage on the Battlefield." What must we assume in the background for such a headline to make sense without profaning a nation's highest award of valor? Minimally, fortitude and discipline, intention to act while undergoing the experience of fear, some notion of sacrifice with regard to one's own life, and so forth, for what is courage without these things? That a robot might simulate them is surely not enough to warrant the attribution of virtue, unless we change the meaning of some terms.

At bottom, to bestow respect on someone or something for their (its?) actions is to deem agents "responsible" for them. Mixed in with the many definitions of the term "responsible" is the matter of accountability. Sometimes this term refers to an agent of cause, as when a fireman might explain to me that the toaster was responsible for my house burning down. But I cannot hold the toaster accountable for its actions, though I might its manufacturer. *Moral* responsibility travels with such accountability. To return to the robot soldier once more, the robot can be the precipitating cause of an action, and hence responsible in the same sense as a toaster; what must we add to it to make it accountable, and hence also morally responsible, for its actions? From the engineering point of view, we have no way to say. Indeed, MTT and the criterion of implementability make such a distinction between causal and moral responsibility impossible in the first place. This is because stipulating the means of implementation is precisely to have determined the causal properties responsible for moral responsibility and, indeed, for the virtues themselves, if we should choose to implement a virtue ethics. So, the fact that the robot soldier was designed to be courageous either undermines its ability to be so, though certainly not to act so, or we invert the strategy and say that its ability to act so is precise proof that it is so.

Even explicit awareness of the inverted strategy as such will not stop us from bestowing moral esteem on machines, any more than knowing that my Ragdoll kitten was genetically bred to bond with human beings stops me from feeling the warmth of its affection. ("Ragdoll" here represents a feline breed that was controversially engineered to be passive and amiable.) Indeed, if our moral admiration can be raised by the behavior of fictitious characters simulated by actors—Captain Picard in the TV program *Star Trek*, for instance—then all the easier it will be to extend it to real machines that look, think, and act like us. This psychological propensity (and epistemic necessity) to judge internals on the basis of external behavior is not the main concern, however, as it may first appear, precisely because we are not dealing here with a matter of misplaced attribution. Rather, on the contrary, MTT and the criterion of implementability suggest that such attribution is quite properly placed. Success in this arena would thus seem to raise even deeper concerns about the nature of human morality, our moral objectivity, and our right to implement a human-centered ethics in machines.

If, for instance, implementability is a requirement for a valid approach to morals (thereby resituating full moral agency as a sufficient, though not necessary, condition for moral behavior, as previously noted), then the details of how, when, and why a moral agent acts the way it does is partly explained by its implementation. To the extent that human beings are moral, then, we must wonder how much of our own sense of morals is tied to its implementation in our biology. We are ourselves, in other words, biologically instantiated moral machines. To those working in neuroethics and the biology of morality more generally, there is nothing surprising about this. Ruse (1995), for instance, has already noted that our values may be tied implicitly to our biology. If so, then human virtues are *our virtues* partly because we are mammals. Is there any reason to think that human virtues are those that we *should* implement in machines? If so, on what grounds? Why mammalian virtues as opposed to reptilian, or perhaps, even better, virtues suited to the viability and survival advantages of the machines themselves?

The question of an objectively valid account of morality is once again on the table, this time complicated by details of implementation. Even though questions of biological, genetic, neurological, and technological determinism are still hotly debated today (yet another indication of the difficulty of the hard problem of ethics), we are nonetheless left wondering whether soon the notion of accountability may be jettisoned by the necessity of scientific and technological discovery. If so, moral responsibility would seem to vanish with it, leaving only causal responsibility to remain. Research in building moral machines, it would seem, adds yet another challenge to a conventional notion of moral responsibility that is already under attack on other fronts.

In 2007, Anderson and Anderson wrote:

Ethics, by its very nature, is the most practical branch of philosophy. It is concerned with how agents ought to behave when faced with ethical dilemmas. Despite the obvious applied nature of the field of ethics, however, too often work in ethical theory is done with little thought to real world application. When examples are discussed, they are typically artificial examples. Research in machine ethics, which of necessity is concerned with application to specific domains where machines could function, forces scrutiny of the details involved in actually applying ethical principles to particular real life cases. As Daniel Dennett [2006] recently stated, AI "makes philosophy honest." Ethics must be made computable in order to make it clear exactly how agents ought to behave in ethical dilemmas. (2007, 16)

At the very least, we must agree that the criterion of implementability suggested here makes ethics honest, and herein lies the problem. For present purposes, I define "ethical nihilism" as the doctrine that states that morality needs no internal sanctions, that ethics can get by without moral "weight," that is, without some type of psychological force that restrains the satisfaction of our desire and that makes us care about our moral condition in the first place. So what, then, if the trajectory I have sketched should turn out to be correct and that internal sanctions are merely sufficient conditions for moral behavior? Will future conceptions of ethics be forced to make do without traditionally cherished notions, such as conscience, responsibility, and accountability? If so, have we then come at last to the end of ethics? No doubt, if the answer is no, it may be so only by embracing a very different conception of ethics than traditional ones like those mentioned earlier (for possibilities, see Floridi and Sanders 2004 and Coeckelbergh 2010b).

Acknowledgments

I wish to acknowledge Colin Allen, Susan Anderson, Larry Colter, Dick Connolly, Deborah Johnson, Jim Moor, Dianne Oliver, and Wendell Wallach for past conversations on ethics that have led me to the views expressed here. I would particularly like to thank Colin Allen, Luciano Floridi, Christopher Harrison, Patrick Lin, Mark Valenzuela, and Wendell Wallach for their comments on earlier drafts of this chapter.

References

Allen, C., G. Varner, and J. Zinser. 2000. Prolegomena to any future artificial moral agent. *Journal of Experimental & Theoretical Artificial Intelligence* 12 (3): 251–261.

Anderson, S. 2011. How machines might help us to achieve breakthroughs in ethical theory and inspire us to behave better. In *Machine Ethics*, ed. Michael Anderson and Susan Anderson, 524–530. New York: Cambridge University Press.

Anderson, M., and S. Anderson. 2007. Machine ethics: Creating an ethical intelligent agent. *AI Magazine* 28 (4): 15–26.

Beavers, A. 2001. Kant and the problem of ethical metaphysics. *Philosophy in the Contemporary World* 7 (2): 47–56.

Beavers, A. 2009. Between angels and animals: The question of robot ethics, or is Kantian moral agency desirable. Paper presented at the Eighteenth Annual Meeting of the Association for Practical and Professional Ethics, Cincinnati, Ohio, March 5–8.

Beavers, A., ed. 2010. Robot ethics and human ethics. Special issue of *Ethics and Information Technology* 12 (3).

Coeckelbergh, M. 2010a. Moral appearances: Emotions, robots, and human morality. *Ethics and Information Technology* 12 (3): 235–241.

Coeckelbergh, M. 2010b. Robot rights? Toward a social-relational justification of moral consideration. *Ethics and Information Technology* 12 (3): 209–221.

Coleman, K. 2001. Android arete: Toward a virtue ethic for computational agents. *Ethics and Information Technology* 3 (4): 247–265.

Dennett, D. 1984. Cognitive wheels: The frame problem in artificial intelligence. In *Minds, machines, and evolution: Philosophical studies*, ed. C. Hookway, 129–151. New York: Cambridge University Press.

Dennett, D. 2006. Computers as prostheses for the imagination. Invited talk presented at the International Computers and Philosophy Conference, May 3, Laval, France.

Floridi, L., and J. Sanders. 2004. On the morality of artificial agents. *Minds and Machines* 14 (3): 349–379.

Gips, J. 1995. Towards the ethical robot. In *Android Epistemtology*, ed. K. Ford, C. Glymour, and P. Hayes, 243–252. Cambridge, MA: MIT Press.

Kant, I. [1785] 1981. *Grounding for the Metaphysics of Morals*, trans. J. W. Ellington. Indianapolis, IN: Hackett Publishing Company.

Kurzweil, R. 1990. *The Age of Intelligent Machines*. Cambridge, MA: MIT Press.

Mill, J. S. [1861] 1979. *Utilitarianism*. Indianapolis, IN: Hackett Publishing Company.

Moor, J. 2006. The nature, importance, and difficulty of machine ethics. *IEEE Intelligent Systems* 1541–1672: 18–21.

Plato. 1993. *Republic*, trans. R. Waterfield. Oxford, UK: Oxford University Press.

Powers, T. 2006. Prospects for a Kantian machine. *IEEE Intelligent Systems* 1541–1672: 46–51.

Ruse, M. 1995. *Evolutionary Naturalism*. New York: Routledge.

Turing, A. 1950. Computing machinery and intelligence. *Mind* 59 (236): 433–460.

Wallach, W. 2010. Robot minds and human ethics: The need for a comprehensive model of moral decision making. *Ethics and Information Technology* 12 (3): 243–250.

Wallach, W., and C. Allen. 2009. *Moral Machines: Teaching Robots Right from Wrong*. New York: Oxford University Press.

VIII   Epilogue

22 Roboethics: The Applied Ethics for a New Science

Gianmarco Veruggio and Keith Abney

The previous chapters in this book have covered a multitude of ethical issues raised by the new science of robotics, from issues about the use of robots in policing and the military, to assist in various social activities, including entertainment and even sex, to discussion of the possibility that robots will one day have rights and be moral agents themselves. This possibility highlights an important ambiguity in the use of the term "robot ethics," as the phrase has at least three distinct meanings.

First, it applies to the philosophical studies and researches about the ethical issues arising from the effects of the application of robotics products on our society. In this sense, *roboethics* suggests the development of a very broad "applied ethics," which, similarly to the ethical studies related to bioethics, deals with the universal, fundamental ethical issues. These are related to the need to protect and enhance human dignity and personal integrity; to secure the rights of the weakest; and to limit the "robotics divide" in all those instances in which robotics products could either worsen the existing inequalities, or create some new ones. In this meaning, roboethics pertains to all the issues deriving from the relationship among science, technology, and society, and it benefits from the related studies in psychology, sociology, law, comparative religions, and so on.

Second, robot ethics could refer to the moral code to which the robots themselves are supposed to adhere (presumably a morality somehow programmed into them). For any level of robotic autonomy, there will be some code the programmers create that the robot must follow in order to do what it ought; in effect, that will be a moral code for the robots, in this second sense. This will enable humans to make the judgment that the robot acted morally, in obeying its programmed moral code and doing what it ought to do, or that the robot acted immorally, in doing something that it wasn't supposed to (that it ought not to have done), whether due to a electromechanical glitch, or a bug in the software, or lack of foresight about the conditions of its use, or otherwise incompetent programming. But the robot itself is unaware of its own programmed-in morality; it is "just following orders," whether it does so badly or well.

The last consideration leads us to yet a third sense of robot ethics: it could refer to the self-conscious ability of the robots themselves to do ethical reasoning, to understand from a first-person perspective their choices and responsibilities, and to freely, self-consciously choose their course of action. Such an ability would make robots full moral agents, themselves (and not their programmers, designers, or builders) personally responsible for their actions. This third sense of robot ethics would imply that robots have a morality they choose for themselves, not merely one they slavishly, mindlessly must follow; they would share the human trait of self-conscious, rational choice, or *freedom*.

To help disambiguate and explore the first sense of robot ethics as described here, one of the coauthors of this epilogue (Veruggio) has coined the term "roboethics" to indicate an applied ethics whose objective is to develop scientific, cultural, and technical tools that can be shared by different social groups and beliefs. These tools aim at promoting and encouraging the development of robotics for the advancement of human society and of the individual, and to help to prevent its misuse. As per this definition, it is clear that roboethics is a human-centered ethics: it is not the—"artificial" or "natural"—ethics of the robots, but the ethics of the robotics researchers, of the producers, and of users of the robots.

The exploration of those professional responsibilities underlies the development of the Roboethics Program and Roboethics Roadmap (Veruggio 2007), as follows. Following the First International Symposium on Roboethics in 2004, many leading roboticists determined to work in collaboration with scholars of humanities. The aim of this common endeavor was to roadmap the ethical issues surrounding the emerging science of robotics in order to create a cross-cultural and interdisciplinary consciousness of these new social challenges. The results of these common and synergic efforts should be (a) a general cultural (ethical, social, and legal) framework for robotics; (b) a professional ethics for the roboticists; and (c) the technical standards, regulatory rules, and the legal apparatus for the robotics market products.

But discussion of this first sense of robot ethics, or roboethics ineluctably leads to considerations of the second sense: as robots gradually become more autonomous, what moral codes shall we program into them? How can we guarantee that the robots we create will do little to no unintended harm, that they will commit no immoral actions? What moral codes shall we program in: deontological, utilitarian, virtue, or just war theory? (Selmer Bringsjord and Joshua Taylor in chapter 6 in this book even investigate programming a divine-command ethics into military robots!) And what will this mean for personal, moral, and legal responsibility? As robots increase in autonomy and complexity, and their use becomes ever more pervasive in society, will the robotic programmer, builder, user, or the robot itself be the proper locus of moral evaluation and legal responsibility?

The chapters in this book have examined all three senses of robot ethics. The design and programming of robotic ethics and ethical issues concerning the military use of robots, and varieties of human–robot interaction (from sex to health care) all primarily involve senses one and two of robot ethics. As the discussion of moral behavior by robots advances, we may eventually have to face the possible third sense: the (as yet distant) specter that the robotics community may one day be responsible for creating something that SETI has so far failed to discover—a new race of intelligent beings capable of doing ethics, beings that raise difficult questions about the nature and extent of morality, questions that have been obfuscated as long as the only moral agents were members of *Homo sapiens*. This third sense is also investigated in part I (especially by Keith Abney in chapter 3), and in part VII, on robot rights and ethics.

But the creation of fully autonomous artificial moral agents is still off in the distance, if it is even possible; and in the meantime, roboticists have a serious responsibility to examine sensitive issues about their work in the first and second senses of robot ethics, as their creations gradually become more complex, more autonomous, more pervasive, and more enmeshed in the activities of everyday life. As such, roboethics currently involves key issues of regulation, including issues of safety and responsible use and development, even while robots remain mere human tools and not (yet) moral agents. Ideally, philosophers and roboticists (and even lawyers!) should work together on this project (as demonstrated in many of the chapters), as applied ethics works best when experts in both ethics and its applied field have mutually fertile conversations and reach plausible positions, ideally forming a consensus that informs action.

This chapter, the epilogue of this edited collection, is intended as a snapshot of some current developments in the field of robot ethics, or roboethics, in all three senses. As such, we will attempt to explain some important and unifying themes of this text, and, of robot ethics more generally, clear up some common misconceptions, and gesture toward the future of the field. To begin, we need clarification on the discipline that informs robot ethics, that is, the field of robotics.

22.1   Robotics, a New Science?

Robotics, of course, deals with robots; so what exactly is a robot? One definition: a robot is "a machine, situated in the world, that senses, thinks, and acts" (Bekey 2005, and chapter 2 of this volume). A typical robot uses sensors to detect aspects of an external world, software to reason about it, and actuators to interact with it; as such, all proper robots have at least some degree of autonomy and, hence, a sort of intelligence. So, we can define robotics as a branch of engineering that deals with autonomous machines—that is, robots. Robotics is but a nascent discipline, yet in contemporary robots we can already see glimpses of the fulfillment of the human dream of designing an artificial intelligence embodied in an autonomous entity, whether it be a friendly companion or pet (like AIBO) or a terrifying weapon of war (like the Predator drone).

Some have called the rise of robotics a "Third Industrial Revolution" (Thurow 1999) as machines progress from mere tools into something that potentially has "a mind of its own"; as robotics advances, investigations into complex notions like autonomy, learning, (self-)consciousness, evaluation and judgment, free will, emotions, and the like, formerly the province of philosophers, shall become part and parcel of engineering practice. As the previous chapters demonstrate, the ever-expanding capabilities of robots will pose multiple new ethical challenges (given "ought implies can"), as will the various modes of their deployment: there will be biorobots, military applications of robotics, nanny robots in children's rooms, socially assistive robots taking care of the elderly, and many more. Each of these applications will create new quandaries as a new kind of machine intelligence interacts with humans, sometimes taking human jobs, but even more often usurping traditional human roles and creating tension, as usually happens when new ways challenge venerable traditions.

Robotics thus forms a new science (and related emerging technologies) at an early stage; as philosophers of science such as Thomas Kuhn (1970) or Larry Laudan (1984) point out, new sciences are born from both the rational quest to solve problems and test solutions, and the nonrational thrust of societal forces and *gestalt* shifts in one's worldview. The future developments of robotics will likely require scrutiny, if not full-scale revision, of some of our contrastive concepts, such as person (moral being) versus mere machine, freedom versus determinism, or intelligently autonomous versus merely algorithmic. Such possible revisions in our basic concepts may well result in a *gestalt* shift in our worldview, and lead to a radically new science. The emerging science of robotics thus has far-reaching implications, and likewise itself depends upon a syncretic melding of disciplines involving knowledge from many fields, as is clearly demonstrated by scanning the entries in the huge *Handbook of Robotics* (Siciliano and Khatib 2008).

Robotics holds another promise, one not shared by all emerging sciences—the possibility of major development by way of a potentially immense number (and value) of applications, which in turn is controlled by the so-called forces of the market. Governments have made huge investments into robotics applications, from Japan's METI (Ministry of Economy, Trade and Industry) promise of 4 billion yen or more in the humanoids challenge (Robertson 2007), to the 160 billion dollars in the U.S. Future Combat Systems Program. Such applications raise new possibilities and, hence, new worries as well. Given "ought implies can," novel moral issues arise when science and technology give us new capabilities and new possibilities—but not before. So, popular opinion and expression has made doing roboethics far more difficult, as rampant confusion reigns over what robots can and cannot do, and over how they are similar to (and different from) human beings. These popular misconceptions about robotics largely stem not from its being a new scientific discipline, but from its status as an ideology.

22.2 The Robotics Ideology

An *ideology* evokes belief in certain ideas that transcend a mere evanescent opinion; instead, to qualify as an ideology, there must exist a major concretion of symbols or memes that are passed down over generations and shape the thoughts of many. A helpful contrast is with knowledge (which, unlike ideology, never occurs in the plural). To qualify as public knowledge, as opposed to a mere ideology, there needs be some set of reasons for belief that approaches a settled consensus by experts. Many popular beliefs about robots do not reflect widespread knowledge of robotics, but instead fit the criteria for an ideology. These beliefs are shaped by myths, legends, and the imagination of fiction writers and the public at large, rather than by facts and reasonable (to experts!) possibilities.

In the eighteenth century, one of the main missions of scientists in the field of electromagnetism was to remove magic from physical phenomena, turning (for example) lightning from being seen as the work of the gods or the "black arts" into something naturalistically explicable. Roboethics, in order to advance, currently needs to perform such a demystification, freeing robotics from the magical conception still dominant today in the popular imagination. Roboethicists need to help design plausible visions of the future and the options it may hold (and choices we must make), based on fact and informed speculation, not fancies and atavistic fears borne of science-fiction movies.

One example of the power of ideology in roboethics is the legend we term the "Rebellion of the Automata," in which robots rise up and overthrow their human masters, a theme so common in the literature about robots as to seem almost trite. Yet for now (and for the foreseeable future), robots are simply not self-conscious, and so while a complex robot can malfunction or break or engage in behavior that surprises its programmer, it can never consciously rebel! (Put differently, robot ethics in the third sense is as yet impossible—and may always be). Yet much of the popular fear of robots stems from the belief that they will rise up against their human masters and engage in murderous revolt. Perhaps this myth originated for reasons related to the development of Western civilization, going back to ancient Egypt and classical Greece, if not even further. In this history, we see many cultures dominated by authoritarian kingdoms whose ultimate authority was based on religious understandings, often where subdued sons surrounded a god-like king, and the king's constant fear was revolt by his family/slaves/subjects. Perhaps the worries over the so-called rebelling automata are because we think of them not as artificial tools, but instead as human slaves, illegitimately treated as a mere tool for their master's use—that is, they are treated as mere *automata*, but we fancifully believe they are capable of more! Perhaps this recurring myth is driven by our collective guilt over the history of slavery and a need for reassurance in the face of uncertainty over our robotic future? (Or perhaps it is driven by our own theory of mind and our overwillingness to attribute agency to mindless creatures—more on this follows).

But in reality, our robots are not (for now, anyway) our "slaves" in any robust sense, as they have no will of their own; and the historical origins of robots do not actually include such fictions as a Golem or a Frankenstein's monster that could rebel against its master. For current roboethics, continuing to take such tales seriously seems as silly as believing that our ancestors were the Flintstones, and our grandchildren will be the Jetsons! These tales arouse highly unrealistic expectations among the public about the near future of robotics, while simultaneously helping mask public recognition of actual near-term developments and their moral implications. Real technological advance often progresses far slower than the public is aware; and actual revolutionary technological advance is often undreamt of, even by science-fiction novelists.

To take but two examples, Arthur C. Clarke (an engineer and a scientist, as well as novelist) forecast that in 2001 we should have arrived on Jupiter, taking off from a lunar base, piloted by an autonomous robotic spacecraft of murderous intelligence (HAL), whose murders were based on its own moral reasoning. Or take the novelist Philip K. Dick, whose 1968 novel *Do Androids Dream of Electric Sheep?* was made into the movie *Blade Runner* (a source of innumerable images in the robotics literature). The novel/movie is set at the year 2019, and has autonomous biological robotic androids (replicants) with superhuman powers who wish to rise up against their enforced servitude and gain their freedom as persons. The timeframe to attain robotic moral personhood, or robot ethics in the third sense, that was assumed by these artists, was definitely more than a bit optimistic! And the robots they envisage cause great harm and destructiveness to the ordinary human protagonists—as befits the attempt to create literary and narrative drama, but not the attempt to engage soberly with the real implications of robotics.

Literature and novels are primary human arts; but reality is not a mere social construction or a novel. So to do roboethics responsibly, we need to redefine the *liaison dangereuse* between literature and robotics. We need other myths, images, and metaphors, which are more proper to the practice of robotics, and not to the anthropology of the human/automaton tragedy and legend. A different cultural history may make a society's ideology and myth less prone to such distortions and fears. For instance, the Japanese mythology does not include such fears of the evil robots overthrowing their human yoke. On the contrary, Japanese depictions of robots are largely beneficial and friendly to humanity, and popular opinion in Japan is much more sanguine about human–robot interaction. Perhaps the Japanese view of robots as beneficent helpers—not the violent, rebellious machines of Western science fiction—is rooted in the Shinto religion, which blurs the boundaries between animate and inanimate objects. The Shinto mythology may help the Japanese avoid undue fear of robots, and perhaps even avoid the "Uncanny Valley" of creepiness that seems to afflict those from other cultures when viewing humanoid robots.

Another myth that forms part of a related pernicious robotic ideology we could call the "Pinocchio Syndrome": the idea that humanoid robots could evolve into humans. Pinocchio is the main character of a novel for children by Italian author Carlo Collodi, made into the animated film by Walt Disney. It is a naughty, pinewood marionette that gains wisdom through a series of misadventures, which lead it/him to become a real human as reward for his good deeds (Collodi [1883] 2009). Implicit in this myth is the idea that reproducing human functions ever more perfectly coincides with producing a human being. This Pinocchio Syndrome commits an acknowledged flaw of reasoning, the fallacy of composition; for even if we could design and manufacture a robot endowed with reasoning powers about symbolic properties (i.e., language) analogous to those of humans, the former would belong to another, different kind of entity, another *species* (albeit nonbiological). Passing some version of the Turing Test may or may not be enough to become a "person" in some sense to be defined (as discussed by Rob Sparrow, in chapter 19 of this volume); but it certainly would not make one *human*. Our nature as humans is not merely the ability to express symbolic properties, but also the result of our biophysical powers and properties, as well as the human relationships that we develop and mature from birth until death—"human" is, in part, a relational concept. So, human nature inevitably contains both socio-cultural and biological components, and robots may gain capacities that make them our equals or betters in certain ways, but (trivially) they can never be *Homo sapiens*.

22.3 Robots and Moral Agency

But the third sense of robot ethics may not be so quickly dismissed. The future possibilities of cyborgs bring up the possibility that what may be morally crucial may not be unique to our biology. Kevin Warwick (2002, and chapter 20, this volume) explores some of these issues, as he himself has become a cyborg, and investigates the possibilities of machines with human neural cells. We can extend his thought experiment: if (admittedly, a *very large* if) we could gradually replace all of our higher brain biological functions with mechanical, robotic replacements, until we had a completely robotic brain, with no interruption in first-person self-consciousness, why would the result not qualify as a moral person, even if no longer a completely biological human? And, if so, why would biological arms or legs or digestive tract be morally crucial? Surely those humans with artificial arms, legs, and so forth, are full moral persons. So, what of robots' moral status? With the appropriate abilities, why could they not be moral persons? Do we not have to face the possibility of robot ethics in the third sense?

The foregoing considerations suggest that biological humanity is not morally crucial if robots could attain first-person self-consciousness and deliberative agency, the usual requirement for moral responsibility and the hallmark of moral personhood. But could robots ever attain agency, the ability that philosophers have long claimed set us apart from the other animals? You or I can be held responsible for our actions; we can be tried in a court of law, and found guilty or innocent, in a way that makes no sense (thus far) for any other species here on Earth (genetic engineering or discovery of extraterrestrial intelligences on other worlds pending!). But will this remain true for robots? Now, unlike the other animals we have thus far encountered here on Earth, robots have the promise of being excellent (indeed, superhuman) logical reasoners, and the prospect of such sophisticated machine reasoning has no doubt contributed to the Pinocchio Syndrome. But to be an agent, plausibly one needs more than mere mechanical reasoning; there are several possibilities (and much active research) on what more is needed.

For one possibility, perhaps one needs what Kant ([1781/1787] 1997) termed the "transcendental unity of apperception" (hereafter abbreviated TUA), in which conceptual reasoning and the appearances of objects due to sensation are tied together in a single, self-aware consciousness, able to experience a unified first-person self-consciousness—to experience (and not merely "say") the thought "I choose to do X, not Y." Mere machines (like some bank ATMs or socially assistive robots) can already speak, but they (presumably) have no self, no awareness that they are speaking—they *mean* nothing by what they say; the only *meaning* is in the (human) mind of the hearer, not in the utterance itself, or in the robot that utters it. Is TUA what is missing? If so, can an increasing complexity of programming cause TUA to emerge, or is it separated from mere algorithmic programming by some unbridgeable divide? Is it that such complex programming can simulate the syntax of human language, but a program (even a very complex program) can never have a mind that understands its *meaning*? Philosophers and neuroscientists such as John Searle (1984) with his "Chinese Room argument" and Paul and Patricia Churchland (Churchland and Churchland 1990) have hotly debated such issues and the debate rages on today.

Relatedly, the Catholic philosopher José Galván wrote: "The symbolic capacity of man takes us back to a fundamental concept which is that of free will. Free will is a condition of man, which transcends time and space. Any activity that cannot be measured in terms of time and space cannot be imitated by a machine, because it lacks free will as the basis for the symbolic capacity" (Galván 2004). So what robots may necessarily lack for agency is freedom (the freedom needed for TUA?), not mere instrumental reason. If Galván is correct, then it will continue to be the case that whenever a machine makes a statement or even displays an emotion, this doesn't mean that it feels that emotion, but only that it is using an emotional language to interact with the humans. There is more to agency than mere behavior—there is an interiority, a self who knows what it is like to be someone, in a subjective sense still unexplained by science.

Perhaps, however, one could ask: why is moral agency so important? If we merely evaluate the morality of actions by their consequences, rather than by the intentions behind the act, moral agency may not be crucial for moral practice. Perhaps as long as robots obey moral codes (in the second sense of robot ethics), difficult questions about their ability to become moral agents are irrelevant. Is it the results of actions, and not self-conscious intentions, that ultimately matter?

In human terms, most ethics presumes agency matters, because of our theory of mind. The ability to detect agents within the human community was a key to our evolution as a social species, and we are so hardwired for it that we attribute agency promiscuously, naively attributing the human ability to choose on the basis of reasons and goals to dogs and cats, cars and trains, even trees and clouds and volcanoes and the weather and . . . well, just about everything we interact with. Because this "intentional stance" works so well in understanding other humans, we have a tendency to use it to explain everything: so the Hawaiians explained volcanic eruptions by the agency of a displeased goddess Pele, and the Greeks explained shipwrecks as due to a similar rage of their god Poseidon.

The history of science comprises the long and difficult attempt to remove such teleological thinking from being applied to the natural world, so much so that some scientists and philosophers (like the Churchlands) attempt to remove it from humans themselves. But the fallacy of the intentional stance as applied to robots and the resulting Pinocchio Syndrome comes from the older and more typical human tendency to ascribe a theory of mind like our own to things that act in relevantly similar ways—and so we attribute emotions to the robot that we see speaking, precisely because of our own human emotions and mind. Robots may come to simulate many human abilities, but any simulation always lacks some of the reality of that which it simulates—or else it would not be a simulation, but identity. A related complication arises because of the nature of the different decision-making systems within the human brain.

Neuropsychological research has overwhelming support for the theory that human cognition actually involves not one but two primary systems, the one reflexive, and the other deliberative. The deliberative, fully self-aware "rational" system is an evolutionary newcomer, but perhaps as a result is often overridden by the older, usually subconscious reflexive system. As the speed required for decision making increases (whenever we must decide "in a hurry"), the fast, ancestral, emotional system continues on as usual, while the more modern deliberative frontal cortex system gets left behind. As a result, we become more prone to stereotyping, more vulnerable to emotional reactions, wishful thinking or confirmation bias, or various other "weaknesses of the will" in which we choose something that, upon deliberation, we would think is bad for us. Our moral judgments and beliefs are influenced by our cognitive limitations and evolved methods of dealing with the breakdown of rational control. The result is that human agency resembles not a finely tuned machine, but a "kluge" (Marcus 2008), a Rube Goldberg-esque construction that leaves us with a sense of reason and deliberation that is both temporally behind and somewhat subservient to our sense of impulse and reflex.

Taking such concerns about the evolutionary background of moral agency seriously, some theorists advance an alternative: perhaps it is not TUA or freedom or computational complexity that enables moral consideration, but embodiment. On this view, robots, equipped with mechanical bodies, sensors, and actuators, as well as computational abilities, would have minds, but not human minds—because they lack human *bodies*. The research program known as Embodied Cognition (EC) (Brooks 1999; Lakoff and Johnson 1999) rejects the strong AI view that all cognition consists in computational, representational symbol manipulation. EC's account of conscious (and subconscious) cognition therefore emphasizes the embodied experiences of organisms as opposed to abstract symbol manipulation, and aims to explicate how such embodiment shapes knowledge. Common to EC accounts is the idea that normal everyday human interactions consist, not in algorithmic mental computing, but in nonmentalist embodied engagements. If this approach is correct, then perhaps it is not freedom or TUA that are needed for a self-consciousness, but a body; and the type of body will determine the type of mind that inhabits it.

So, conceptual clarification is needed in order to advance this debate: does being a moral person merely require the ability to engage in symbolic representations (so any computer could qualify?!), or embodiment with freedom of action in an external world, or TUA, or . . .? The questions and confusion over robotic (self-)consciousness, robotic emotions, and robot rights and responsibilities are often based on the confusion generated by the use of the same words for intrinsically different items, and by further unclarity or equivocation over the abilities and resources necessary to have emotions, consciousness, rights, and responsibilities.

One attempt to solve at least the representational and equivocation problems is to express potential ontological differences through a specific notation. We might indicate with an "R dot" (R.) the properties of our presumably mindless robotic artifacts, to distinguish them from the capabilities known to be held by self-conscious human beings. So:

- Humans have intelligence (and agency)
- Robots have R. intelligence (and no self-conscious agency—so far)

This notation could help keep us aware of these ontological differences, and so also help avoid flaws in our moral reasoning. It is worth recalling this device began with Isaac Asimov, inventor of the Three Laws of Robotics (Asimov [1942] 1968). One character in Asimov's novels was the robot-detective R. Daniel Olivaw, so named because humanoid robots in the novel's futuristic society are virtually indistinguishable from human beings; so to avoid any confusion between humans and robots, all robots should have their name preceded by an R (Asimov 1954).

22.4   Roboethics, a Work in Progress

As Anthony Beavers (this volume, chapter 21) points out, given *"ought* implies *can,"* implementability is a requirement for any plausible approach to morals; and he suggests that our own morality is ineluctably tied to its implementation in our biology—humans are "biologically instantiated moral machines." He then asks a crucial question for the future of roboethics: "Is there any reason to think that human virtues are those that we *should* implement in machines? If so, on what grounds? Why mammalian virtues as opposed to reptilian, or perhaps, even better, virtues suited to the viability and survival advantages of the machines themselves?"

In other words, in the development of roboethics, must human engineers place their own (biologically inspired) ethics into robots, or will we gradually develop a kind of "alien" ethics, suitable for robots with very different bodies and capacities, but perhaps unsuitable for *Homo sapiens*? Given "ought implies can," how could we think it would be otherwise?

What becomes clear is that, far from the received biological nature we humans have, a robotic nature will be a choice its engineers make for it. The issues that pervade the ethics of human enhancement and the possibility of genetic engineering, and particularly the issues of "playing God" in fashioning a new human nature, thus apply with even greater force to robots—as roboticists and ethicists will be deciding the moral code of machines with novel capabilities, until (and unless) the day comes that they choose their moral code for themselves. From such considerations, the task of the robotics community must include becoming master of our own destiny, and anticipating future developments and social needs about the ethical, legal, and societal aspects of such research and its potential applications.

At the same time, given robotics' status as an ideology, it is necessary that those not involved in robotics keep themselves up to date on the field's real and scientifically predictable developments, in order to base the discussions on data supported by technical and scientific reality, and not on appearances or emotions generated by legends. To achieve this goal, we need an internationally open debate. Currently we are living in the Age of Globalization, and robotics will have a global market, just like computers, video games, cars, or cameras. This also means that roboethics is the daughter of our globalized world. It is an ethics that should be shared by most of the world's cultures, and capable of being translated into international laws that could be adopted by most of the nations of the world.

But, given that there are significant differences in the way the human–robot relationship is considered in the various cultures and religions, only a large and lengthy international debate will be able to produce useful philosophical, technical, and legal tools. At a technical level, we need a huge effort by the standard committees of the various international organizations, to achieve safety standards, just like for any other machine or appliance. In the case of robots, this task is more complex, due to the potential unpredictability of autonomous learning machines. Most obviously, this means that, in accordance with the precautionary principle, for now we will have to impose limits on the autonomy of the robots, especially in sensitive circumstances, when the robot could be harmful. At a legal level, we will need a whole new set of laws, regulating, for instance, the mobility of robots in work places or public spaces, setting clear rules about the liability and accountability of their operations. At a philosophical level, we need to discuss in depth the serious problem of the lethality of robots, for instance, in military applications. Such is precisely the mission that led to starting the Roboethics Program, and developing the Roboethics Roadmap (Veruggio 2007). The basic idea was to build the ethics of robotics in parallel with the construction of robotics itself. The goal was not only to prevent problems or equip society with cultural tools with enough time to tackle them, but also to pursue a much more ambitious aim. Indeed, it seems that robotics' development is not so much driven by abstract laws of progress, but more so by complex relations with the driving forces of the economic, political, and social system. And therefore dealing with roboethics means influencing the route of robotics.

It is certainly a great responsibility, which cannot however be avoided. Indeed, in society there cannot be a "non-choice" stance; to avoid regulation is itself a choice. Abstention ultimately ends up favoring the strongest, and in our case, in the current political, social, and economic system of the world, this means one thing only: a development policy largely driven by the interests of multinational corporations. As Philippe Coiffet said: "A development in conformity with a Humanist vision is possible but initiatives must be taken because 'natural' development driven by the market does not match with the desired humanist project" (2004).

Roboethics is precisely one of these initiatives. A crucial step is the dissemination of accurate information on robotics and its applications. The first task is to inform society and try to remedy the delusions borne from the robotics ideology. Education activities are crucial, and they should target in particular our younger citizens. It is also important to inform and, indeed, educate policy makers at a national and international level. Guidelines for the ethical application of robotics to society should come out from deep transdisciplinary and multidisciplinary discussions held by scientists and scholars of humanities (law, philosophy, sociology, psychology, and so on).

Different cultures, religions, and approaches should be taken into account. International roundtables should be organized, sponsored by alliances of states like UNESCO, the European Union, and so on, with the assistance of professional orders and associations.

22.5 The Primacy of Principles over Regulations: The Example of Military Robots

Roboethics thus should be the result of deep discussions about general ethical principles that bear on pressing practical concerns, not merely far-off scenarios. Yet, discussions should not be hidden behind any technical issue—when the guiding moral principles are not clearly defined, the pace of discovery and innovation is too fast for that: no regulation could match the speed of innovations. In doing roboethics, then, we adopt the methodology of triage, which teaches us to select the most urgent subjects and, once clear about them, see what comes next. In light of this methodology, we would like to gesture at some crucial principles for analyzing one of the most critical robotics applications—military robotics—which is certainly one of the most difficult challenges for roboethics, as already surveyed by coauthor Keith Abney in "Robots in war: Issues of risk and ethics" (Lin, Bekey, and Abney 2009).

Dealing with ethical principles when robotics weapon systems are deployed implies a close examination, among other subjects, of the doctrine of just war through history; of the details of modern, industrial warfare, and how its new possibilities problematize traditional concepts; and of the influence of military politics as well as new technology. Consideration must also be paid to the various agreements humans have already made to limit the nature of warfare, such as the Geneva and Hague Conventions and other treaties related to technological warfare, including dual use and export control agreements and other treaties (Altmann 2009).

The import of technological evolution in warfare is hard to overstate: dramatic turning points in human history occurred when development of novel weapons systems guaranteed military advantages and political power to the side that employed them. History further offers us numerous cases in which technological military superiority was used to make wars even crueler. To lessen the inhumanity of war, societies have agreed on ethical codes, codified in *jus in bello* restrictions on the ways war may morally be waged. The principles crucially include the requirement that one must exercise both discrimination and proportionality in attacks, never intentionally targeting civilians.

In this vein, it is worth reading from the "Declaration Renouncing the Use, in Time of War, of Explosive Projectiles Under 400 Grammes Weight":

Considering: that the progress of civilization should have the effect of alleviating as much as possible the calamities of war; that the only legitimate object, which states should endeavor to accomplish during war is to weaken the military forces of the enemy; that for this purpose it is sufficient to disable the greatest possible number of men; that this object would be exceeded by the employment of arms, which uselessly aggravate the sufferings of disabled men, or render their death inevitable; that the employment of such arms would, therefore, be contrary to the laws of humanity; the Contracting Parties engage mutually to renounce, in case of war amongst themselves, the employment by their military or naval troops of any projectile of a weight below 400 grams, which is either explosive or charged with fulminating or inflammable substances. (Declaration of Saint Petersburg 1868)

The preceding words were signed by most of the world powers to renounce precisely those "inflammable substances" which formed the novel technological development of chemical warfare. Unfortunately, despite the treaty, such weapons were extensively used in the following hundred years in many war theaters. More recently, facing threats of "weapons of mass destruction" (WMD), there have been new arms control conventions to reduce their proliferation. But worries persist about their use.

Robots offer a new kind of weapon. The explicit aim of much military robotics research is to develop autonomous robots that substitute for human soldiers; to create, in effect, a new army, "manned" (what will the new word be?) by untiring and near-invincible robotic soldiers that can defeat the enemy without cruelty and with discriminating selection of the military targets. Robots are also proclaimed to be able to solve some of the crucial problems of human warfare, in the three (now four) Ds: *Dull, Dirty, Dangerous,* and (as suggested in conversations with Patrick Lin, coeditor of this volume) *Dispassionate.*

Real warfare, unlike the movies, is often *Dull*; but robots can engage in extended reconnaissance and patrol, well beyond limits of human endurance, and can stand guard over perimeters in ways impossible for humans. Warfare is often also *Dirty*—but robots can work with hazardous materials, or after nuclear/biochemical attacks, or in environments unsuitable for humans, for example, underwater or in space. Warfare, of course, is also *Dangerous*—but robots can tunnel in terrorist caves, or control hostile crowds, or clear improvised explosive devices (IEDs), and save the lives and limbs of human soldiers. Finally, robotic warfare could be *Dispassionate*: in the heat of battle, seeing brothers in arms wounded, or bored and homesick, or fearful and seeing the enemy as subhuman, soldiers often let their emotions get the best of them and commit atrocities on the battlefield, to say nothing of the all too common crimes of the rape and pillaging of innocent civilians. Robots in war need have no emotions, no fears, no homesickness, no passions to satisfy, no bloodlust to quench.

Does that mean robot soldiers are automatically a good idea, and morally permissible to deploy? Not so fast. The prospect of a robotic army also has troubling ethical implications, especially given the push to further autonomy for military robots. It evokes a (for now) fanciful belief in the logical culmination of this trend, wars waged without human bloodshed at all—only machines fighting other machines. But there are many problems with the probable sequence of events that would lead to such an outcome, as pointed out by Jutta Weber (2009). First of all, a robotic army is often depicted as the zenith of technological perfection: fully autonomous robots, linked in clusters by superefficient networks, endowed with learning capabilities, perhaps even with self-conscious powers. Second, the perfect and "emotionally correct" (dispassionate) robotic warrior is lethally equipped, and could kill combatants (other human beings) with total autonomy, that is, without any *human* control or responsibility. It is implied, then, that those robots can be so perfectly programmed and so high in intelligence that they can analyze the situation "objectively," unfailingly obeying the laws of war and rules of engagement.

Such is but a dream, at least for now. Any professional involved in the fields of computer science and robotics knows the impossibility of guaranteeing both the performance and safety of a complex technological product such as a robot. If this is true in civilian situations, it is ever more difficult in a military theater, where avoiding "friendly fire" and making correct (non-)combatant discrimination is morally and practically crucial. In view of current limitations of robotic technologies, robots cannot yet achieve the performances of human-level perceptual recognition that are required to distinguish friends or bystanders from foes. The same argument can apply to the performance of networks (gluing together the robot soldier's clusters): disruption by weather conditions, technological imperfection, the heightened speed of warfare, and enemy hacking all constitute risks that could disable robots' communications. Furthermore, robot-soldiers furnished with learning capabilities able to generate a behavioral evolution according to the learning algorithms could generate unforeseeable consequences, unpredictable even by their designers. In short, given currently foreseeable technology, it is probable that autonomous robotic soldiers could go terribly wrong.

Bearing in mind the candor of the 1868 Declaration on the Explosive Projectiles compared to the hundred years of war tragedies that followed, one could regard the like, well-rounded words representing the robot soldiers—loyal to the various international conventions' regulations; respectful of civilians, the defenseless, and those who surrender; programmed to be humane; endowed with ethical firing rules—as fairy tales that, at least in the short term, no one could seriously believe. Until fully autonomous robots demonstrate (in realistic simulations) that they are no more likely to commit war crimes than human soldiers, it seems immoral to deploy them.

Third, fully autonomous systems thereby gain the status of subject of responsibility, as they are the decision makers in the war theater. If a robot commits a war crime, who is to blame—the commanding officer, the designer, the engineer/builder, the company selling it, or the robot itself? In reality, this provides autonomous robots with a license to kill. To allow such robots to exist is extremely serious, and it should not be taken for granted without informed debate and consent by humankind. This calls into question a fundamental principle: before discussing "how," we should decide "if" a fully autonomous robot can be allowed to kill a human.

22.6 Conclusion

This volume and this chapter hope to have clarified the definition, scope, and at least some of the aims of robot ethics. The subject is a difficult one, given the complexity of robotics—a new science highly interconnected with almost all technological fields, whose products can, in turn, be applied to almost every field of human activity. For these reasons, the ethical, legal, and societal issues of robotics share many common elements with another field of knowledge and practice—medicine—and its associated applied ethics, bioethics.

Roboethics also borrows from many other applied ethics, including computer and military ethics, which helps account for the slowness in disentangling old from new issues. In order to communicate crucial aspects of roboethics, it is also important to remember that the mere uttering of the word "robot" opens up a Pandora's box of images, myths, wishes, illusions, and hopes, which humanity has, over centuries, applied to automata. Tales, novels, science-fiction stories, movies—and also some roboticists who "jazz up" their papers to shock the layman—have loaded robotics with many improper conceptions.

Further, the development of robotics is driven not only by the curiosity of the researcher, but also by the turbulent forces of the global market, forces more responsive to profit than to ethics and the well-being of humanity and of our ecosystem. These forces usually count ethics as an annoying constraint or, at best, they reckon with it only to "avoid ethical issues becoming barriers to market."

That is why it is important to clear from the field the many incorrect notions about robots—a machine that is so complex that it often becomes unintelligible, even to its designer, but always an artificial product of technology, ontologically and irreparably different from a human being. And that is why it is crucial to tackle not the mythical worries due to ideologies and utopian hopes or dystopian fears, but the real issues facing robotics in the larger society—before it's too late.

References

Altmann, Jürgen. 2009. Preventive arms control for uninhabited military vehicles. In *Ethics and Robotics*, ed. Rafael Capurro and Michael Nagenborg, 69–82. Amsterdam: IOS Press; Heidelberg: AKA Verlag.

Asimov, Isaac. [1942] 1968. Runaround. In *I, Robot*, 33–51. London: Grafton Books.

Asimov, Isaac. 1954. *The Caves of Steel*. Garden City, NY: Doubleday.

Bekey, George. 2005. *Autonomous Robots: From Biological Inspiration to Implementation and Control*. Cambridge, MA: MIT Press.

Brooks, Rodney. 1999. *Cambrian Intelligence: The Early History of the New AI*. Cambridge, MA: The MIT Press.

Churchland, Paul, and Patricia Churchland. 1990. Could a machine think? *Scientific American* 262 (1): 32–37.

Coiffet, Phillipe. 2004. Machines and robots: A questionable invasion in regard to humankind development. Conference speech, International Symposium on Roboethics, January 30–31, Villa Nobel, Sanremo, Italy.

Collodi, Carlo (aka Carlo Lorenzini). [1883] 2009. *The Adventures of Pinocchio*, trans. Geoffrey Brock. New York: New York Review of Books.

Declaration of Saint Petersburg. 1868. <http://www.icrc.org/ihl.nsf/0/3c02baf088a50f61c12563c d002d663b?OpenDocument> (accessed March 23, 2011).

Galván, José Maria. 2004. On technoethics. *IEEE-Robotics and Automation* 10 (4): 58–63.

Kant, Immanuel. [1781/1787] 1997. *Critique of Pure Reason*, trans. P. Guyer and A. Wood. New York: Cambridge University Press.

Kuhn, Thomas. 1970. *The Structure of Scientific Revolutions*, 2nd ed. Chicago: University of Chicago Press.

Lakoff, G., and M. Johnson. 1999. *Philosophy in the Flesh: The Embodied Mind and Its Challenge to Western Thought*. New York: Basic Books.

Laudan, Larry. 1984. *Science and Values*. Berkeley: University of California Press.

Lin, Patrick, George Bekey, and Keith Abney. 2009. Robots in war: Issues of risk and ethics. In *Ethics and Robotics*, ed. Rafael Capurro and Michael Nagenborg, 49–68. Amsterdam: ISO Press; Heidelberg: AKA Verlag.

Marcus, Gary. 2008. *Kluge*. New York: Houghton Mifflin.

Robertson, Jennifer. 2007. Robo Sapiens Japanicus: Humanoid robots and the posthuman family. *Critical Asian Studies* 39 (3): 369–398.

Searle, John. 1984. *Minds, Brains, and Science*. Cambridge, MA: Harvard University Press.

Siciliano, Bruno, and Oussama Khatib, eds. 2008. *Springer Handbook of Robotics*. Berlin: Springer.

Thurow, Lester. 1999. *Building Wealth*. New York: HarperBusiness.

Veruggio, Gianmarco. 2007. *The EURON Roboethics Roadmap, European Robotics Research Network, Atelier on Roboethics, 2005–2007*. <http://www.roboethics.org> (accessed November 22, 2010).

Warwick, Kevin. 2002. *I, Cyborg*. London: Century.

Weber, Jutta. 2009. Robotic warfare, human rights and the rhetorics of ethical machines. In *Ethics and Robotics*, ed. Rafael Capurro and Michael Nagenborg, 83–104. Amsterdam: IOS Press; Heidelberg: AKA Verlag.

List of Contributors

Keith Abney is a philosopher of science and senior lecturer at California Polytechnic State University, San Luis Obispo. With Patrick Lin and George Bekey (coeditors of this volume), he has coauthored the grant-funded report *Autonomous Military Robotics: Risk, Ethics, and Design* (2008) and other papers on robot ethics. Abney serves on the ethics committee at a major local hospital and teaches courses on environmental ethics, social ethics, business ethics, bioethics, philosophy of religion, and more. He earned his BA from Emory University and his ABD at University of Notre Dame. He also has authored publications and numerous conference papers on naturalism, the natural–artificial distinction, and environmental ethics, on issues concerning sustainability and future rights and issues of existential risk, as well as on the ethics of enhancement.

Colin Allen is a professor of history and philosophy of science and professor of cognitive science in the College of Arts and Sciences at Indiana University (IU), Bloomington, where he has been a faculty member since 2004. He also holds an adjunct appointment in the Department of Philosophy and is a faculty member of IU's Center for the Integrative Study of Animal Behavior. His main area of research is the philosophical foundations of cognitive science, particularly with respect to nonhuman animals. He is interested in the scientific debates between ethology and comparative psychology and in current issues arising in cognitive ethology. Allen has also published on other topics in the philosophy of mind and philosophy of biology, and artificial intelligence. His most recent book is *Moral Machines: Teaching Robots Right from Wrong* (2009), coauthored with Wendell Wallach.

Peter M. Asaro is a philosopher of science, technology, and media and is an assistant professor of media studies and film at the New School University in New York. His work examines the interfaces among social relations, human minds, bodies, and digital media. His current project focuses on the social, cultural, political, legal, and ethical dimensions of military robotics and unmanned aerial drones, from a perspective that combines media theory with science and technology studies. Asaro's research has been published in international peer-reviewed journals and edited volumes, and he is writing a book that examines the intersections among military technology, interface design practices, and media culture. He earned his PhD in the history, philosophy, and sociology of science, and master of computer science degree from the University of Illinois at Urbana-Champaign; has held fellowships at the Austrian Academy of Sciences in

Vienna, the Digital Humanities HUMlab at Umeå University in Sweden, and the Center for Cultural Analysis at Rutgers University; and has designed human–computer interfaces, machine learning algorithms, robot vision systems, and natural language interfaces at the National Center for Supercomputer Applications (NCSA), the Beckman Institute for Advanced Science and Technology, Iguana Robotics, and Wolfram Research.

Anthony F. Beavers is a professor of philosophy at the University of Evansville, Indiana, where he directs the Cognitive Science Program and the Digital Humanities Laboratory. His interests largely concern issues in the intersection of computing and philosophy, particularly regarding artificial intelligence, machine ethics, information ethics, and computer modeling. He has also worked for more than twenty years on issues connected with ethical metaphysics. Beavers serves as the president of the International Association for Computing and Philosophy. His latest editorial projects include special issues for *Synthese* (with Colin Allen), *Ethics and Information Technology*, *The Journal of Experimental and Theoretical Artificial Intelligence*, and *The IEEE Transactions on Affective Computing*.

George A. Bekey is professor emeritus of computer science at the University of Southern California (USC) and distinguished professor of engineering at California Polytechnic State University in San Luis Obispo. He has worked in robotics for about thirty years and is the founder of the Robotics Research Laboratory at USC, author or coauthor of over one hundred published technical papers in robotics and several books, including the text *Autonomous Robots* (2005). He has received a number of awards for his work and served as president of the Robotics and Automation Society of the Institute of Electrical and Electronics Engineers (IEEE). In recent years, he and the coeditors of this volume (Patrick Lin and Keith Abney) have collaborated on several projects in robot ethics and published several papers on the subject. Bekey is a fellow of the IEEE, the American Association for Artificial Intelligence (AAAI), and the American Association for the Advancement of Science (AAAS); he is also a member of the National Academy of Engineering.

Paul Bello received his bachelor of science in computer and systems engineering with a dual major in philosophy from Rensselaer Polytechnic Institute in 1999. He stayed on at RPI and completed an MS in computer science in 2001 and received a PhD in cognitive science in 2005 under the supervision of Selmer Bringsjord. In 2002, Bello was hired as a computer scientist at the Air Force Research Laboratory's Information Directorate, where he completed his dissertation on reasoning about conditional obligations. In May 2007, he became program officer at the Office of Naval Research where he now directs the cognitive science program, which focuses on extending and developing computational cognitive architectures in the form of unmanned platforms and intelligent displays. Bello's personal research interests are in the computational foundations of human social cognition, with an emphasis on computational cognitive models of mental-state attribution and moral judgment.

Jason Borenstein is the director of Graduate Research Ethics Programs and codirector of the Center for Ethics and Technology at Georgia Tech. He has taught graduate courses on the subject of the responsible conduct of research (RCR) and undergraduate courses such as biotechnology and ethics, ethics and the technical professions, philosophy of science, and science and values in the policy process. Borenstein is also a coeditor of the *Stanford Encyclopedia of Philosophy*'s Ethics and Information Technology section. His research interests include engineering ethics, robotic ethics, human subjects research, genetic ethics, and ethics assessment. His work has appeared in *Science and Engineering Ethics, AI & Society, Communications of the ACM, Journal of Academic Ethics, IEEE Technology & Society, Accountability in Research*, and *Studies in Ethics, Law, and Technology*, and other journals.

Selmer Bringsjord specializes in the logico-mathematical and philosophical foundations of artificial intelligence (AI) and cognitive science, as well as collaboratively building AI systems on the basis of formal reasoning. Since 1987, he has been on faculty in the Departments of Cognitive Science and Computer Science at Rensselaer Polytechnic Institute (RPI) in Troy, New York, where as a full professor he teaches AI, formal logic, human and machine reasoning, and philosophy of AI. Funding for his work has come from the Luce Foundation, National Science Foundation, AT&T, IBM, Apple, AFRL, ARDA/DTO/IARPA, DARPA, AFOSR, and other sponsors. Bringsjord is author of the critically acclaimed *What Robots Can and Can't Be* (1992), *Superminds: People Harness Hypercomputation, and More* (2003), and other books. His papers range in approach from the mathematical to the informal, covering such areas as AI, logic, gaming, philosophy of mind, and ethics. He received a bachelor's degree from the University of Pennsylvania and a PhD from Brown University.

M. Ryan Calo is a director and lecturer at the Stanford Law School Center for Internet and Society. His work has appeared in *The New York Times, Associated Press*, and other local and national media. Prior to joining the law school, Calo was an associate in the Washington, DC, office of Covington & Burling, LLP, where he advised companies on issues of data security, privacy, and telecommunications. He holds a JD from the University of Michigan Law School and a BA in philosophy from Dartmouth College, and he served as a law clerk to the Honorable R. Guy Cole Jr. of the United States Court of Appeals for the Sixth Circuit. Calo is on the planning committee of National Robotics Week and cochairs the American Bar Association committee on Robotics and Artificial Intelligence. He blogs about robotics and the law on the Stanford Law School website.

Marcello Guarini holds a PhD from the University of Western Ontario. He is an associate professor in the Philosophy Department at the University of Windsor, where he holds a research leadership chair. He is a 2009–2010 holder of a Digital Humanities Fellowship from the Shared Hierarchical Academic Research Computing Network. He has done work in artificial neural network modeling of moral case classification, and his general research interests are in philosophy of mind (including philosophy of artificial intelligence and cognitive science) and epistemology, especially work at the intersection of these two general areas. Analogical reasoning is another focus in his research. His work in machine ethics has grown out of these more general interests. Guarini has published in *IEEE Intelligent Systems, Journal for Experimental and Theoretical AI, Synthese, Minds and Machines, Philosophy of Science*, and other journals.

James Hughes is the executive director of the Institute for Ethics and Emerging Technologies, as well as lecturer in public policy and director of institutional research and planning at Trinity College in Hartford, Connecticut. He holds a doctorate in sociology from the University of

Chicago, where he taught bioethics at the MacLean Center for Clinical Medical Ethics. Hughes is author of *Citizen Cyborg: Why Democratic Societies Must Respond to the Redesigned Human of the Future* (2004) and is working on a second book tentatively titled *Cyborg Buddha*. Since 1999, he has produced a syndicated weekly radio program, *Changesurfer Radio*. Ordained as a Buddhist monk while working in Sri Lanka in the 1980s, Hughes has written on the relationship of Buddhism and bioethics.

David Levy graduated from St. Andrews University, Scotland, in 1967. He taught classes in computer programming at Glasgow University for four years before moving into the world of business and professional chess playing and writing. He wrote more than thirty books on chess, won the Scottish Championship, and was awarded the International Master title by FIDE, the World Chess Federation, in 1969. In 1968, Levy bet four artificial-intelligence professors that he would not lose a chess match against a computer program within ten years; he won that bet. Since 1977, he has been involved in the development of many chess-playing and other programs for consumer electronic products. Levy's interest in artificial intelligence expanded into other areas of AI, including human-computer conversation, and in 1997 he led the team that won the Loebner Prize competition in New York; he won the Loebner Prize again in 2009. His fiftieth book, *Love and Sex with Robots*, was published in November 2007, shortly after he was awarded a PhD by the University of Maastricht for his thesis entitled "Intimate Relationships with Artificial Partners." Levy is president of the International Computer Games Association and CEO of the London-based company Intelligent Toys Ltd. His hobbies include classical music and playing poker.

Patrick Lin is the director of the Ethics + Emerging Sciences Group, based at California Polytechnic State University, San Luis Obispo. He has published several books and papers in the field of technology ethics, including coauthoring *What Is Nanotechnology and Why Does It Matter?: From Science to Ethics* (2010) as well as the grant-funded reports *Autonomous Military Robotics: Risk, Ethics, and Design* (2008) and *Ethics of Human Enhancement: 25 Questions & Answers* (2009). On robotics, Lin has appeared in international media such as *BBC Focus*, *BBC Radio*, *Forbes*, *National Public Radio (US)*, *Popular Mechanics*, *Popular Science*, *Reuters*, *Science Channel*, *The Christian Science Monitor*, *The Times* (UK), and others. Lin earned his BA from the University of California at Berkeley, and his MA and PhD from the University of California at Santa Barbara. He is an associate professor in Cal Poly's philosophy department, an affiliate scholar at Stanford Law School's Center for Internet and Society, and an adjunct senior research fellow at Centre for Applied Philosophy and Public Ethics (CAPPE, Australia). He was previously an ethics fellow at the U.S. Naval Academy and a postdoctoral associate at Dartmouth College.

Gert-Jan Lokhorst studied medicine and philosophy at Erasmus University Rotterdam, The Netherlands (MMedSci 1980, MA 1985, PhD 1992). His publications span and link diverse areas including logic, artificial intelligence, philosophy of mind, philosophy of technology, and neuroethics; and he leads a number of research projects in these areas in the Philosophy Department at Delft University of Technology.

Richard M. O'Meara is a professor of global affairs at Rutgers University, Newark. He is a retired U.S. Brigadier General and has worked as the assistant to the Judge Advocate General for

Operations and as an assistant to the Army General Counsel. He is a combat veteran with tours in Vietnam and Panama. As an adjunct at the Defense Institute for International Legal Studies, he has taught rule of law issues in such diverse locations as the Ukraine, Moldova, Rwanda, Sierra Leone, Guiana, Thailand, Philippines, Cambodia, Peru, El Salvador, and Iraq. He holds a JD from Fordham University and MAs in history and international affairs, and he is completing his dissertation on emerging military technologies at Rutgers. He recently completed a fellowship at the Stockdale Center for Ethical Leadership, U.S. Naval Academy, and has written extensively and presented on the issue of emerging technologies and their impact on the ethics of military leadership.

Yvette Pearson   is an associate professor in the Department of Philosophy and Religious Studies at Old Dominion University (ODU). She is also one of the directors of ODU's Institute for Ethics and Public Affairs. Before joining the ODU faculty in 2002 as a visiting assistant professor, she earned her PhD in philosophy from the University of Miami. While she teaches primarily undergraduate philosophy courses, she has also taught graduate-level courses in business ethics and public health ethics. Her research interests include ethical and social policy issues related to human procreation, direct-to-consumer marketing of genetic tests, embryonic stem cell research, and the use of robot caregivers.

Steve Petersen   earned his bachelor's degree in philosophy and mathematics from Harvard and his doctorate in philosophy from the University of Michigan. He is now an assistant professor of philosophy at Niagara University. His research mostly pursues an algorithmic approach to "good thinking," and thus lies somewhere in the intersection of traditional epistemology (what is it to think well?), computational epistemology (how might a machine think?), philosophy of mind (what is thinking anyway?), and philosophy of science (what are the simplest explanations, and why believe them?). He also sometimes gets paid to act in plays.

Matthias Scheutz   received degrees in philosophy (MA 1989, PhD 1995) and formal logic (MS 1993) from the University of Vienna and in computer engineering (MS 1993) from the Vienna University of Technology in Austria. He also received the joint PhD in cognitive science and computer science from Indiana University in 1999. Scheutz is an associate professor of computer and cognitive science in the Department of Computer Science at Tufts University. He has over one hundred peer-reviewed publications in artificial intelligence, artificial life, agent-based computing, natural language processing, cognitive modeling, robotics, human–robot interaction, and foundations of cognitive science. His research and teaching interests include multiscale agent-based models of social behavior as well as complex cognitive and affective robots with natural-language capabilities for natural human–robot interaction.

Amanda Sharkey   has an interdisciplinary background that began with a first degree in psychology, followed by a variety of research positions at the University of Exeter, MRC Cognitive Development Unit, and Yale and Stanford universities. After completing her PhD in psycholinguistics in 1989 at the University of Essex, she conducted research in neural computing at the university before moving to the University of Sheffield, where she is now a senior lecturer in the Department of Computer Science and researches human–robot interaction and associated ethical issues, swarm robotics, and combining neural nets and other estimators. Sharkey has over seventy publications, is a founding member of the scientific committee for the international series of workshops on multiple classifier systems, and is editor of the journal *Connection Science*.

Noel Sharkey is a professor of AI and robotics and a professor of public engagement at the University of Sheffield. He has held a number of research and teaching positions in the United Kingdom (Essex, Exeter, Sheffield) and the United States (Yale and Stanford). Sharkey has moved freely across academic disciplines, lecturing in departments of engineering, philosophy, psychology, cognitive science, linguistics, artificial intelligence, and computer science. He holds a doctorate in experimental psychology and a doctorate of science. He is a chartered electrical engineer, a chartered information technology professional, and a member of both the Experimental Psychology Society and Equity (the actor's union). He has published over a hundred academic articles and books, as well as national newspaper and magazine articles. In addition to editing several journal special issues on modern robotics, he has been editor-in-chief of the journal *Connection Science* for twenty-two years and an editor of both *Robotics and Autonomous Systems* and *Artificial Intelligence Review*. Sharkey's research interests include biologically inspired robotics, cognitive processes, history of automata/robots (from ancient to modern), human–robot interaction and communication, representations of emotion, and machine learning; but his current research passion is for the ethics of robot applications. He was an EPSRC Senior Media Fellow (2004–2010) and is now a Leverhulme Research Fellow (2010–2012) on the ethical and technical appraisal of robots on the battlefield.

Rob Sparrow is a senior lecturer at the Centre for Human Bioethics at Monash University, where he teaches and researches on ethical issues raised by new technologies. In addition to researching the ethics of robotics, Sparrow also writes about the ethics of human enhancement and publishes on topics in political philosophy.

Joshua Taylor is a PhD student at Rensselaer Polytechnic Institute in computer science. His research interests include artificial intelligence and formal logic. He is the primary developer of the Slate system and works in the Rensselaer AI & Reasoning (RAIR) Lab.

Jeroen van den Hoven is professor of ethics and technology at Delft University of Technology. He is scientific director of the Centre of Excellence of the Three Technical Universities in The Netherlands in The Hague. He is editor-in-chief of the journal *Ethics and Information Technology* and recently published an edited volume, *Information Technology and Moral Philosophy*, with John Weckert (2009). He is winner of the 2009 World Technology Award in the Ethics category.

Gianmarco Veruggio is responsible for the Operational Unit of Genoa of CNR-IEIIT. In 1980 he received a degree in electronic engineering from the University of Genoa, Italy. His research interests encompass robot mission control, real-time human–machine interfaces, control system architectures for telerobotics, and Internet robotics. In 1989, he founded the CNR-IAN Robotlab, which he headed until 2003, to carry out research and missions in experimental robotics in extreme environments. He led several marine robotics campaigns in Antarctica and in the Arctic. In 2000, he founded Scuola di Robotica (School of Robotics), a nonprofit association, to promote this new science among young people and society. His research on the complex relationship between robotics and society led him to coin the term—and propose the concept—of "roboethics" and to dedicate increasing resources to the development of this new applicative field of ethics. He serves as the corresponding cochair of the IEEE Robotics and Automation Society's Technical Committee on Roboethics and as a distinguished lecturer. In 2009, he was presented with the title of Commander of the Order of Merit of the Italian Republic.

Wendell Wallach is consultant, lecturer, and scholar at Yale University's Interdisciplinary Center for Bioethics. For the past six years he has chaired the center's research study group on technology and ethics, and is also a member of research groups on animal ethics, end-of-life issues, neuroethics, and post-traumatic stress disorder (PTSD). He coauthored, with Colin Allen, *Moral Machines: Teaching Robots Right from Wrong* (2009). Formerly, he was the cofounder and managing partner of two computer consulting companies: Farpoint Solutions (a regional consultancy located in Connecticut) and Omnia Consulting Inc. (an international consultancy and software developer). Wallach's research interests include the societal, ethical, and policy challenges posed by emerging technologies, the prospects for implementing moral decision-making capabilities in computers and robots, and the cognitive mechanisms that support moral decision making.

Kevin Warwick is a professor of cybernetics at the University of Reading, England, where he carries out research in artificial intelligence, control, robotics, and biomedical engineering. Warwick took his first degree at Aston University, followed by a PhD and a research post at Imperial College, London. He subsequently held positions at Oxford, Newcastle, and Warwick universities before being offered the chair at Reading. He has been awarded higher doctorates (DScs) both by Imperial College and the Czech Academy of Sciences, Prague, in addition to honorary doctorates from Aston, Coventry, and Bradford universities. He was presented with The Future of Health Technology Award from MIT, was made an Honorary Member of the Academy of Sciences, St. Petersburg, and received the IEE Senior Achievement Medal, the Mountbatten Medal, and the Ellison-Cliffe Medal. In 2000, Warwick presented the Royal Institution Christmas Lectures. He is perhaps best known for carrying out a pioneering set of experiments involving the implant of multielectrodes into his own nervous system. With this in place, he carried out the world's first experiment involving electronic communication directly between the nervous systems of two humans.

Blay Whitby is a philosopher and ethicist concerned with the social impact of new and emerging technologies. His publications include "Oversold, Unregulated, and Unethical: Why We Need to Respond to Robot Nannies," "On Computable Morality," and "Sometimes It's Hard to be a Robot: A Call for Action on the Ethics of Abusing Artificial Agents." His books include *Reflections on Artificial Intelligence: The Legal, Moral and Ethical Dimensions* and *Artificial Intelligence, A Handbook of Professionalism* (1996). Whitby is a member of the Ethics Group of BCS, The Chartered Institute of IT, and an ethical advisor to Royal Academy of Engineering. He is a regular speaker in academic, commercial, military, and community settings and has participated in several high-impact science/art collaborations. Whitby received his doctorate in the social implications of artificial intelligence in 2003 and holds degrees in philosophy, politics, and economics (BA, Oxford), philosophy (MA, Sussex), and intelligent systems (MSc, Sussex). Whitby lectures at the University of Sussex, where he chairs the University Ethics Committee for Science and Technology.

Index